SUSTAINABLE BRIDGE STRUCTURES

Sustainable Bridge Structures

Editor

Khaled M. Mahmoud

Bridge Technology Consulting (BTC), New York City, USA

CRC Press
Taylor & Francis Group
Boca Raton   London   New York   Leiden

CRC Press is an imprint of the
Taylor & Francis Group, an **informa** business

A BALKEMA BOOK

Front Cover:
San Francisco-Oakland Bay Bridge Eastern Span, California, USA
Photo courtesy of Steve K. McClanahan

Back Cover:
San Francisco-Oakland Bay Bridge Eastern Span, California, USA
Photo courtesy of Richard Shewmaker

Cover Design:
Khaled M. Mahmoud
Bridge Technology Consulting (BTC)
New York City, USA

Published by: CRC Press/Balkema
P.O. Box 11320, 2301 EH Leiden, The Netherlands
e-mail: Pub.NL@taylorandfrancis.com
www.crcpress.com – www.taylorandfrancis.com

First issued in paperback 2020

© 2015 by Taylor & Francis Group, LLC
CRC Press/Balkema is an imprint of the Taylor & Francis Group, an informa business

No claim to original U.S. Government works

ISBN 13: 978-0-367-73793-1 (pbk)
ISBN 13: 978-1-138-02878-4 (hbk)

Visit the Taylor & Francis Web site at
http://www.taylorandfrancis.com

and the CRC Press Web site at
http://www.crcpress.com

Typeset by MPS Limited, Chennai, India

Table of contents

Seismic analysis of bridges

Bridge rehabilitation & strengthening

Bridge bearings

Bridge history & aesthetics

Preface

The ever-increasing traffic demands, coupled with deteriorating condition of bridge structures, present great challenges for maintaining a healthy transportation network. The challenges encompass a wide range of economic, environmental, and social constraints that go beyond the technical boundaries of bridge engineering. Those constraints compound the complexity of bridge projects and motivate innovations in bridge engineering technology towards the design and construction of sustainable bridges. The sustainability aims at minimizing the cost of bridge construction projects and the associated environmental impact on the society, while maintaining healthy economic development.

On August 24–25, 2015, bridge engineers from all over the world gathered at the 8th New York City Bridge Conference to discuss and share experiences on the construction and maintenance of sustainable bridge structures. This volume contains select papers that were presented at the conference. The peer-reviewed papers are valuable contributions and of archival quality in bridge engineering.

Main cable dehumidification aims at reducing moisture content inside bridge cables. The technique was first applied to the new Akashi-Kaikyo Bridge in Japan, which opened to traffic in 1998. Cable dehumidification aims at active control of the relative humidity within the cable microenvironment to reduce the moisture content, the principle cause for wire degradation and the reduction in cable carrying capacity. Dehumidification offers a cost-competitive and effective alternative to previous cable management strategies such as painting, wrapping or oiling. The objective of this preservation strategy is not only to reduce the trapped moisture within the cable, but also to minimize the cumulative wire breaks over time. The dehumidification technique has been applied in Europe on several suspension bridges. Dehumidification of the main suspension cables of the William Preston Lane Jr. Memorial (Bay) Bridge in Stevensville, Maryland is the first main cable dehumidification project in the United States. The proceedings lead off with two papers that discuss details of this first application of the dehumidification on an American suspension bridge. The first paper is "Suspension bridge main cable dehumidification – an active system for cable preservation," by Beabes et al.; and the second paper "Chesapeake Bay Bridge dehumidification design," authored by Nader et al. Many cable-stayed bridges include a deck composed of a steel grillage made composite with concrete panels. These deck systems are subject to requirements for fracture critical or structurally redundant members, sometimes including the guidelines published by the U.S. Federal Highway Administration (FHWA). However, this type of bridge includes inherent redundancy, as there are multiple load paths to redistribute the effects of member failure. Given the redundant nature of these bridges, a blanket designation of fracture critical member (FCM), or System Redundant Member (SRM) may be a conservative approach that affects the economy of the structure. In "Fracture analysis of steel cable-stayed bridges," Soule and Tappen demonstrate an approach to determining where FCM fabrication is necessary utilizing the East End Bridge Crossing, currently under construction near Louisville, Kentucky, USA. The authors discuss the development of performance criteria suitable to the site, as well as the analyses that were performed to evaluate the structure after a member rupture against those criteria. The paper suggests general guidelines appropriate for assessing this type of construction. Opened to traffic in 2005, the Binh Bridge, located in Haiphong City, Vietnam, is a 3-span cable-stayed bridge with composite girders and a center span of 260 m. In 2010, 3 ships swept upstream by a typhoon and collided with the bridge, resulting in serious damage to the edge girder and several cables. Tokuchi and Kaifuku provide in "Repair of Binh Bridge damaged by ship collision," details of the repair works which included cutting of the damaged steel girders and on-site welding of new plates and stay cable replacement utilizing the adjacent stay cables.

The Route 46 over Musconetcong River Bridge connects the town of Hackettstown with Mount Olive and Washington Townships between New Jersey's Morris and Warren counties. The 127-foot long bridge, which carries 13,200 vehicles per day via one 15-foot lane in the east and west directions, is of significant importance to commuters and the surrounding residents and business communities. Built in 1924 and spanning a Class 1 waterway, the structure lies immediately adjacent to the remains of a historically significant gristmill in the northwest quadrant, and borders local businesses in the other three quadrants. In "Reconstructing a bridge in ten days: New Jersey Route 46 over Musconetcong River accelerated replacement" Adams and Deeck discuss the challenges faced during the design and construction of the project during a ten-day road closure using Prefabricated Bridge Elements and the State of New Jersey's first application of Ultra-High Performance Concrete. Steel bridge surfaces exposed to aggressive environment must be protected to preserve structural integrity and provide longevity. Metallization is a thermal spray solution commonly used in steel bridge fabrication. Highway bridge design standards in North America specify values for slip coefficients to be used in slip-critical connections for various faying surfaces. Currently, these standards do not prescribe a slip coefficient value for metallized faying surfaces used with slip-critical bolted connections. Thus, bridge fabricators are compelled to mask off joint faying surfaces before metalizing, which is time-consuming and expensive. In "Slip and creep performance for metallized connection faying surfaces used in steel bridge construction," Ampleman et al. present results from two different research work carried by Université Laval, Quebec City, Canada, and the U.S. Federal Highway Administration in Virginia, on the slip performance of metalized faying surfaces. Both short-duration slip tests under static load and long-term sustained creep tests, performed at the laboratories of the two partner institutions, are reported in this paper. The slip resistance is then characterized based on the Canadian Highway bridge design code and the AASHTO LRFD bridge design code. Skewed and/or horizontally curved steel I-girder bridges make up a significant portion of the steel bridge population in the United States. The structural behavior of such bridges is more complicated than their non-skewed, tangent counterparts due to additional effects from skew/curvature. Therefore, supplemental guidance is recommended for both the design and construction phases. Figures 1–2 are examples of a severely skewed and a highly curved bridge. The New Jersey Department of Transportation (NJDOT) Research Bureau retained Cambridge Systematics (CS) and Greenman-Pedersen Inc. (GPI) to develop design and construction engineering guidelines and checklists to instruct designers on how to properly address out-of-plumb issues for skewed and curved steel I-Girder bridges during the design phase of the project. The guidelines were developed based on current AASHTO design specifications, available research papers and reports, and GPI's past project experience. In "Design and construction guidelines for skewed/curved steel I-girder bridges," Liang et al. present the design portion of the research project, and construction engineering guidelines and notes to be included in the contract. The Nhat Tan Bridge is located on the new route from Noi Bai new international airport to downtown in Hanoi, Vietnam. The bridge opened to traffic in January 2015. The main bridge is a 1500 m long, 6-span cable stayed bridge with 8 traffic lanes. This scale of multiple span cable stayed bridge is the first application in Southeast Asia and also very rare type of bridges in the world. In "Construction of the Nhat Tan Bridge superstructure," Matsuno and Taki describe the challenges encountered in the development and application of construction methods.

The analysis of ultimate limit state (ULS) of a structure requires a stability study until failure. This is complex mechanical behavior to compute with standard tools. Cracking, damage, and elastic-plastic law are among the phenomena, which often lead to numerical problem of convergence and interpretation of results. It is therefore often advised to use codes instead (Eurocodes, AASHTO, etc.), but this solution comes at the expense of accurate analysis of the physical behavior of failure. In "Limit analysis for steel beams connection nodes," Arquier and Cespedes present an alternate solution, which combines two parallel and complementary methods. Used in a finite element mesh for rigid-plastic calculations, these two methods lead to a full determination of the physical failure: mechanism, stresses distribution and safety factor. No standard approach for the analysis of integral abutment bridges appears in the AASHTO LRFD Bridge Design Specifications or other international codes. In "Integral abutment bridges and the modeling of soil-structure interaction," Rhodes and Cakebread present the approaches most suitable for modeling common integral abutment bridge forms, expanding upon recent UK guidance regarding soil-structure interaction approaches. The authors discuss material properties, initial stress state and the incorporation of the effects of soil ratcheting and both continuum and spring-type ('subgrade modulus') finite element models. Many steel bridges built decades ago have redundancy issues since redundancy was not accommodated in the design. These major non-redundant steel bridges are in various forms, such as two-girder bridges, tied arch bridges with tension ties, and truss bridges. With the lack of redundancy, failure of one member of the bridge would lead to the failure of the entire bridge. Serious attention is necessary for this structural performance, structural reliability, and, most importantly, public safety issue. In "Improving structural reliability using a post-tensioned concrete floor system for major non-redundant steel bridges," Chang and Lawrie discuss the structural reliability improvement using a post-tensioned concrete floor system for major non-redundant steel bridges rehabilitation. A non-redundant structure can be represented as a series system, in the reliability engineering aspect, in which when one of the system components fails, the entire system fails. A structure with redundancy, on the other hand, is considered as a combination system made of series and parallel configurations, where a parallel configuration is one that does not fail unless all the components fail. The authors provide illustrative examples for further demonstration of the structural reliability improvement.

Laboratory testing of bridge components has an important role in verifying their long-term performance and thus minimizing their life-cycle costs. The life-cycle cost of a bridge's expansion joints are likely to be many times higher than the initial supply and installation costs. The long-term performance of these critical bridge components, and their fatigue performance in particular, should thus be a key factor in the design. While the long-term performance of a particular type of expansion joint can in many cases be evaluated on the basis of the performance to date of expansion joints that have been in service for many years, it is often desirable to require evidence in the form of standardized laboratory testing. In "AASHTO fatigue testing of modular expansion joints – setting new standards," Moor et al. discuss the fatigue performance of bridge expansion joints. The authors present recent fatigue testing, in the "infinite life regime", of a modular joint in accordance with the AASHTO. The design of "depth-critical" steel superstructures for even simple span bridges is an emerging design concept for many engineers today and is inadequately explained in the AASHTO LRFD Bridge Design or Construction Specifications. Traditional empirical design assumptions as presented in the AASHTO Specifications do not necessarily apply to slender beams, which require large cambers and specialized sequences of construction. Shallow girder design will only continue to rise in popularity for situations in which designers are forced to provide additional roadway underclearance where functionally obsolete bridges are replaced, or where construction of a new bridge requires spanning longer over widened roadways without the addition of a pier. In "Design of depth critical steel bridge superstructures," Schaefer and Ricks address the unique concerns of designing steel bridges for minimum superstructure depth through the discussion of a shallow single span bridge. An extradosed prestressed bridge is a girder bridge that is externally prestressed, using stay cables over a portion of the span. Extradosed prestressed bridges can provide an economical bridge solution for spans in the transition range from conventional girder bridges and cable stayed bridges. In "Proportioning and design considerations for Extradosed prestressed bridges," Stroh discusses initial proportioning guidelines for this bridge type, based on work by the author in developing the design for the first extradosed prestressed bridge in the US, the Pearl Harbor Memorial Bridge in New Haven Connecticut, and from reviewing over 60 extradosed prestressed bridge designs worldwide. Good maintenance of a bridge requires the information on traffic loads. A method of estimating the traffic loads is bridge-weigh-in-motion (BWIM). The conventional BWIM is based on the strains of main girders. To obtain supplemental information of truck velocity, the strains of transverse stiffeners are measured additionally. The approach involves the integration of time-history response of strain and is called BWIM by Integration Method with Transverse Stiffeners (BWIM-IT).

In "Bridge-weigh-in-motion for axle-load estimation," Yamaguchi and Kibe extend BWIM-IT to the estimation of axle loads.

The Gerald Desmond Bridge is a vital link and a major commuter corridor, which connects Long Beach with Terminal Island in Southern California. The Port of Long Beach intends to replace the existing deteriorating bridge with a 2000 ft long cable stayed bridge with 1000 ft main span and two 500 ft side spans. The 515 ft tall towers, which provide primary means of vertical support to the cable-stayed bridge, are relatively slender tall hollow reinforced concrete sections thus requiring assessment of possible buckling. The buckling resistance of the towers comes from the global structural system with the stay cables providing restraint to the top of the towers and the viscous dampers providing restraint at deck level. The octagonal tower shape at the connection to the pile cap tapers to a diamond in the upper part of the tower. The bridge is located in an area of extreme seismic hazard. The non-linear time history analysis of the bridge includes simultaneous tri-axial earthquake accelerations as well as gravitational acceleration. In "Post-earthquake stability of Gerald Desmond Bridge," Banibayat et al. discuss the explicit nonlinear time history analysis performed for tower stability during an earthquake and show the tower remain stable and elastic after Functional Evaluation Earthquake (FEE) and 1000 year return period Safety Evaluation Earthquake (SEE) event. When completed, the Izmit Bay Bridge, with a 1550 m main span will be the world's 4th longest suspension bridge. The bridge crosses the Izmit Bay in western Turkey from North to South and situated at very close premises of the North Anatolian Fault. Therefore, from the early stages of design, detailed geophysical and geotechnical surveys have been conducted to find out the faulting, and the consequent earthquake risk in the region to lay the required basis for detailed design. In "Impact of secondary fault findings on the design of Izmit Bay Bridge in Turkey," Cetinkaya et al. discuss the conceptual bridge design, the detailed geophysical studies and the implication of the faults revealed by those geophysical studies on the conceptual design. The existing Poplar Street Interchange at I-55/I-64 in downtown St. Louis, Missouri, adjoins the approach viaduct to the Poplar Street Mississippi River Bridge. A major seismic retrofit initiative was completed on the approach viaduct within the last 15 years. The ramps were not part of the retrofit project and are currently being replaced. The eastbound approach viaduct will also be widened to accommodate an additional lane. In "Poplar Street Interchange replacement – seismic design," Rolwes presents the design response spectrum, which is developed through the use of site-specific probabilistic rock accelerations in conjunction with at-depth analysis of ground motion. Using the refined spectrum, it was demonstrated, without time-consuming modeling, that the widening and new ramps would not appreciably affect the existing viaduct structure and associated retrofit details. Mexico City is vulnerable to earthquakes. In "Seismic behavior of a long viaduct in Mexico DF: a combined FEM and SHM approach," Simon-Talero et al. propose a methodology for tackling seismic challenges, based on the combined use of Finite Element Modeling and Structural Health Monitoring practices. This methodology involves the implementation of precise FEM models combined with real-time, web-based remote monitoring systems providing dynamic data obtained from a sensor networks.

A visually distressed vintage conventionally reinforced concrete deck girder bridge was identified by routine inspection. Subsequent investigation showed the distress was due to a poorly detailed splice location for the flexural steel and the ratings determined the girders to be significantly understrength. The bridge was shored to allow it to remain in service until it could be strengthened. The bridge was strengthened using near-surface mounted titanium alloy bars. Round titanium alloy bars with a unique deformation pattern were specially developed for this application. Experimental research was conducted to evaluate the behavior of the as-built poorly detailed girder and then to evaluate the performance of the strengthening approach. Realistic full-scale girder specimens were constructed, instrumented, and tested to failure. The specimens mimicked the in-situ materials, loading interactions, and geometry. In "Titanium alloy bars for strengthening a reinforced concrete bridge," Higgins et al. discuss details of the proposed approach. The Girard Point Bridge, carrying Interstate highway 95 (I-95) over the Schuylkill River, is located in Philadelphia, Pennsylvania. The bridge is an 18-span, double-deck, through-truss steel structure. The main spans are comprised of two 353-ft cantilevered anchor spans and a 700-ft center span including a 390-ft suspended portion. Construction of the truss was completed in 1973 and the entire bridge opened to traffic in 1976. The upper deck carries the southbound traffic and the lower deck carries the northbound traffic of I-95. Fatigue cracks were first reported in 1993 in some of the floor beam end connections in the three-span cantilever-suspended unit of the bridge. The cracks occurred in the floor beam web and the triangular knee brace, and initiated from the horizontal web-to-flange connection welds at the end of the floor beam top flange. The fatigue cracks were found to have grown in length and location overtime, and spread over nearly all the floor beam end connections of the same construction detail by the late 1990s. In "Fatigue investigation and retrofit of double-deck cantilevered truss I-95 Girard Point Bridge," Zhou and Guzda discuss the retrofit design and construction for extensive fatigue cracks in the end connections of floor beams on a double-deck, cantilever-suspended steel truss bridge. The leaking expansion joints are a major source of multi-span bridge deteriorations in Canada and North America. Flexible link slabs made of Engineered Cementitious Composite (ECC) forming a joint free bridge can replace expansion joints. ECC is a special type of high performance fiber reinforced cementitious composite with high strain hardening characteristic and multiple micro-cracking behavior under tension and flexure. The locally available aggregates and supplementary cementitious material (SCM) have been used to produce sustainable and cost effective ECC mix for the link slab application. The use of flexible ECC link slab in joint free bridge deck has been an emerging technology. Limited research has been conducted on the fatigue performance of such ECC link slabs. In "Flexural fatigue performance of ECC link slabs for bridge deck applications," Hossain et al. present the results of experimental investigation on ECC link slabs subjected to flexural fatigue loading at stress levels of 40% and 55% for 400,000 cycles. The authors compare the structural performance of ECC link slabs with their self-consolidating concrete (SCC) counterparts based on load-deformation/moment-rotation responses, residual strength, strain developments, cracking patterns, ductility index and energy absorption.

With recent advancements in bridge design technology, bridge bearings are required to address significant further challenges in addition to their primary functions of resisting loads and accommodating movements and rotations. In "Modern bearings for key bridges – special functions & type selection," Kutumbale and Moor present developments in the design of bridge bearings, focusing on innovative solutions such as uplift-restraining bearings subjected to fatigue loading, temporary locking of bearings to resist construction loads, bearings with adjustable height and easily replaceable bearings. The Pearl Harbor Bridge known locally as the Q Bridge carries I-95 traffic over the Quinnipiac River in New Haven, Connecticut. The original plate girder structure was built in 1958 and was designed to accommodate 40,000 vehicles per day. When that total approached 140,000 vehicles per day the Connecticut Department of Transportation decided a new structure was needed. The new twin $635 million cable stayed extradosed bridges are now nearly complete and feature some high load multirotational bearings that were designed for a vertical capacity in excess of 44,400 kN (10,000 kips) which makes them some of the largest bridge bearings ever fabricated in the world. In "High load multirotational bearings for an Extradosed bridge," Watson and Conklin discuss the issues surrounding the design, manufacture and testing of these devices. The authors provide details of the testing conducted at the University of California at San Diego's SRMD facility on these bearings. Curved highway bridges are widely used in modern highway systems, often being the most viable option at complicated interchanges or other locations where geometric restrictions apply. Among the great variety of seismic isolation systems available, the lead rubber bearing (LRB), in particular, has found wide application in highway bridge structures. However, conventional LRBs, which are manufactured from standard natural rubber and lead, display a significant vulnerability to low temperatures. In "Seismic isolation of highway bridges: effective performance of LRBs at low temperatures," Mendez Galindo et al. describe the challenge faced in the seismic isolation using LRBs of a curved highway viaduct where low temperatures must be considered in the design. Specifically, the LRBs must be able to withstand temperatures as low as $-30°C$ for up to 72 hours, while displaying acceptable variations in their effective stiffness. This extreme condition required the development of a new rubber mixture, and the optimization of the general design of the isolators.

Charles Ellet, Jr. (1810–1862) was a multi-talented engineer who was far ahead of his time and who made important contributions in the fields of long span suspension bridge-building; river training and flood controls in western rivers; transportation planning and economics; canal and railroad building; and demonstrating merits of iron-clad steam rams in naval warfare. Ellet built the first permanent wire suspension bridge in the U.S. over the Schuylkill River in 1842, the first suspension bridge across the Niagara Gorge in 1848, and the first suspension bridge with a span over 1,000 feet at Wheeling, Virginia in 1849. In "Charles Ellet, Jr., the pioneer American suspension bridge builder," Gandhi highlights Ellet's contributions in building and promoting suspension bridges in the U.S. Another great figure in bridge engineering is Gustav Lindenthal (1850–1935). Upon his death, Lindenthal was referred to by some journals as "The Dean of American Bridge Builders". He was born in Bruun, Austria in 1850, and immigrated to the United States in 1874. He started his own consulting engineering firm in 1881. In 1888 he initiated the pursuit of building a major suspension bridge across the Hudson River connecting New Jersey with Manhattan, a pursuit, which continued for the following 45 years. In "Lindenthal and his pursuit of a bridge across the Hudson River," Gandhi examines the various schemes developed by Lindenthal, and the circumstances, which prevented Lindenthal from achieving his lifelong dream of building a bridge across the Hudson River. Throughout the twentieth century, the West Broadway Bridge, over the Passaic River, provided vehicles and pedestrians a main transportation link to the center of Paterson, New Jersey. Built in 1897–1898, the three-span concrete and steel Melan arch bridge was technologically innovative in its early years. Growing traffic demands, poor physical condition, insufficient capacity, scour vulnerability, inadequate safety features, and lost/altered architectural elements were the driving needs for this rehabilitation and preservation project for this historic bridge. In "Rehabilitation of the West Broadway Bridge over the Passaic River, Paterson, New Jersey," Zamiskie and Chiara provide details of the rehabilitation project, including improvement of the crossing in order to continue its service into the twenty first century and preserving this unique structure that has merited listing on the National Register of Historic Places.

The editor is grateful for the efforts of the authors and reviewers, who produced the archival quality of these proceedings.

Khaled M. Mahmoud, PhD, PE
Chairman of Bridge Engineering Association (BEA)
www.bridgeengineer.org
President of Bridge Technology Consulting (BTC)
www.kmbtc.com
New York City, USA

New York City, August 2015

Cable-supported bridges

Chapter 1

Suspension bridge main cable dehumidification – an active system for cable preservation

S. Beabes
AECOM, Baltimore, MD, USA

D. Faust
AECOM, Philadelphia, PA, USA

C. Cocksedge
AECOM Ltd., London, UK

ABSTRACT: Main cable dehumidification was first applied to the new Akashi-Kaikyo Bridge in Japan. Since its inception, it has been successfully adapted as a main-cable preservation strategy for existing bridges. Cable dehumidification actively controls the relative humidity within the cable microenvironment and removes moisture, the principle cause for wire corrosion and the reduction in cable reliability. Dehumidification offers a cost-competitive and effective alternative to previous cable management strategies such as painting, wrapping or oiling. It has been demonstrated that this preservation strategy not only successfully removes the trapped moisture within the cable, but also reduces cumulative wire breaks over time. Main cable dehumidification has recently been implemented on the William Preston Lane Jr. Memorial (Bay) Bridge in Stevensville, Maryland, the first main cable dehumidification project in the United States, and is under evaluation on other US suspension bridges.

1   INTRODUCTION

According to the National Bridge Inventory, the United States has approximately eighty-five suspension bridges on the national highway system of which nearly 30% are structurally deficient. Challenged with an aging infrastructure, competing priorities and limited funding, bridge owners must focus efforts on maintaining their assets through system preservation projects. For suspension bridges, this demands a proactive approach to main cable preservation strategies.

Internal cable investigations on many older suspension bridges revealed that wire corrosion is a common problem. The causes of the corrosion are attributed to many factors including zinc depletion and hydrogen embrittlement, typically caused by the presence of moisture trapped within the interior cable microenvironment. If not effectively addressed, these factors combine to cause wire breakage leading to a reduction in cable reliability and eventually the need to replace the main cables – a last resort for most suspension bridge owners.

The findings from cable investigations have demonstrated that traditional passive wire protection systems, such as red lead paste, wire, and paint or elastomeric wrapping systems are ineffective in preventing moisture intrusion into the cable. Main cable dehumidification provides bridge owners with an active protection system that alters the cable microenvironment by reducing relative humidity to non-corrosive levels. Studies of acoustic monitoring data demonstrate that this method will reduce wire breakage over time. Furthermore, the system operating and maintenance costs of

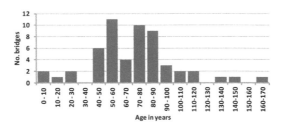

Figure 1.   Age distribution of major US suspension bridges.

cable dehumidification compare favorably to those of other systems. For these reasons, the installation of main cable dehumidification is now seen as a reliable and cost effective technique for the preservation of main cables on existing suspension bridges.

2   SUSPENSION BRIDGES IN THE UNITED STATES

2.1   *Bridge inventory*

The United States was a major contributor to the construction of suspension bridges during the 19th and 20th centuries. Beginning with the Wheeling Bridge in 1849 and ending with the Verrazano Narrows Bridge in 1964, the US was continuously pushing the limits on the maximum span lengths for suspension bridges. During this time, notable US bridges such as the Brooklyn Bridge (1883), George Washington Bridge (1931), Golden Gate Bridge (1937) and the Verrazano Narrows Bridge (1964) were the longest-span bridges built at the time of their construction.

Today, according to the 2013 National Bridge Inventory, there are approximately eighty-five suspension bridges in the United States on the national highway system. Forty-five of these bridges have a span length greater than 213 m (700ft). Many of these bridges are providing a vital link for motorists and interstate commerce and are experiencing large volumes of traffic in-and-around our nation's major cities.

About fifty-percent of the major suspension bridges in the United States are over 75 years old (Figure 1). Coupled with an average bridge sufficiency rating of approximately 46 out of 100, suspension bridge owners are challenged with repairing, rehabilitating and preserving these critical assets.

2.2   *Main cable construction*

Suspension bridges typically have two cables in tension transferring the weight of the deck and imposed dead and live loads to the towers and anchorages. The cables are primary load carrying members whose failure will cause the collapse of the bridge. The cables are constructed of high-strength wire with a typical diameter of 5 mm (0.196 in) and generally have a zinc coating for corrosion protection. The wires have a high carbon composition and are manufactured to achieve the high tensile strength required for their application; however, this comes with a trade-off in ductility. US suspension bridge cables include aerially spun cables, prefabricated parallel wire strand cables (PPWS), and helical strand cables. Both aerial spinning and PPWS processes involve erecting thousands of miles of wire in a harsh environment of extreme wind, rain and temperature variation. This exposure to the elements and the mechanical abrasion that occurs due to the erection process inevitably damages the galvanized coating on the wires, which leads to holidays and premature corrosion.

Figure 2.   Helical strand cable (left); aluminum fillers between exterior strands (right).

2.2.1   *Aerially spun cables*
The process of aerial spinning involves drawing two or four loops of wire across the bridge using wheels supported by a hauling tramway. The wires are checked to ensure that all wire tensions are equal by evaluating the sag in the wires. The wires are then compacted and banded into strands which are adjusted to the correct profile within the main cable and ultimately compacted together to form a circular cross-section. Today, the traditional wire-by-wire method of establishing the correct tension in the wires has been replaced by the controlled tension method. Most of the larger US suspension bridges, such as the Golden Gate Bridge, have aerially spun (AS) cables.

2.2.2   *Prefabricated parallel wire strand cables*
Prefabricated parallel wire strand cables consist of shop-fabricated parallel wire strands with a hexagonal cross-section and socketed-ends that are hauled across the bridge in their shop-fabricated bundle supported on guide-rollers mounted on temporary catwalks. This type of cable will typically have fewer crossing wires than aerially spun cables and no wire splices. This is primarily due to the strands being shop fabricated where the process of placing the wires within the strand is more easily managed. The Westbound William Preston Lane Jr. Memorial (Bay) Bridge in the US was the first suspension bridge where PPWS was used to construct the cables.

2.2.3   *Helical strand cables*
Helical strand cables are constructed of parallel strands, each comprising helical layers wrapped around a core wire with each layer typically alternating in the opposite direction. Helical strand cables were primarily used on shorter span bridges and possessed the inherent benefit of self-compacting under tension. This benefit eliminated the need to compact and apply strapping around the wires to keep them together. Once the helical wire strands were placed, the outer spaces (voids) between the helical strands were filled with wood or metal fillers shaped to fit within the voids and provide a circular cross-section to the cable. Several bridges in the United States were constructed using helical strand cables including the Eastbound William Preston Lane Jr. Memorial (Bay) Bridge (Figure 2).

2.2.4   *Main cable corrosion protection*
Main cables may be seen as highly susceptible to corrosion. Depending on the diameter of the cables, there may be tens-of-thousands of individual wires closely compacted together with numerous voids and concealed wire surfaces to trap moisture and promote corrosion. Considering this inherent problem in combination with high-strength wires that can be susceptible to stress corrosion, effective protection of the cables is critical to the safety and reliability of the bridge. To address these concerns, multiple corrosion protection methods have historically been applied in an attempt to protect main cables against corrosion.

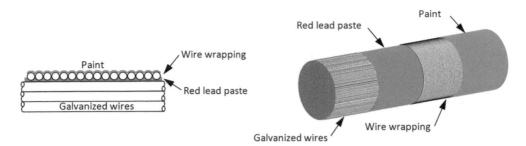

Figure 3.   Traditional main cable corrosion protection system.

The individual high-strength steel wires within the cable are generally hot-dipped galvanized. After the cables are compacted into a circular cross-section, red lead paste was traditionally placed over the outer wires of the cable, followed by soft-annealed galvanized wire wrapping applied under tension, and paint (Figure 3). More recently, the red lead paste has been replaced with zinc paste due to health and environmental concerns. The paint provides a reasonably durable coating to exclude moisture, but it can also trap moisture inside the cable.

Various cable corrosion protection systems have been used on different bridges in the US over time. Bright wire instead of galvanized wire was used on a select few bridges including the Williams-burg Bridge; wire wrapping was not installed on some bridges; and, different exterior coatings or wrapping were used instead of paint, such as neoprene-hypalon wrapping on the Westbound William Preston Lane Jr. Memorial (Bay) Bridge. Regardless of what system was used, water has inevitably found its way into the cables and corrosion has occurred.

2.3   *Main cable corrosion*

The fundamental cause of corrosion in main cable wires is generally understood to be water. If the water is removed, corrosion does not occur. There are many sources contributing to the presence of water within the cable microenvironment. Typical causes may include the entrapment of water during cable construction, water intrusion from cracks in the external paint or wrapping system, and gaps at cable bands or hood seals. Another less obvious source of water is condensate from the variations in temperature and relative humidity inside the cable. Regardless of the source, the presence of water contributes to corrosion mechanisms including zinc depletion and hydrogen embrittlement. If the water or moisture within the cable is not addressed, corrosion in the galvanized wires will eventual occur leading to fracture of the wires.

2.3.1   *The micro-climate within the cable*
Since the main cable is typically painted or wrapped, an enclosed microenvironment is created that exhibits its own micro-climate. With daily and seasonal variations in ambient conditions, the temperature and relative humidity (RH) within the cable varies.

From empirical investigations and findings from internal cable inspections, the presence of water within the cable makes the internal cable environment highly humid. Due to the fluctuation in ambient temperatures, different wet-dry zones can be developed within the cable.

2.3.2   *Zinc depletion*
The water inside the cable will react with atmospheric pollutants such as carbon dioxide and sulfur dioxide from sources such as automobiles to contribute to zinc depletion – the degradation of the galvanized coating on the cable wires. The depletion rate is dependent on the quantity of pollutant and the duration, frequency and quantity of moisture within the cable microenvironment. The corrosion process will be accelerated in areas where holidays or other damage has occurred in the

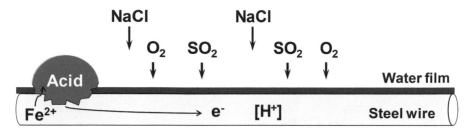

Figure 4.   Corrosion – an electrochemical process.

Figure 5.   Wire fracture due to corrosion.

galvanized coating, often attributed to the installation process or the crossing of wires predominately seen in parallel wire cables – either AS or PPWS.

Once the zinc protection has been depleted, corrosion of the steel wires will occur. The cable microenvironment typically contains the necessary components of water and oxygen in addition to atmospheric pollutants, such as sulfur dioxide and sodium chloride, to promote corrosion through an electrochemical reaction (Figure 4).

2.3.3   *Hydrogen embrittlement and surface irregularities*
After zinc depletion has occurred, other corrosion mechanisms typically follow that involve surface corrosion, pitting, and wire fracture of which hydrogen embrittlement is suspected to contribute. During the corrosion process, a hydrogen rich environment is formed within the cable microenvironment. The availability of hydrogen and its absorption into hardened steel will tend to reduce the ductility of the steel leading to fracture at a lower strain – a phenomenon referred to as hydrogen embrittlement. Research has also pointed to the corrosion process and the resulting surface irregularities on the wires as creating a profile conducive to local stress concentrations contributing to the brittle fracture of the wires (Figure 5).

2.4   *Main cable condition inspections*

With the inevitable intrusion of water into the main cables and its detrimental effects, routine visual and hands-on biennial inspections supplemented with programmed internal cable investigations are warranted. NCHRP Report 534 (TRB, 2004) and FHWA Report No. FHWA-IF-11-045 (FHWA, 2012) offer guidelines on the inspection and analysis of suspension bridge cables using the NCHRP Report 534 or the BTC method.

The inspection of main cables is a difficult and expensive process. Many of the suspension bridges in the United States are located on major transportation facilities with high volumes of traffic making sustained access for internal cable investigations a challenge. Typically, the inspection process requires collaborating with a contractor to construct a work platform or others means of access at multiple panel points along the cable to support the internal cable investigation.

Figure 6.   Internal cable investigation.

The interval for internal cable investigations is based on the findings from the first internal cable investigation, which is recommended to occur at approximately 30 years from when the bridge was built, or sooner depending on the findings from the hands-on biennial inspection of the cable exterior. Based on the age of the bridge and the level of corrosion in the wires, subsequent internal investigations may vary from 5- to 30-year intervals, with more frequent investigations occurring as the age of the bridge and level of deterioration in the wires increases.

The primary method for the internal investigation of suspension cables is a visual inspection supplemented by laboratory testing of specimen wires obtained during the investigation (Figure 6).

For the first internal investigation, the visual inspection of the exterior of the cable and the presence of signs of internal deterioration guide the selection of the cable panels to be unwrapped, wedged open, and visually inspected. If no deterioration is evident from the external inspection, then a minimum of three panel locations on each cable is recommended for internal investigation. Subsequent locations for internal investigation will depend on the findings from the previous investigations; the greater the level of corrosion, the more panels that are recommended for internal investigation. This may result in 4 panels, 6 panels or up to 16–20% of the panels on each cable to be investigated.

To assist in developing a strategy for the location of internal investigations, acoustic monitoring results can be used. Acoustic monitoring is not intended to replace the internal investigation, but the results of the monitoring may assist in identifying panels with potential active wire breaks.

2.5   *Main Cable Preservation Strategies*

2.5.1   *Paint*
In the absence of doing nothing, one option to consider in preserving the cable is the application of a new paint system. Depending on the age of the bridge, an investment in an updated paint system may be a viable alternative. Over time, paint systems have improved from alkyds to moisture cured urethanes to today's water-based elastomeric paints. However, the paint system will require regular maintenance including over-coating. The application of a new paint system does have drawbacks, as the paint may not only minimize water intrusion but also trap existing water in the cable promoting a corrosive environment.

2.5.2   *Oiling*
Cable oiling is an intrusive process requiring the sequential unwrapping and wedging of panels from their high-point to their low-point, while installing specially formulated oil within the cable. The intention is to provide a protective barrier of oil on-and-around each wire to stop corrosion – an objective that may not be attainable. Oiling is a difficult and expensive process requiring sustained access to the cable, as well as re-compacting and re-wrapping the entire length of the cable.

Figure 7.   Bulge and leak in cable wrapping from oiling.

Casualties of oiling include the inevitable leaking of the oil from the cable and likely the necessity to repeat the process in the future (Figure 7). Additionally, future re-oiling may be problematic if the cable becomes obstructed from previous oiling.

2.5.3   *Wrapping*
An elastomeric membrane wrap may be considered in place of the existing paint system or an older wrapping system. Similar to paint, membrane wraps have improved over time and provide a durable UV and ozone resistant wrap. However, just like the elastomeric paint, the membranes will not only assist in keeping water out, but will also trap existing moisture in the cable. It is important to maintain the existing wrapping wire as this provides good resistance to any accidental damage that might breach the membrane over time.

The installation of cable wrapping requires extensive effort. Traveling gantries or main cable access platforms will be required for access. In addition to the installation of the wrapping itself, sealing of the cable bands and hoods will also need to be performed. With the access other cable rehabilitation work may be prudent such as tightening the cable band bolts.

With the capital investment and significant effort involved in the installation of a new protective wrapping, it can be seen as prudent to consider main cable dehumidification. The same elastomeric wrapping that is installed to provide a weather-resistant seal can also provide an air-tight seal for the installation of a dehumidification system. The installation of the wrapping in combination with dehumidification will not only address the immediate protection of the cable, but also address the presence of trapped or future moisture in the cable.

2.5.4   *Dehumidification*
Unlike painting, oiling or wrapping, main cable dehumidification is an active system for cable preservation. Dehumidification addresses the primary issue for cable corrosion by removing moisture from within the cable microenvironment by monitoring and controlling the relative humidity (RH) to a targeted threshold where corrosion ceases. For existing bridges, the RH target can be set low to dry the cable out then adjusted upwards to save on operating costs while maintaining the RH below the threshold to minimize further corrosion.

Main cable dehumidification was first developed for use on a new suspension bridge in Japan. Subsequent to its development, the system has been adapted and successfully implemented on several existing suspension bridges. AECOM has previously designed and implemented main cable dehumidification on three major suspension bridges in the UK (Figure 8) to include the Forth Road, M48 Severn (Cocksedge & Bulmer, 2009), and Humber Bridges (Cocksedge et al, 2011).

Most recently, AECOM has developed the main cable dehumidification system for the William Preston Lane Jr. Memorial (Bay) Bridge in Maryland, the first-of-its-kind project in the

Figure 8.   Forth Road Bridge (left), M48 Severn Bridge (middle) and Humber Bridge (right).

Figure 9.   Chesapeake Bay Bridges (left), Westbound Chesapeake Bay Bridge Cable Access (right).

United States (Figure 9). As the owner's engineer, AECOM developed the design of the cable dehumidification system, developed the contract documents including performance specifications for the work, as well as assisted with construction support services including field support, witnessing of wrapping trials, air trials, and commissioning of the systems. The Westbound Bay Bridge cable dehumidification was commissioned in early 2014 and the Eastbound Bay Bridge is anticipated to be commissioned by mid-2015 (Beabes et al, 2015).

3   APPLICATION OF DEHUMIDIFICATION TO SUSPENSION BRIDGE CABLES

3.1   *Principles of dehumidification*

Protection of steel through the control of humidity is a long-proven technique dating back to the first half of the twentieth century. Extensive investigation work was carried out by Vernon and others producing some very revealing results (Vernon, 1935). Key amongst these was the discovery of a critical humidity below which corrosion of steel did not take place. Figure 10 illustrates some typical results. At very high humidities (99%), corrosion of a specimen was very rapid, but below 60% there was no visible rusting and the gain in weight, as a result of rusting, became nearly zero after 20-days exposure. After 50 days at 60% RH the increase in weight was one-hundredth of the specimen held at 99% RH.

It should be noted that at a given relative humidity the actual amount of water that can be held by air varies with temperature; the higher the temperature, the greater the amount of water that can be retained in air without condensing. The relationship, known as a Psychrometric Chart, is shown in Figure 11. It can be seen that fully saturated air at 38°C (100°F) will hold about 10 times the quantity of water than it is able to at −1°C (30°F). This means that when a cable is being dried, much more water can be removed during the summer than the winter.

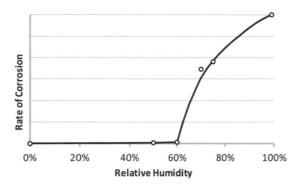

Figure 10.   Variation of corrosion rate with RH.

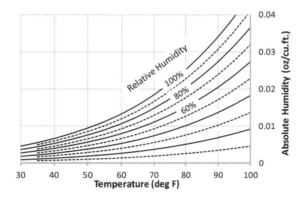

Figure 11.   Psychrometric chart.

3.2   *Parallel wire versus helical strand cables*

Although aerial spun and prefabricated parallel wire strands cables are constructed differently, once the cable is compacted and the wire wrapping is installed, the cable cross-section is nearly the same. The individual wires will generally be parallel to each other with small voids formed due to the circular shape of the wires and the inability to fill all interstitial spaces between one and the other (Figure 12).

Helical strands will generally experience the same geometric phenomenon as parallel wires except the helical strands will be comprised of multiple wires resulting in larger diameter cylindrical wire-bundles. However, the voids between the helical strands will be larger since the larger diameter strands cannot be compacted nearly as close as the smaller diameter wires. There are voids within each strand, but these are neglected. Additionally, depending on the number and size of the helical strands in comparison to the overall diameter of the cable, voids will also occur around the outer circumference of the cable requiring fillers to be used to create a nearly circular cable to receive the wire wrapping (Figure 13).

The typical void ratio for a parallel wire cable is around 20% due to crossing wires and imperfect compaction; however, since helical strands can be better aligned with each other, a void ratio approaching 10% is achieved.

Since main cable dehumidification relies on the voids in the cable for the dry air to be injected and exhausted along the length of the cable, both types of cable are conducive to dehumidification. While helical strand cables have a smaller void area, it is nevertheless easier to blow air along them because the pressure loss is much less, as it is related to the void area divided by the void perimeter.

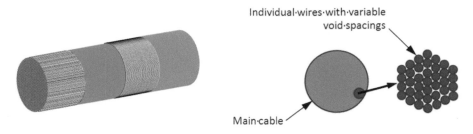

Figure 12.   Parallel wire cable – isometric view (left), cable cross-section (right).

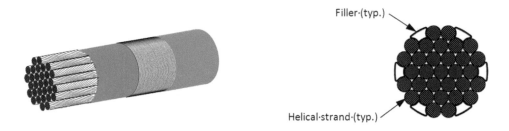

Figure 13.   Helical strand cable – isometric view (left), cable cross section (right).

This means that the layout of injection and exhaust points, injection pressures, and air exchanges within the cable will differ for parallel wire and helical strand cables.

3.3   *Cable sealing*

As discussed above, normal round wrapping wire and conventional paint provide poor air-tightness in the cables, so it is typically necessary to enhance the protection. The correct choice is crucial as it is likely to represent a large cost-component of the whole system due to the requirement to access the entire length of the cables.

3.4   *Flow trials*

There is now a good body of evidence that air can be successfully blown along the inside of suspension bridge cables, and this means that in general, it is not necessary to carry out any flow trials in advance of implementing a cable dehumidification system. However, there may be cases when it is useful to carry out flow trials. One example was for the Philip Murray (South 10th Street) Bridge in Pittsburgh, which has two atypical cable features. The first was the use of a smaller wire to form the cable; instead of the usual 5 mm (0.196 in) wire, it has 4 mm (0.165 in) wire meaning that the gaps between wires for the air to flow are smaller than usual resulting in greater flow resistance. The other feature was the cable had been wrapped in a neoprene wrap as part of an earlier rehabilitation project when the wrapping wire had been removed. It was possible that the neoprene wrap might have been sufficiently airtight to eliminate the need to replace it. Therefore a series of simple air-tests was implemented using a controllable blower mounted at mid main-span (Figure 14) and tubes inserted into the cable bands to provide exhausts. The air-tests proved that even with the smaller gaps between wires, air could successfully pass along the cable. In addition, the air-tests proved that the wrap was very leaky and will need to be replaced. It is worth noting that this is an example of the application of cable dehumidification to a smaller suspension bridge – 275 m–725 m–275 m (900 ft–2,380 ft–900 ft).

Figure 14.   Flow trials on Philip Murray Bridge – temporary blower installed at mid main-span (left); view of main cable (right).

3.5   *System layout*

There are a number of factors to be considered when determining the layout of the system. These include the location and distance between the injection and exhaust points. For a parallel wire cable the maximum blowing length should not exceed about 150 m to 180 m (500 ft to 600 ft) and for a helical cable the blowing length can typically be doubled. The location of the plant rooms will vary from bridge-to-bridge depending on suitable locations and the source of the power supply.

3.6   *M&E plant and control system*

Plant rooms are normally designed to act as plenum chambers with the dehumidifier creating a buffer volume of dried air that is gradually taken by the fans and blown into the cables. Fresh air is continuously drawn into the plant room to make up for the air blown to the injection points.

The dehumidifier and other equipment are typically off-the-shelf components to reduce capital costs, simplify future maintenance, and minimize long-term operating costs.

If possible, there are significant advantages in manufacturing, assembling and testing the complete plant rooms off-site prior to field installation. The off-site assembly will assist in debugging the system versus addressing issues on the bridge, as well as minimizing field time and traffic interruptions during the onsite installation and setting to work of the system.

The injection points on the cables are supplied from centrifugal fans located in the plant rooms. Each fan is fitted with a high efficiency motor and is controlled by an inverter to consistently maintain the required injection pressure.

Two stages of filtration are usually employed; the first being a coarse filter on the intake air and the second being a HEPA filter at the fans. These are essential to prevent dirt and debris from being injected and potentially clogging the voids between the cable wires.

A supervisory control and data acquisition (SCADA) system is used to monitor and control the systems via a dedicated website using industry standard software. A robust monitoring system is important because it indicates whether the system is running efficiently and effectively reducing the humidity inside the cables. Monitoring data also provides a quantifiable record of the drying progress.

4   PROCUREMENT APPROACH TO MAIN CABLE DEHUMIDIFICATION PROJECTS

It is considered that the method of procurement for main cable dehumidification is as important as the design itself. While cable dehumidification installations are built using proven materials, established construction techniques and standard off-the-shelf components, the integration of these

Figure 15.   Procurement process.

components does not lend itself to traditional low bid procurement. Bridge owners should be prepared to adopt a best-value procurement strategy with an emphasis on competitive pricing and long-term system reliability.

A successful procurement begins by selecting an engineer experienced in the design of cable dehumidification systems and the development of procurement documents that have been used to build at least one working dehumidification system. In general terms, the bridge owner's design engineer should have extensive knowledge of suspension system analysis and design supplemented with expertise in mechanical, electrical and control systems.

The owners engineer's scope of work should include the development of procurement documents incorporating both prescribed construction and performance based specifications. Prescriptive designs should be developed for all civil works including materials, details and workmanship standards required for constructing the dry air injection and exhaust system, as well as cable wrapping and sealing to achieve air-tightness. All infrastructure improvements required to upgrade existing power supply systems should also be prescribed, as should any structural work required to support mechanical plant rooms. Performance specifications should be developed for mechanical and control systems themselves. This will permit the contractor to select components and equipment suppliers that they are most familiar with and will ultimately lead to price competitiveness for these elements. Under this scenario, final integration and detailing of these components is performed by a system integrator/engineer hired by the contractor.

Experience has shown that for specialty construction, such as cable dehumidification, the use of best-value contractor selection – as opposed to strict low price bidding – will provide bridge owners with the right combination of price competitiveness and technical capability (Figure 15).

The cornerstone of such an approach as best-value selection is the implementation of a process which will limit the competitive field to contractor teams which can demonstrate competency in the requisite work. This is best achieved through a rigorous prequalification of teams, typically comprised of a general bridge contractor, specialty contractors for mechanical and electrical installations, and a system integrator and/or engineer. Recognizing that the scarcity of cable dehumidification installations will limit the number of contractors with directly relevant experience, prequalification should focus on the core elements of work including: general bridge structural work; cable unwrapping and rewrapping; cable access and work platform design/installation; cable wire splicing; and electrical and mechanical dehumidification system installation. As indicated, contractor teams must also include professional engineers with the experience and qualifications to provide final systems that comply with contractual performance specifications for mechanical and control systems.

The importance of demonstrating technical competence must be carried through to construction and should include contractual provisions requiring the contractor to perform proof-tests utilizing offsite test rigs before advancing into full production. Such demonstrations should be clearly articulated in the procurement documents and should address requirements for achieving specified tension, lapping and bond strength of the wrapping material and air-tightness, as demonstrated through pressure tests (Figure 16).

To facilitate a complete understanding of the procurement documents and to help with necessary trouble-shooting during construction, it is important for the owners engineer to have a full-time field presence throughout construction. The owners engineer should also participate in and be a signatory to final system acceptance along with the contractor and contractor's integrator/engineer.

Bridge owners should also contemplate short- and long-term system maintenance needs and consider how best to incorporate these into the procurement document. Given the reliance upon

Figure 16.   Test rig for wrapping demonstration (left); pressure test on test rig – soap bubbles indicate leaks in wrapping (right).

Table 1.   Relative costs of cable protection options.

Option	Cost	Application	Protection
Paint	$$$	✗ 10–15 years	✗ Low
Oil	$$$$	✓ 20 years	✓ Medium
Wrap	$$	✓✓ 40 years	✗ Low
Dehumidification	$$$	✓✓ 40 years	✓✓ High

Dehumidification is nearly as cost-competitive as wrapping, but offers a much greater level of protection.

standard equipment and control packages, many owners may find it cost-effective to utilize existing in-house staff for routine maintenance functions. Alternatively, owners may find it more appropriate to include long-term maintenance in the procurement document, in effect transferring system life-cycle costs to the contractor. At a minimum, procurement documents should require the contractor to provide short-term maintenance for at least one year beyond the normal warranty period with training and owner hand-off provisions clearly specified and tied to final payment.

5   LIFE CYCLE CONSIDERATIONS

5.1   *Overall life cycle costs*

Major bridges have become a vital part of the nation's infrastructure and represent a very high dollar value investment. If a major bridge becomes so deteriorated that the only option is replacement, then the cost is not only the cost of rebuilding, but also the cost of the disruption and delays during reconstruction. It is therefore incumbent on the bridge owner to provide interventions to prolong the life of the asset for as long as possible. Suspension bridge cables are by their nature very costly to replace, as well as any work on the cables tends to be expensive.

Section 2.5 has discussed the potential alternatives that are available to the bridge owner to provide protection to the cables, and this section examines the economics of each of the alternatives. The costs of the alternative strategies need to be compared with the value of the asset. Table 1 assesses the relative costs of the four options.

While paint is relatively cheap to apply, it has to be re-coated frequently and experience has shown that it is not an effective barrier against moisture intrusion. Oiling a cable is an expensive operation as the whole length needs to be unwrapped and then re-wrapped. While it might provide a good level of protection, it is unlikely that the oil will coat every wire and over time the oil will dry out requiring re-oiling. It also makes future dehumidification much more challenging as it reduces the air pathways through the cable. Wrapping of the cable is moderately expensive as the whole length must be accessed; however, while it may keep water out of the cable it will also trap moisture inside and potentially be very detrimental. The additional cost of dehumidification is small compared with the cost of full-length wrapping, but it provides the best cable protection currently available.

One potential advantage of dehumidification that can be factored in to the overall costs is the frequency and extent of internal cable inspections can be reduced to the extent of just verifying that the dehumidification is performing correctly, particularly when acoustic monitoring is coupled with dehumidification.

5.2   *Maintenance costs*

The potential maintenance liability of dehumidification systems may be of concern to bridge owners. However, a well designed and built system should not pose any serious concerns. The system will be an assembly of standard off-the-shelf components that will require no more maintenance than that of an HVAC system in a medium-size building. Of course, one of the differences is that the components on the bridge are likely to be located in very exposed locations; therefore, measures should be taken in the design to protect sensitive instruments and equipment.

The dehumidification system will require regular servicing and maintenance to ensure the system performs efficiently and effectively. Planned preventative maintenance is carried out to ensure key items such as dehumidifiers, fans, heat pumps, filters, and electrical systems receive the necessary attention. The layout of the plant room requires careful planning during the design-development stage to ensure adequate access is provided around each item to allow for maintenance of serviceable components such as filters and complete disassembly or replacement of main plant components. The cable dehumidification system should comprise standard plant and components which are readily available and can be quickly sourced in the event of failure. This enables a simple system to be adopted which is economical to maintain. Regularly replaced items such as filters should be kept in spare-parts storage so they are immediately available when required.

To ensure the plant continues to maintain the design conditions, it is necessary to check the calibration of sensors and monitoring and control devices as part of the planned preventative maintenance regime. The maintenance required on the majority of the dehumidification control equipment can be carried out by qualified electricians who are familiar with conventional HVAC plant and components.

After commissioning and testing the dehumidification system, the software requires minimal intervention, as it automatically modulates to compensate for any changes in ambient conditions. As the dehumidification software can be accessed via the internet, the system is easy to monitor and any changes required can be undertaken by authorized operators remotely.

Experience on other bridges has led to the conclusion that it is better to minimize the use of copper cable in favor of fiber optics for the control system. The copper cable is susceptible to interference resulting in signal degradation. The fiber optic cables provide increased signal reliability, capacity for future expansion, and easy fault location compared to a long run of copper cable.

As maintenance of cable dehumidification equipment is likely to be a new activity for a bridge owner, a maintenance requirement lasting several years is frequently written into the installation contract. This essentially provides a longer guarantee period and can be in the form of a fixed period of two years with an optional extension of a further two years, or any interval to suit the individual owner's preference. This extended warranty period provides the owner sufficient time to incorporate the running of the system into his day-to-day activities but it also incentivizes the installation contractor to achieve a high quality installation so that future maintenance during the warranty period is minimized. It may suit the installation contractor to sub-let this work to an electrical maintenance specialist. In the UK, several bridge owners use the same maintenance contractor and this has provided excellent continuity of knowledge and successful continuing operation. For bridges in the UK, the annual cost of maintenance contracts is between USD $50,000 to $100,000.

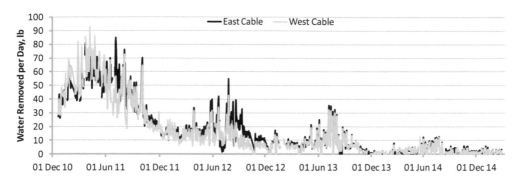

Figure 17.   Daily removal of water from the cables.

5.3   *Operational costs*

Cable dehumidification systems operate 24 hours a day, 7 days per week; therefore, one of the main operating costs is electricity consumption. During the initial drying phase, the cable dehumidification plant will be operating at a relatively high capacity to accelerate the main cable drying process. While the operation of the dehumidification plant remains continuous, changes in ambient condition throughout the year vary the heating, cooling and moisture removal process. Although this causes the electrical demand to vary throughout the year, the plant operation does maintain a steady base load.

After the initial drying process is completed and the relative humidity level in the exhaust is consistently below 40%, the relative humidity of the air supplied into the cable can be increased, and the injection pressure and air flow reduced to conserve electricity. Changes to the air flow rate or humidity should only be carried out incrementally and will require continual review of the exhaust RH levels to avoid a detrimental increase in relative humidity within the cable.

As an example, the cable dehumidification systems installed on the three UK suspension bridges, with spans between 1,000 m to 1,400 m (3,300 ft to 4,600 ft), use between about 600 and 1200 kWh per day. Taking a unit cost of electricity as USD \$0.10, this equates to between USD \$25,000 and \$50,000 per year.

6   EVIDENCE OF EFFECTIVENESS OF DEHUMIDIFICATION

In the UK, internal inspections have been carried out on the Severn Bridge and Forth Road Bridges after the cable dehumidification systems were installed. On the Severn Bridge, two previously inspected panels were re-inspected and it was found that there had been only a small loss in theoretical strength, less than had been projected previously. On the Forth Road Bridge it was found that the factor of safety on the cables has not materially diminished, and that it was expected that the factor of safety of the main cables will not diminish significantly in the future as long as the dehumidification system continues to function effectively.

It is possible to estimate the quantity of water removed from the cables through integration of the air flow and change in humidity as the air passes through the cable. An illustration of this for one of the UK bridges is given in Figure 17. This plots the daily quantity of water removed from each cable. It is apparent that the majority of the initial drying occurred during the first summer and that in each subsequent summer the additional drying has reduced to negligible levels.

A more tangible demonstration of the benefits of dehumidification can be seen in the reduction in the number of wire breaks recorded by the acoustic monitoring system on one of these bridges. The correlation between reduction in cable humidity and the reduction in cumulative wire breaks is clearly apparent in Figure 18. Note that the average cable relative humidity has been simplified.

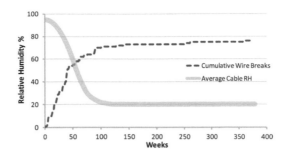

Figure 18.   Reduction in wire breaks as cable humidity reduces.

7   CONCLUSIONS

An analysis of the National Bridge Inventory reveals an aging infrastructure requiring bridge owners to maintain their assets through system preservation projects. For owners of suspension bridges, high on the agenda is the need to proactively maintain the cables. Suspension bridge wire is susceptible to hydrogen embrittlement which places an increased risk and the need to provide very effective protection.

The main driver of deterioration in cables is the presence of moisture which penetrates through gaps in the protective system. Conventional paints have proved to provide insufficient protection and oiling is not the panacea it was hoped it would be.

The principle of dehumidification has been established worldwide, initially for new bridges, but the technology has now been successfully applied to older and deteriorated cables.

The procurement of the contractor to install the dehumidification system is a key part of the process and a low-bid approach is less likely to achieve the quality product required. It is recommended that a best-value contractor selection process is adopted that brings together the right combination of price competitiveness and technical ability of the contractor's team.

Ongoing maintenance of the systems is straightforward provided standard plant and components are used and well designed and accessible plant rooms have been provided. The monitoring system requires the use of high quality exterior grade components, suitably protected, with network cabling based on fiber optics rather than copper cable.

Maintenance and operating costs have been found to be reasonable and good value in comparison with the economic and political value of the major infrastructure assets they are protecting.

REFERENCES

Beabes et al, 2015. Preserving the Suspension System of Maryland's WPL (Bay) Bridge – The First Main Cable Dehumidification Project in the United States, *2015 International Bridge Conference, to be published.*

Cocksedge, C.P.E & Bulmer, M.J, 2009. Extending the life of the main cables of two major UK suspension bridges through dehumidification, *Bridge Structures*, 5: 4, 159–172.

Cocksedge, C.P.E et al, 2011, Humber Bridge Main Cable Dehumidification and Acoustic Monitoring – The World's Largest Retrofitted Systems, *Bridge Structures 7 (2011)* 103–114.

FHWA (Federal Highway Administration), 2012. Primer for the Inspection and Strength Evaluation of Suspension Bridges, Publication No. FHWA-IF-11-045. Washington, DC: FHWA.

TRB (Transportation Research Board), 2004. NCHRP Report 534 Guidelines for Inspection and Strength Evaluation of Suspension Bridge Parallel Wire Cables. Washington, DC: TRB.

Vernon, W.H.J, 1935. A Laboratory Study of the Atmospheric Corrosion of Metals, *Trans. Faraday Soc.* 1935, 31, 1668–1700.

Chapter 2

Chesapeake Bay Bridge dehumidification design

M. Nader, G. Baker, J. Duxbury & C. Choi
T.Y. Lin International, San Francisco, CA, USA

E. Gundel
Kiewit Infrastructure Co., Woodcliff Lake, NJ, USA

A. Tamrat
Maryland Transportation Authority, Baltimore, MD, USA

ABSTRACT: The cable dehumidification system on the Chesapeake Bay (William Preston Lane) Bridges in Maryland, USA, is the first such installation in North America. Historically, the main cables of the suspension bridges were not fully sealed from moisture. By applying waterproof wrapping to the cables and injecting dehumidified air, the cables are dried of retained water, and the relative humidity is reduced to a level that will reduce future corrosion and extend the service life of these structures.

This paper presents theory and practice of cable dehumidification, including principles of cable drying, the calculation of effective air flow within the cable, based on the cables' construction and condition, and the leakage rate of the wrapping system. Maintaining adequate air flow for timely cable dry-out and humidity control is achieved by effective sealing, optimal placement of injection and exhaust points, calibrated instrumentation, and effective mechanical, monitoring and control systems.

1 INTRODUCTION

The cable dehumidification system on the Chesapeake Bay (William Preston Lane) Bridges in Maryland, USA, is the first such installation in North America. In July 2011, the Maryland Transportation Authority, with technical support from AECOM and Ammann & Whitney, released a contract solicitation to design and build this system for the bridges. T.Y. Lin International (TYLI) was retained by Kiewit Infrastructure Co. to design the dehumidification systems for the parallel wire and helical strands of the main cables.

The major suspension bridges consist of two parallel structures, opened in 1952 and 1973, and have main spans of 1600 ft. See Figure 1.

As for many historic suspension bridges, the main cables of the bridge were not fully sealed from moisture intrusion, which resulted in some corrosion. By applying waterproof wrapping to the cables and injecting dehumidified air, the cables are dried of their retained water, and the interior relative humidity (RH) is reduced to a level below 40%, which will reduce future corrosion and extend the service life of these structures. The special conditions of older cables, such as corrosion products and the presence of preservative oil around the wires, inhibit the flow of air, and engineering the system for these conditions requires the consideration of many factors. This paper describes the methodology TYLI used to develop the successful systems that have been installed on the Chesapeake Bay Bridges.

Figure 1.   The Chesapeake Bay Bridges.

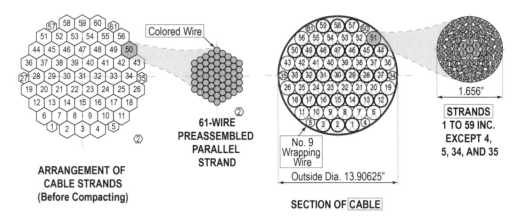

Figure 2.   Parallel wire strands (left) and helical wire strands (right).

2   DEHUMIDIFICATION DESIGN OVERVIEW

The crossing consists of two bridges with two different cable systems. The first bridge, completed in 1952, has a cable composed of 61 helical wire strands. The second bridge, completed in 1973, is one the first bridges with cables composed of parallel wire strands. See Figure 2. Due to the different distribution of voids in the two cables, two different dehumidification systems were developed.

Each dehumidification system consists of injection and exhaust sleeves, and a cable sealing system, consisting of the cable wrapping, cable bands, anchorages and cable saddles, as well as the mechanical, electrical, operation and control designs of the dehumidification units based on project demands.

See Figure 3.

Note that at each injection sleeve the air flows into the cable in both directions from the sleeve, and each length to the nearest exhaust sleeve is referred to as the "flow length".

The designs included the design and layout of the dehumidification plants (for the main cables and anchorages); design of the cable band injection sleeves; design of the cable band exhaust sleeves, along with the layout of all ductwork and other miscellaneous support and anchorage

Figure 3.   Injection sleeve and cable wrapping.

Figure 4.   Mock-up for cable band sealing and cable wrapping.

details. A series of laboratory and in situ mock-ups were performed to verify the performance of the system. Figure 4 show the leakage mock-up. The overall project was multi-faceted, and included the rewrapping, dehumidification and installation of security shielding at the main cables and anchorages of the Eastbound and Westbound Bay Bridges.

Every suspension bridge cable is unique, due to the conditions of erection, compaction, wrapping, painting, and the possibility of corrosion products. Each of these variables will affect the resistance of the cable to air flow, the optimal spacing of injection points, and the system requirements for dehumidification and air flow capacity.

3   CABLE AIR FLOW

The starting point in the dehumidification design is to determine the properties of the cable voids and to determine the amount of flow that the cable can accommodate. The hypalon wrapping system

(Kaczinski, 1999) retains a working pressure in the order of 3000 Pa, and the pressure drops along the cable length, according to the Darcy-Weisbach equation:

$$\Delta p = f_D \frac{L}{D_H} \frac{\rho V^2}{2} \tag{1}$$

where L = flow length inside the cable; V = air flow velocity inside the cable; f_D = friction factor; and $D_H = 4\frac{A}{D}$ = the hydraulic diameter

For laminar flow with a Reynolds number less than 2000, $f_D = \frac{64}{Re}$, and

$$\text{Re} = \frac{v}{D_H} V$$

where v = kinematic viscosity, μ/ρ.

If A is the void area of the cable cross-section and Q is the flow rate, then the air velocity may be expressed as

$$V = \frac{Q}{A}$$

Combining these four equations, the following relationship results:

$$\Delta p = \frac{32\mu}{D_H{}^2} \times \frac{L}{A} \times Q = G \times \frac{L}{A} \times Q, \text{where } G = 32\frac{\mu}{D_H{}^2} \tag{2}$$

This means that for a given cable the head loss between injection and exhaust sleeves, neglecting any leakage, is proportional to the flow rate, Q.

All cable sealing systems have some leakage, and the flow rate will vary along the cable length. For a well-constructed wrapping system the variation in flow due to leakage may be represented as:

$$Q(x) = Q_0 e^{-kx} \tag{3}$$

Where Q_0 is the flow rate at the injection point, x is the distance along the cable and k is the rate of leakage, for example .004 per meter. The pressure is then expressed as:

$$\frac{dp}{dx} = -\frac{G}{A} Q_0 e^{-kx} \tag{4}$$

Letting L represent the length of cable between the injection and exhaust points, integrating the above expression and setting the boundary condition that $p(L) = 0$, the pressure function becomes

$$p(x) = \frac{G}{A}\frac{Q_0}{k}[e^{-kx} - e^{-kL}] \tag{5}$$

Note that the initial pressure P_0 is generally chosen by design, based on the pressure capacity of the cable wrapping system. Setting this as a boundary condition when $x = 0$, the initial value of Q_0 may be solved as

$$Q_0 = \frac{P_0 A k}{G[1 - e^{-kL}]} \tag{6}$$

And the pressure becomes

$$p(x) = \frac{P_0[e^{-kx} - e^{-kL}]}{[1 - e^{-kL}]} \tag{7}$$

Note that for small values of kL,

$$Q_0 \cong \frac{P_0 A}{GL} \text{ and} \tag{8}$$

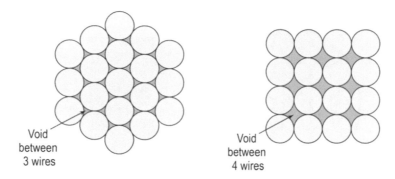

Figure 5.   Wire packing.

$$p(x) \cong \frac{P_0\,(L-x)}{L} \tag{9}$$

Equations 8 and 9 lead to the following conclusions. Neglecting the small head losses at the injection and exhaust sleeves, the distribution of pressure along the cable is primarily a linear function of the flow length between the injection and exhaust points. Regardless of the condition of the cable itself, for a very good wrapping system the pressure varies almost linearly from the initial pressure P_0 to zero at the exhaust point. *This distribution of pressure is not a function of the cable's resistance to flow.*

The flow, however, is inversely proportional to the flow length, even at the injection point. This means that for a given cable in a given condition the choice of the placement of the injection and exhaust sleeves determines how much air flow will eventually be able to enter the cable. Note that the flow rate is not a function of the climatic conditions of temperature or humidity.

Through the factor G (see Equation 2), the flow rate Q_0 is also proportional to the square of the hydraulic diameter, which is defined as four times the ratio of the flow area and the perimeter of the area.

Figure 5 shows examples of voids for the closest wire packing, where the void is bounded by only three wires, as well as one with a less dense packing, bounded by four wires.

In the three-wire case, for wires of 5.03 diameter, such as are used in the westbound Chesapeake Bay Bridge, D_H has a value of 0.52, while the four-wire case has a value of 1.37. It is also possible to compute the hydraulic diameters of voids surrounded by five or six wires. Given that the flow varies as the square of this diameter, the flow through the four-wire void is seven times that through the three-wire void. Note that this increase is in the velocity of flow, and not simply due to the greater area of the individual void.

To further illustrate this,

Figure 6 shows the layout of the 42 mm diameter strands of the eastbound Chesapeake Bay Bridge. While the strands of the westbound bridge are compacted together, the strands are helical and retain their identity. There are voids between the strands as well as between individual wires. While the smaller voids are similar in nature to the wire voids on the westbound bridge, the strand voids of the eastbound bridge have a hydraulic diameter of 4.3 mm, eight times the size of the smallest wire voids. This means that flow rates 64 times higher are possible in such a cable.

In the actual cable the distribution of voids will be random, and the total void ratio can determined from the compacted cable diameter and the total wire area. Based on the actual void ratio is possible to infer the average void size between wires, and thus the average flow velocity in the cable. For the westbound Chesapeake Bay Bridge the average hydraulic diameter predicted by the overall cable void ratio and the total wire perimeter would be 1.3 mm.

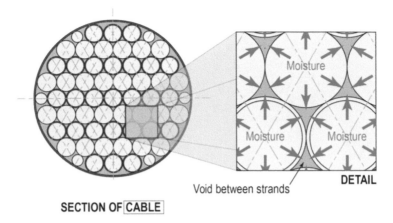

SECTION OF CABLE

Figure 6.   Eastbound bridge strand arrangement, showing moisture moving from voids between wires to voids between strands.

Ultimately, the real flow rate is verified by in situ measurements, either in a pre-construction mock-up, or by measuring flow rates in the completed installation. For a cable that has experienced wire corrosion, or whose wires have been treated with oil coatings or oil injection, the total void area is reduced and the hydraulic diameter of individual voids may be reduced by corrosion products or oil. In such cases both the flow area and flow velocity will be reduced, and the impact to flow capacity may be significant. In situ flow testing is advisable prior to final design of the layout of injection sleeves.

For the westbound Chesapeake Bay Bridge, in situ flow measurements of the westbound bridge vary with time and location. The highest flow rate averages about $0.2\,\mathrm{m^3/min}$ at an injection sleeve serving two flow lengths of 115 m in length. This means a flow of $0.1\,\mathrm{m^3/min}$ over a length of 115 m with a total head loss of 2900 Pa.

Equation 8 may be solved to determine the effective hydraulic diameter, considering:

$A = 0.01793\,\mathrm{m^2}$
$P_0 = 2900\,\mathrm{Pa}$
$Q_0 = 0.1\,\mathrm{m^3/min}$
$L = 115\,\mathrm{m}$, therefore $G = 260\,\mathrm{kg/m^3/sec}$ from equation 8

For air $\mu = 1.83 \times 10^{-5}\,\mathrm{kg/m/sec}$, therefore $D_H = 1.5\,\mathrm{mm}$, since $G = 32\frac{\mu}{D_H^2}$.

At another injection sleeve the flow is 0.085 m3/min, serving two flow lengths of 125 m. By a similar calculation the effective hydraulic diameter $D_H = 1.0\,\mathrm{mm}$.

The variation of flow rates results from varying conditions of the cable interior. For parallel wire cables such as used in the westbound Chesapeake Bay Bridge, the void dimensions are small and the hydraulic diameter is easily affected by the degree of cable compaction and by any corrosion products.

In contrast, for the eastbound bridge the cable's flow resistance was very low and the maximum design flow rates were achieved at a reduced pressure.

The placement of injection sleeves has sometimes been done based on judgment and experience, and flow measurements are not always correlated to the selected flow length. Head loss is reported in terms of Pa/m, and a value of 10 Pa/m has been suggested (Bloomstine, 2011) as a typical value for flow resistance of parallel wire cables. This value has been cited for information in some Requests for Proposal for cable dehumidification. As noted above, the pressure drop is simply a function of the chosen length between injection and exhaust points, and is not related to adequate air flow rates or effective dehumidification.

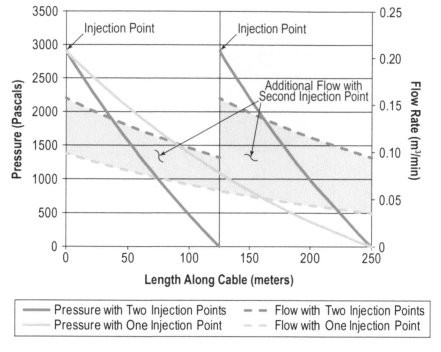

Figure 7.   Comparison of total flow with two injection points vs. one injection point along the same cable length.

Figure 7 illustrates the importance of flow length in cable dehumidification, considering a realistic cable with a hydraulic radius of 1.6, a leakage rate of 0.4% per meter, and applying Equations 5, Equation 6, and Equation 3 above. Two cases are considered. One case assumes a distance of 125 m between injection and exhaust sleeves, and the second case assumes a distance of 250 m. For simplicity the first case is shown twice, to indicate that twice as many injection sleeves would be required. Note that the injected pressure of 2900 Pa is the same in both cases and the pressure is zero at each exhaust point. It is a physical requirement that the pressure reduce to zero over the flow length, since the air exhausts to the atmosphere.

Note that the rate of pressure drop is 23 Pa/meter in the first case and 12 Pa/m in the second case. The rate of flow, however, is significantly higher in the first case than in the second. This result is contrary to the description of pressure drop per meter as a measure of flow resistance. This effect is apparent in the original Darcy-Weisbach equation. If Δp and L are fixed by the installed geometry, then whatever the resistance of the cable, the flow velocity, V, must reduce in response to the level of resistance.

We propose the use of the *effective hydraulic diameter, $D_{H,eff}$, as an objective means of defining the actual flow resistance of a bridge cable*. This parameter may be field-measured over any test length, based on Equation 2 above, and may thereafter be used for the engineering design of air flow as part of the cable dehumidification strategy. It may also be used to compare measurements made on different cables.

As illustrated in Figure 7, for very long flow lengths or cables resistant to flow, the flow will simply reduce to the level consistent with the installed pressure drop. The highlighted area is the increase in total flow that is gained by dividing the flow length into two parts. In this case the flow is doubled using two injection sleeves. The importance of total flow is that it determines the cable dry-out period.

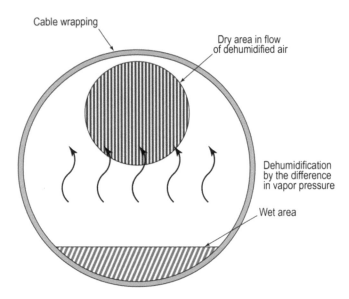

Figure 8.   Dehumidification by the difference in vapor pressure within a cable section.

4   CABLE DRY-OUT

When water is present in a sealed cable, the vapor pressure in the air within the cable equals the vapor pressure of the liquid water. When dry air is injected into the cable there arises a difference in vapor pressure, causing the evaporation of the water in the cable. See Figure 8.

The basic relationship is expressed as:

$$W_e = \frac{H \times A \,(VP_s - VP_a)}{H_L} \tag{10}$$

where W_e = rate of evaporation; H = latent heat transfer rate; A = total wetted surface area; VP_s = vapor pressure of saturated air at the water's temperature; VP_a = vapor pressure of the air over the water; and H_L = latent heat of vaporization.

This equation may be used to estimate an upper limit on the rate of evaporation of water in the cable, or in open areas such as the anchorage. It may be used to compute the maintenance dehumidification performance once the cable is effectively dry.

For parallel wire stands the void space inside the cable is about 20% of the total cable volume and the water contained in the cable is typically estimated to be 5% of the void volume. For the Chesapeake Bay Bridges this results in about 900 gm of water in each meter of length. Air at 20°C has a saturated moisture capacity of about 15 grams of water per kg of air and a cubic meter of air has a density of 1.17 kg. This equates to 17.5 grams of water per cubic meter. Air with an RH of 20% can absorb 80% of this amount or 14 g/m3.

Due to the long length of the cable with respect to its cross-section, the dry air entering at the injection point will soon become saturated as it contacts the water inside the cable and Equation 10 does not apply. When the system is first switched on, the first meter of cable, with 900 grams of water, will require about 64 m3 of dry air to remove the water. Noting peak flow rates in Figure 7 above, which are 0.15 and 0.09 m³ /min., the first meter of cable would take 7 hours and 12 hours respectively to dry. Noting that the flow rate reduces over the length due to leakage, the time to dry each subsequent meter of cable is a little longer.

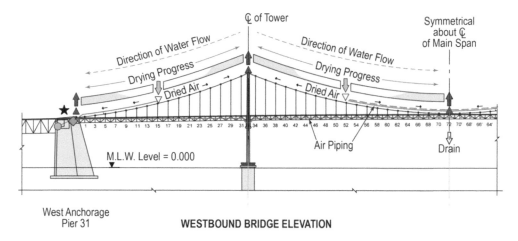

Figure 9.   The drying process.

If W is the weight of water per meter and m is the moisture capacity of the dehumidification air, then the total drying time can be computed as:

$$T = \int_0^L \frac{W}{Q(x)m}\,dx \tag{11a}$$

or,

$$T = \frac{W}{k\,Q_0 m}\left(e^{kL} - 1\right) \tag{11b}$$

given Equation (3).

For the example given in Figure 7, the dry-out time for two lengths of 125 m is about 50 days, while for the 250 m length it is about 210 days, more than four times longer. Not only is the flow rate much higher in the two shorter segments, but the drying takes place simultaneously in both segments. In contrast the last half of the 250 m length does not start to dry until the first half is completely dry.

Figure 9 below illustrates how the drying process proceeds from the injection point to the exhaust point. The air injected into the cable becomes saturated with moisture prior to reaching the exhaust point. At the exhaust point no drying takes place until the entire flow length is dry. This has been demonstrated in field measurement of installed systems (Yanaka et al. 2002 and Cocksedge & Bulmer, 2009). Being exposed to the most saturated air and the lowest rate of flow, the cable wires at the exhaust point will always be the last to become dried after the system is initiated. The selection of too great a length between injection and exhaust sleeves may result in a dry-out period measured in years. For this reason it is important to consider the placement of the exhaust points, to ensure that the longer drying times are acceptable in those locations.

Figure 10 shows a commonly used arrangement, in which the exhaust of the cable is vented into the anchorage splay chamber. This is a practical solution, since it passes dry air through the splay saddle. However, as noted above, the benefit to the splay saddle is postponed until the entire cable is dry. An undesirable consequence is that the relative humidity in the splay chamber becomes 100% during the dry-out period.

This condition would likely be worse than the previous state without dehumidification. One solution is to place an injection point close to the cable entry point to the anchorage, so as to avoid a long period when moist air enters the anchorage. Another solution is that used on the Chesapeake Bay Bridges. The supplemental dehumidification unit that dries the strand anchorage chamber has

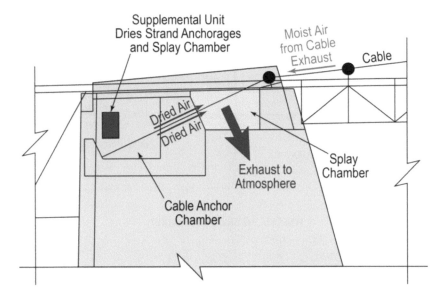

Figure 10.   Commonly used arrangement of exhaust point at anchorage splay chamber.

sufficient capacity that dry air flows upward through the rod bores and enters the splay chamber, clearing the moist exhaust from the cable.

5   OPTIMAL SLEEVE PLACEMENT

In deciding the number and placement of injection sleeves, it is important to consider the desired dry-out period, as well as the cost-benefit of adding sleeves.

The desiccant-wheel dehumidifiers that have been provided for the Chesapeake Bay Bridges can provide a high volume of dry air at 3% relative humidity (RH). The high capacity of the units is used to provide a performance safety factor and a very low RH, so as to minimize the drying-out period that follows inauguration of the system, a short drying-out period being desirable to get the maximum protective benefit for the initial investment in cost. The plant room containing the unit is air-tight and provides a large buffer of dry air. This allows the dehumidifiers to work intermittently at their most efficient level of drying. The very dry output of the dehumidifiers is mixed with ambient air to produce an air supply at the effective RH for injecting into the cables.

The seasons of the mid-Atlantic coast have a wide range of temperature and humidity. The plant room is climate-controlled using heat pumps, and the system is seasonally set for different thermodynamic cycles, so that the drying is effective throughout the year. The dehumidifier works harder in the winter, supported by the heat pumps, to provide air dry enough to remove moisture from the cable in cold conditions, when condensation is most likely. After the drying-out period the whole system will be placed into a maintenance mode, providing a slightly higher RH air supply that is still sufficient to remove any moisture that may over time permeate the sealing measures of the cable. Operating the system below its maximum capacity will reduce long-term running costs, and prolong the life of the equipment, which will result in better value for the Owners of the bridges.

Figure 11 shows the relative capacities of the dehumidifier and the injection sleeves used on the Chesapeake Bay Bridge. The dehumidification units have a reserve capacity, and the cost benefits may favor the use of more rather than fewer injection points. As shown above, the speed of the drying process is effectively proportional to the square of the number of injection sleeves. The longer term

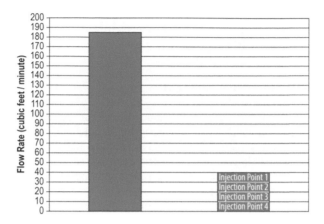

Figure 11. Capacity of the dehumidifier.

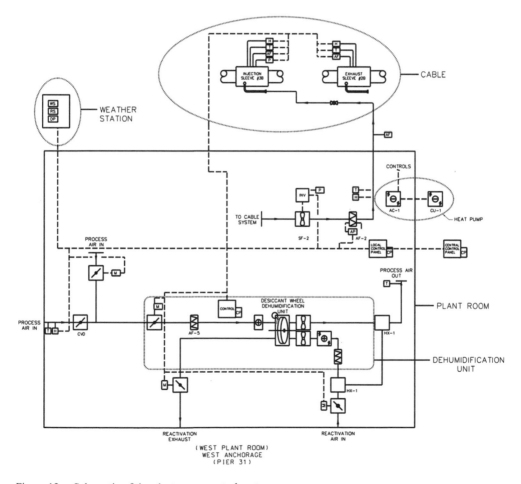

Figure 12. Schematic of the plant room control system.

savings are based not on the initial size of the dehumidification unit, or the number of injection and exhaust sleeves, which have relatively small incremental costs, but on the maintenance operation, using the minimal air flow and dehumidification function that will keep the cable dry.

6   PLANT ROOM AND CONTROL SYSTEM

On the Chesapeake Bay Bridge, the dehumidification plants, heat exchangers, blowers, filters, and other equipment were housed within a carefully designed Plant Room, custom-made to fit the site conditions. For ease of transportation, handling and installation, the plant rooms were designed as modular elements that were delivered in sections and maneuvered into the bridge structure. Each plant room contains a control system to monitor the conditions at each cable injection and exhaust point, and regulate its output accordingly.

Figure 12 shows a schematic of the plant room control system.

7   CONCLUSIONS

This paper presents theory and practice of cable dehumidification, including principles of cable drying, the calculation of effective air flow within the cable, based on the cables' construction and condition, and the leakage rate of the wrapping system. Maintaining adequate air flow for timely cable dry-out and humidity control is achieved by effective sealing, optimal placement of injection and exhaust points, calibrated instrumentation, and effective mechanical, monitoring and control systems.

The dehumidification system on the westbound span is installed and operational. The eastbound span is expected to be completed by the end of 2015. The Chesapeake Bay Bridge cable dehumidification project represents a significant advancement in the areas of bridge rehabilitation and preservation.

REFERENCES

Bloomstine, M. L. 2011. Main Cable Corrosion Protection by Dehumidification – Experience, Optimization and New Development, *Modern Techniques in Bridge Engineering: Proceedings of 6th New York Bridge Conference*: 39–54.

Cocksedge, C. & Bulmer, M.J. 2009. Extending the life of the main cables of two major UK suspension bridges through dehumidification, *Bridge Structures: Assessment, Design and Construction*, 5:4: 159–172.

Kaczinski, M. 1999, Corrosion Protection of Suspension Bridge Cables, *International Conference on Suspension, Cable Supported, and Cable Stayed Bridges: November* 19–21.

Yanaka, Y., Kawakami, Y., Hasegawa, Y., Yamaguchi, K. 2002. Dry-Air Injection System for Existing Suspension Bridge Cable. *The 3rd International Suspension Bridge Operators' Conference.* Awaji Island, Japan.

Chapter 3

Fracture analysis of steel cable-stayed bridges

B. Soule & B. Tappen
International Bridge Technologies, San Diego, CA, USA

ABSTRACT: Many cable-stayed bridges include a deck composed of a steel grillage made composite with concrete panels. These deck systems are subject to requirements for fracture critical or structurally redundant members, sometimes including FHWA guidelines. However, this type of bridge includes inherent redundancy, as there are multiple load paths to redistribute the effects of member failure. Given the redundant nature of these bridges, a blanket designation of FCM (or SRM) may be a conservative approach that affects the economy of the structure.

The example of the East End Bridge Crossing, currently under construction near Louisville, KY is used to demonstrate an approach to determining where FCM fabrication is necessary. The development of performance criteria suitable to the site is discussed, as well as the analyses that were performed to evaluate the structure after a member rupture against those criteria. Suggested general guidelines appropriate for assessing this type of construction are described.

1 INTRODUCTION

1.1 *Fracture and highway bridges*

The potential for a sudden fracture of steel members has long been recognized in the design of bridges. For Highway structures, the potential for fracture and the mitigation measures are described in the AASHTO LRFD, primarily in Section 6.6.2 (AASHTO 2012). This section defines the idea of a fracture critical member, commonly known as an FCM. An FCM is a steel member whose fracture will result in a loss of stability of the structure.

It is the responsibility of the Engineer to designate all FCMs on the plans. An FCM, once identified, will be held to higher standards of procurement and fabrication. This primarily is in the form of more stringent Charpy V-notch requirements, as well as fabrication under a fracture control plan conforming to the requirements of the AASHTO/AWS D1.5 Bridge Welding Code. These additional requirements create a cost premium for steel designated as FCM, so it is prudent to use the designation only when necessary.

Within this section of the AASHTO LRFD, the Engineer is allowed to perform a "rigorous analysis with assumed hypothetical cracked components" to confirm the strength and stability of the fractured structure. This is one potential avenue under this code to demonstrate that a particular member is or is not an FCM. However, there is not a well-established set of guidelines within the code that would address this rigorous analysis. This is recognized within the Commentary, which includes the following text:

"The criteria for a refined analysis used to demonstrate that part of a structure is not fracture-critical has not yet been codified. Therefore, the loading cases to be studied, location of potential cracks, degree to which the dynamic effects associated with a fracture are included in the analysis, and fineness of models and choice of element type should all be agreed upon by the Owner and Engineer. Relief from the full factored loads associated with the Strength 1 Load Combination of Table 3.4.1-1 should be considered, as should the number of loaded design lanes versus the number of striped traffic lanes."

It is clear from this Commentary that a fracture analysis of a complex bridge will inherently be tailored to the needs and requirements of a particular bridge Owner, and further that close collaboration between Owner and Engineer is necessary to define the required analysis.

1.2   *FHWA guidance*

In June of 2012, the Federal Highway Administration (FHWA 2012) issued a Memorandum with the subject "Clarification of Requirements for Fracture Critical Members". This Memorandum discusses the position of the FHWA as it pertains to the fabrication and inspection of FCMs. In particular, it offers guidelines for the classification of various members. The classifications are based on identification of the level of redundancy of the element, either by traditional simplified methods or by a more rigorous analysis. In general, the Memorandum encourages broader use of FCM fabrication, indicating that elements identified as non-FCM via a rigorous analysis should be fabricated as FCM nonetheless.

1.3   *FCM versus SRM*

The FHWA Memorandum also introduces the concept of a System Redundant Member (SRM). An SRM is a member that is fabricated according to a fracture control plan as if it was an FCM, but that need not be classified as FCM for inspection purposes. For fabrication, construction and estimation, the two are identical. There may be a long-term cost reduction to an Owner or Concessionaire, however, by reducing the inspection frequency of a given member.

Classification of a member as SRM as opposed to FCM may be achieved through a rigorous analysis. This solves the apparent contradiction of performing a rigorous analysis to demonstrate the stability of the structure after member failure, only to fabricate the member as FCM – the savings to the Owner is during the life of the structure, not during construction.

2   REDUNDANCY

2.1   *FHWA language and application to cable-stayed bridges*

Any discussion about fracture critical members is essentially a discussion about redundancy. A structure with many alternate load paths can more easily withstand the fracture of any one member. The FHWA Memorandum defines three types of redundancy. First is Load Path Redundancy. This is the traditional understanding of redundancy where there are multiple load paths between supports capable of carrying the required loads. The simplest example is a bridge with multiple longitudinal girders. The second is Structural Redundancy, which may take advantage of load re-distribution in three dimensions, and requires refined analysis to demonstrate. The last type is Internal Member Redundancy. An example of this type is a bolted built-up beam, where fracture in a flange will not propagate into the web. This type of redundancy is not considered further.

The recommendations for the first two categories of redundancy are that Load Path Redundant structures need not be FCM (or SRM), whereas those members that are shown to have Structural Redundancy should be classified as SRM at a minimum. Cable-stayed bridges, however, can be difficult to categorize under this system. It is widely recognized that typical cable-stayed bridges exhibit a high degree of redundancy. This stems from the closely spaced stay cables, each of which is designed to withstand the failure of an adjacent cable. In light of that, there are multiple load paths between supports, and an explicit design for redundancy. However, extending this redundancy to the surrounding deck members requires additional analysis, suggesting Structural Redundancy. This ambiguity is best addressed by an informed discussion of the relevant issues between the Owner, Engineer and other stakeholders.

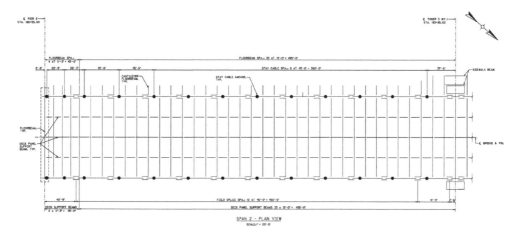

Figure 1. Typical steel grillage.

2.2 *Typical cable-stayed grillage arrangement*

While there are many potential configurations for the deck of a cable-stayed bridge, this paper addresses the most common configuration for short- and medium-range bridges with center spans varying from 700 to 1,500 feet. The concepts presented here could be extended to other structure types.

Most steel cable-stayed bridges in this range have a deck composed of a steel grillage that carries a concrete deck. The steel grillage consists of two or more edge girders that run longitudinally along the entire length of the bridge. These edge girders are supported at regular intervals – generally 30–50 feet – by the stay cables. Transverse floor beams are installed between the edge girders at 10–15 foot intervals to carry the loads transversely to the edge girders. The girders are then made composite with a concrete deck, often composed of precast panels and cast-in-place closure stitches. A typical partial framing plan is shown in Figure 1.

There are many variations on the details of this framing plan, but the essential load path remains the same. Wheel loads are carried through the deck to the nearest floor beam, then transversely to the edge girders, longitudinally along the edge girders and finally through the stay cables into the pylons.

2.3 *Compression*

Cable-stayed bridges have a unique feature that must be considered in any discussion of their redundancy. This is the high compression stresses present in the deck, with peak values in the vicinity of the pylons. This compression is a fundamental consequence of the load-carrying system of any cable-stayed bridge, as can be seen from the schematic in Figure 2.

As the size and inclination of the stays increases, so does the cumulative compression carried by the deck. This compression is necessary for the overall stability of the structure, so any analysis of potentially fractured members needs to evaluate the effect of the fracture on the ability of the deck to carry compression. At a minimum, this should include an accurate representation of the deck compression at the time of fracture and a second-order analysis that will capture the effect of that compression on the distorted structure.

2.4 *Design for stay loss*

Most cable-stayed bridges are designed for cable loss events. For bridges that are designed under the (PTI Guide Specifications 2006), this includes the sudden dynamic loss of any one cable, as

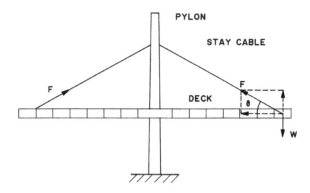

Figure 2.   Deck compression.

well as the planned replacement of each cable. Some projects include additional project specific requirements for additional or more stringent scenarios, but a cable loss event is a standard design consideration. This is an issue that may factor into an analysis of deck fracture, but must be understood in the proper context.

Design requirements for cable loss events clearly speak to the overall redundancy of a cable-stayed bridge. Assuming that a bridge is designed to withstand the loss of any one cable, it is reasonable to conclude that a degree of load path redundancy does exist between the pylons. That is, an inherent feature of the design is that there are multiple load paths between major points of supports, and that independent of any FCM considerations, the structure has been shown to demonstrate the accuracy of that assumption.

This is not sufficient, however, to determine that any or all steel members are not fracture critical. Loss of a cable is not directly equivalent to the fracture of a steel member, and design for cable loss should not automatically relieve a structure of FCM considerations. As cable-stayed bridges are complex structures, additional assessment of individual members or categories of members is necessary, and can lead to a more refined analysis.

3   GENERAL ANALYTICAL APPROACH

3.1   *Establish a criteria*

It is clear from the requirements of the governing codes that a fracture analysis should be undertaken in close collaboration with the Owner. Given that there is not a detailed procedure put into place by the AASHTO LRFD, it is necessary to create a procedure on a project-specific basis. The first decision to be made is whether or not a fracture analysis is appropriate for a particular structure. Several factors may influence this decision, including the expected degree and type of redundancy, the role of the structure, and the cost-benefit profile of reducing the amount of FCM fabrication.

Assuming that a fracture analysis is selected, the next steps are to establish the particular performance requirements that should be met. The overall goal is to establish an analytical framework that will demonstrate that the fracture of a particular element will result in behavior that is within defined performance limits, thereby demonstrating that FCM fabrication is unnecessary. The critical aspects are well summarized by the LRFD commentary quoted above. The key topics to be addressed are:

3.1.1   *Loading*
As described in the AASHTO LRFD commentary, it may be reasonable to consider load cases that vary from the Strength I load case used to evaluate the nominal structure. This idea is consistent with the approach taken in the PTI Guide Specifications for stay loss analysis, which defines load

factors that are reduced from the Strength I load combination. The load case defined in PTI may be considered as a consistent baseline load case for assessing fracture, thereby giving similar levels of reliability for stay loss and member fracture. Structure-specific loadings, such as unique permit or rail loadings should also be considered.

3.1.2 *Locations*

The locations of potential fractures should also be determined, as well as an understanding of what families of elements the locations represent. For example, fracture analysis at two or three edge girder locations may be adequate to characterize "typical" edge girders, but not those at bearings locations, or those that connect to multiple stay anchors. The types of assumed failure and related loading patterns should be described in detail. These will generally correspond to peak effects, such as maximum positive bending at the center of a span, or maximum shear near a support.

3.1.3 *Modeling*

The type of modeling to be performed, specific modeling requirements and model refinement should be defined. As much of the focus is on overall redundancy of the structure, it is likely that a complete model of the bridge is desirable. The modeling should include a full representation of the loaded conditions at the time of assumed fracture, as second-order effects stemming from the total applied load may be important. An accurate representation of the actual dead load distribution is critical.

Other key requirements include the consideration of dynamic effects, either by direct analysis or conservative load factors, the type and magnitude of structural damping, and the extent of geometric and material non-linearity to be considered.

3.1.4 *Performance*

A performance-based criteria should be established for evaluating the response of the fractured structure. The category of the loading combination should be clearly established (i.e., Strength or Extreme), which provides guidance for application of AASHTO capacity provisions. The performance criteria may also include the limits of any localized failure, such as strain limits in steel or concrete, and load-carrying or repair requirements.

3.2 *Identify inherently FCM elements*

A careful review of the structural steel elements on a bridge may identify particular places that are inherently non-redundant. These elements may be determined to be FCM purely on the basis of the structural role that they play, or on the basis of the complexity of assessing the reaction of the structure to their fracture. Examples include tension tie-downs, elements that house multiple stay cable anchors, and the direct load path for bearings or restrainers. These specialized elements may represent locations where the redundancy is unclear, and potential cost savings are minor compared to their overall importance.

3.3 *Evaluate potentially redundant members*

This last step is the responsibility of the Engineer. The analyses in the criteria should be carried out deliberately, with close attention paid to details that may influence the results. Examples may include the assumed damping of the structure or the locations of the failures. The Engineer should also carefully evaluate the analytical results for consistency with the modeling assumptions. For example, if members exceed their elastic limits, it may be necessary to introduce that plasticity into a revised model, complete with an assessment of the local behavior and plastic capacity.

4   EAST END BRIDGE

4.1   *Bridge basics*

The general approach described above was put into practice in the design of the East End Bridge, currently under construction near Louisville, KY. This project is being procured by the Indiana Finance Authority (IFA) as part of the larger Ohio River Bridges project. It is being constructed by Walsh Vinci Construction (WVC), and was designed by Jacobs Engineering. International Bridge Technologies (IBT) performed design services as a sub-consultant to Jacobs Engineering.

The cable-stayed portion of the project consists of a 1200′ main span and two 540′ backspans for a total length of 2280′. An elevation view of the project appears in Figure 3. The bridge carries a total of six design traffic lanes, two 12′ shoulders and an 11′ pedestrian walkway, for a total width of 124′-2″ as shown in Figure 4.

There are two planes of stay cables, anchored into edge girders spaced 107′ apart. Stay cables are anchored every 45′, with closer spacing of approximately 25′ in the vicinity of the end piers. The deck is connected to the end piers via a tie-down, which carries only axial load. The region is ballasted to keep the tie-down under compression, though it will go into tension under variable loads. The stays are anchored to the edge girders above the deck. Transverse restraints are provided at both pylons and end piers, with the only longitudinal restraint at one of the pylons. A rendering of the completed structure appears in Figure 5.

The IFA and the design team jointly agreed upon a set of criteria for analyzing and assessing the bridge for fracture. The remainder of this paper discusses the particulars of that approach, the analytical procedures implemented, and the results of that analysis as they apply to the bridge deck. The bridge pylons also included steel stay anchor boxes that were subject to FCM assessment, but those members are outside the scope of this paper.

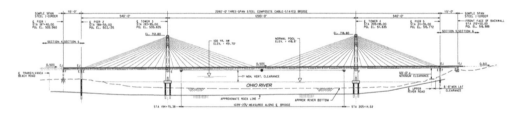

Figure 3.   General elevation.

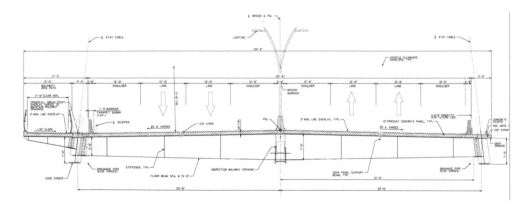

Figure 4.   Typical section.

4.2 Stay loss load case

The stay loss load case given in PTI formed the basis for the fundamental load case to be applied in the fracture analysis. This was seen as a consistent choice, as it would result in a similar level of reliability as the criteria that was already in place. In addition, the dynamic requirements described in PTI were also adopted. Under these requirements, a dynamic analysis may be used to establish the expected amplification after a sudden rupture. However, a minimum dynamic impact factor of 50% would be applied as an alternative static load case.

4.3 Non-redundant members to remain FCM

The first step of the analysis was to rule out members that could be considered non-redundant. There were several members that fell into this category. They were:

1. The tie-downs at the end piers
2. Any floor beams that support ballast loads.
3. Floor beams at the pylons, which resist transverse loads.
4. Edge girder sections above the tie-downs.

These members were designated as FCM without any further analysis. The edge girder section in point 4 merits additional discussion. Typically, edge girders are only 45′ long between bolted field splices, and contain only one stay cable anchorage. Therefore, the fracture of any one edge girder could result in the slackening of only one cable, and would not propagate to the surrounding cables. This is consistent with the stay loss analysis performed as part of the standard design. The edge girder section above, however, has two stay anchors welded to a section between field splices. Therefore a fracture of that section could result in the loss of two cables, which would be inconsistent with the design basis.

4.4 Analytical procedure

The modeling and analysis of the bridge was performed using the program LARSA 4D. The model was fully three-dimensional and used a combination of beam, truss, plate, and spring elements to accurately represent the geometry and stiffness of the bridge deck, floor beams, edge girders, stay cables, pylons, and foundations. A rendering of the full model is shown in Figure 6.

A typical deck section was modeled with beam elements to represent the steel floor beams and edge girders as well as plate elements representing the composite concrete deck slab. Nodes between the individual beam and plate elements were typically spaced at 7.5 foot increments. For select

Figure 5. Bridge rendering.

Figure 6.   Full 3D model of the bridge.

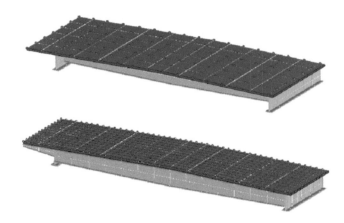

Figure 7.   A typical deck section with a coarse 7.5 foot mesh (top) and refined 2.5 foot mesh (bottom).

models, the mesh was refined and nodes were added such that elements were spaced at 2.5 foot increments in the vicinity of the fracture. This refinement allowed for a more detailed analysis of the forces in the area of concern. A typical deck section with a coarse and refined mesh is shown in Figure 7.

The individual elements were assembled in a full stage-by-stage construction analysis including geometric non-linearity to ensure that an accurate distribution of the loads was achieved. The inclusion of geometric non-linearity was important in the analysis as a fracture holds the potential for large deformations. Load factors were used at each construction step to produce a structural state equivalent to a factored dead load combination. A static live load pattern was then applied to the structure at the end of construction and scaled with an appropriate load factor in order to maximize the intended force effects at the location to be fractured.

To simulate the fracture of a steel member, a short section of the element was removed from the model and a set of equivalent end forces was applied at the same time such that equilibrium was maintained as shown in Figure 8. In all cases the steel element was assumed to have fractured completely (i.e. both flanges and the web were removed). The effects of the fracture were then investigated using a dynamic time-history analysis in order to incorporate the excitation of the structure's vibrational modes.

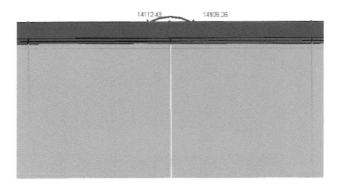

Figure 8. Removal of a short section of a steel member to be fractured and application of equivalent end forces.

For the analysis, a set of equal and opposite forces were applied to the ends of the fractured member. The full load was applied over a timeframe of 0.1 seconds and the resulting dynamic response of the structure was recorded at time intervals of 0.01 seconds until the structure reached a steady-state condition. All forces and demands were then enveloped for the entire response time recorded. As a minimum threshold, a static analysis was also performed. For the static analysis, the set of equal and opposite forces were applied to the ends of the fractured member as a purely static load with a minimum amplification factor of 50%.

4.5 *Studies performed*

The steel members were investigated in several different locations that best represented the areas of peak structural demands. The edge girders were examined in regions of peak shear and bending moments in the back spans, main span, and near the pylons. Peak negative bending moments occurred in the edge girders in the region near the end of the back spans due to the proximity of a tie-down link and the concentration of several stay anchors. In the center of the main span region, the edge girders experienced their peak positive bending moment. Peak shear demands as well as negative bending moments occurred in the edge girders near the pylons due to vertical bearing reactions.

The demands on the floor beams were representative of a simply-supported beam under a uniformly distributed load with peak shear demands at either end and a peak positive bending moment at mid-span. Due to the potential influence of second-order effects, a typical floor beam was investigated in both a region of higher compression near the pylon and a region of lower compression near the centerline of the main span. Images of the model with steel elements fractured in negative bending, positive bending, and shear are shown in Figure 9.

4.6 *Structural assessment*

For the purpose of assessing the structure, the fracture case was treated as an Extreme Event load combination limit state. Therefore all reduction, or phi (Φ), factors were set to 1.0 and localized plasticity was allowed, except for the stay cables as noted below. The assessment followed the following process. In the first step, the model was evaluated for large stresses exceeding prescribed values. These values included yielding of the steel members and concrete stresses higher than 0.6 f_c' in compression or $0.19\sqrt{f_c'}$ in tension which indicated non-linear behavior such as cracking. Yielding of the stay cables was not allowed and stresses were limited to 95% GUTS. Since the model represented a fully factored load combination, the analysis results were assessed directly against these criteria. If no excessive stresses beyond the allowable limits were observed then the structure was assessed to be stable and the fractured member was determined to be not fracture critical.

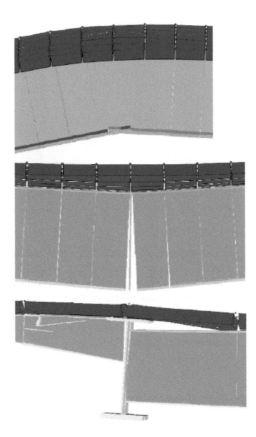

Figure 9.   Modeled steel members fracturing in negative bending (top), positive bending (middle), and shear (bottom).

As a second step, regions with excessive stresses were evaluated in greater detail. The local demands were checked to determine if they fell within the Strength capacity of that element and would therefore be structurally stable. The capacity was checked using an axial force – bending moment (P-M) diagram of the deck slab and the plastic moments of the steel members. For regions of the deck that exceeded this capacity check, the model was modified by reducing the stiffness of the highly stressed plates to approximate the cracking expected to occur in the concrete and the analysis was re-run. This change in stiffness redistributed the forces and decreased the demands in the highly stressed regions. The reduced stiffness of the cracked concrete was determined by performing a moment – curvature (M-Φ) analysis on a cross-section of the deck slab in both the longitudinal and transverse directions with a constant nominal axial load. The analysis incorporated the non-linear material properties of both the unconfined concrete and the steel reinforcement as shown in Figure 10. The effective stiffness of the cracked deck was taken as the slope of a straight line connecting the starting point on the curve with the point marking the first bar yield. The stiffer of either the longitudinal or transverse directions was used in the model as the cracked stiffness of the plate elements. An example of a moment-curvature analysis performed on a cross-section of the deck slab is shown in Figure 11.

If reducing the stiffness of the highly stressed plates proved to be insufficient in decreasing the demands of either the deck or the steel elements to within their respective capacities, the model was again modified by adding non-linear elasto-plastic springs to account for the expected post-yielding behavior. The model was then re-run and the results were assessed for the propagation of yielding or non-ductile behavior (i.e. concrete crushing or steel rupture) in the affected areas.

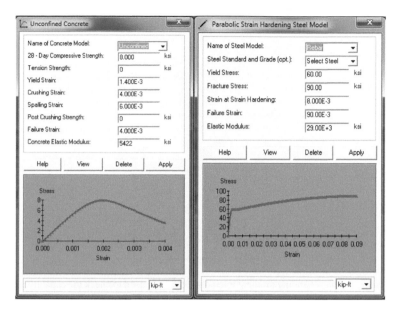

Figure 10.   Non-linear material properties for the unconfined concrete (left) and steel reinforcement (right) used in the moment-curvature analysis.

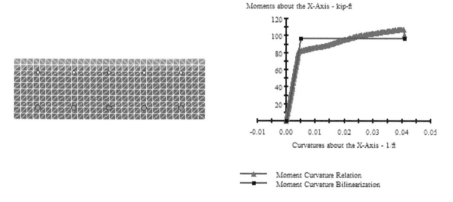

Figure 11.   A cross-section of the deck slab (left) and the moment-curvature analysis results (right).

If yielding remained confined to a localized area and no non-ductile behavior was observed then the structure was determined to be stable. Any instability in the structure indicated that the fractured member would be classified as fracture critical.

4.7   *Results*

The results from the analyses of the edge girder fracture varied considerably due to the differences in the loading and the peak demands at each location. In general, all of the edge girder locations showed highly stressed regions of the deck slab in the vicinity of the fractures in both the longitudinal and transverse directions, but the pattern and severity of those stresses differed. In an initial check, all of the local demands from these regions exceeded the capacity of the deck slab.

After the models were modified by assigning the highly stressed plate elements with the cracked stiffness properties and the analyses were re-run, the demands in the deck near the pylon were reduced enough to fall within the Strength capacity. An example of the capacity check of the deck

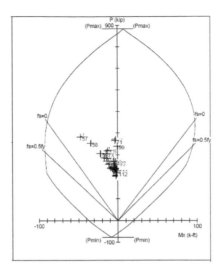

Figure 12.   An axial force – bending moment interaction diagram used to check the capacity of the deck slab.

slab near the pylon is shown in Figure 12. All of the stresses in the steel members and the stay cables near the pylon were also below the prescribed values and thus the edge girder near the pylon was determined to be stable and not fracture critical.

The demands from the cracked plates in the back span region still exceeded the capacity of the deck slab in both the longitudinal and transverse directions. The demands in the longitudinal direction were highly localized to a single plate element. When this demand was averaged with that of any adjoining plate the resultant demand fell to well below the Strength capacity and thus any failure of the deck in the longitudinal direction would be confined to the tributary width of a single plate element and would not propagate. In the transverse direction, the demands in several of the plate elements exceeded the capacity of the deck. These plates were separated in the model with a small gap along a longitudinal line extending over the fracture location. A series of springs was modeled to connect the plates on either side of the gap representing the expected non-linear moment and rotational properties of the deck in the transverse direction. All other degrees of freedom across the gap were tied together. The springs were defined as a bilinear moment – rotational curve with an elastoplastic response. Each spring was unique and the response curve was constructed by performing a moment – curvature analysis on the tributary width of each individual plate element with a constant axial load equal to the maximum tensile force on the plate from the analysis. The resulting curves were idealized with a bilinear approximation such that the initial slope of the line passed through the point of the first bar yield and the overall area under both the actual and idealized curves was equal. A detail of the gap modeled between the plate elements and an example of the non-linear spring properties are shown in Figure 13. The curvatures were translated into rotations by multiplying the values with a calculated plastic hinge length. The model was then re-run and the demands in the spring elements were assessed. Only the spring element located directly over the fracture location reached its effective yield moment. From these results it was determined that any failure of the deck in the transverse direction would not propagate beyond the tributary width of a single plate element. The stresses in the steel members and the stay cables were all below the prescribed values and the edge girder in the back span region was determined to be stable and not fracture critical.

For the model of the edge girder fracture in the main span region, the reduced demands in the deck due to the cracked plates exceeded the Strength capacity only in the longitudinal direction. The steel floor beam immediately adjacent to the fracture location also experienced some large stresses above the yield point. The plate elements which exceeded the capacity of the deck were

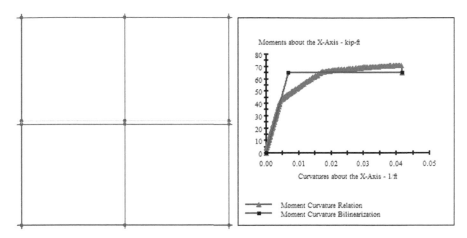

Figure 13.   Detail of the model showing the small gap between plate elements (left) and an example of the non-linear spring properties (right).

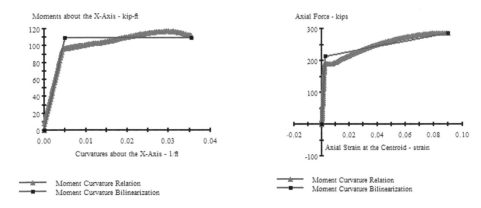

Figure 14.   An example of the moment – rotational (left) and axial force – displacement (right) spring properties.

separated in the model with a small gap similar to the back span region except in this case along a transverse line extending over the fracture location. A series of springs was modeled to connect the plates on either side of the gap that represented the expected non-linear moment and rotational properties of the deck in the longitudinal direction. For the three plates closest to the fracture location, the maximum tensile forces from the analysis exceeded the yield strength of the steel reinforcement in the deck and thus no moment – curvature analysis could be performed. For these locations no moment – rotational spring was defined as the cross-section was assumed to have lost all of its flexural capacity. To ensure that the deck did not rupture from the tension forces a second series of springs was modeled across the gap to reflect the axial stiffness of the cross-section. All other degrees of freedom across the gap were tied together. The second set of springs was defined as a bilinear axial force – displacement curve with an elastic – plastic response. The curve was constructed by analyzing the same cross-section used in the moment – curvature analysis by steadily increasing the axial strain. The resulting curve was idealized as bilinear similarly to the moment – rotational curve except that a secondary post-yield stiffness was determined by setting the ultimate axial force for both the actual and idealized curves equal to each other. An example of both sets of non-linear spring properties is shown in Figure 14. The strains were translated into displacements by multiplying the values with the tributary length of each plate element. After the springs for the

Table 1.   Fracture locations and results.

Fracture Location	Peak Demand	Summary of results	SRM Determination
Edge Girder in Back-Span	Negative Bending	Limited rebar yielding	No
Edge Girder in Main-Span	Positive Bending	Limited rebar yielding	No
Edge Girder near Pylon	Shear	Limited deck cracking	No
Floor Beams	Positive Bending	Limited rebar yielding	No
Floor Beams	Shear	No deck cracking	No

non-linear behavior of the deck were modeled and the analysis was re-run several of the springs reached their effective yield moment over a transverse distance of about 20 feet. However, only the axial force in the spring located directly above the fracture reached its effective yield point. While the yielding of the deck slab did propagate for a short distance from the fracture location it did not result in the instability of the overall structure. Also, the failure of the deck from both axial force and bending moment demands did not propagate more than the tributary with of a single plate. Furthermore, the introduction of the springs redistributed the demands in the steel members such that the stresses were reduced below the yield point. Due to these results, the edge girder in the main span region was determined to be stable and not fracture critical.

In contrast to the edge girders, the analyses of both floor beam locations yielded similar results. For the fracture due to bending near mid-span the initial results from the models indicated no yielding of the steel elements but highly stressed regions in the deck in both the longitudinal and transverse directions that exceeded the Strength capacities. After modifying the plates with the cracked stiffness properties, the demands were reduced such that only the transverse direction exceeded the capacity of the deck. Moment – rotational springs were added to the model similarly to the edge girder fracture in the back span region and the analysis was re-run. With the expected non-linear behavior of the deck modeled, only the spring element located directly over the fracture location reached its effective yield moment and did not propagate beyond the tributary width of a single plate. The fracture location at the end supports due to shear did not produce any highly stressed regions in either the steel or deck elements. The stresses in the stay cables remained well below the prescribed values for all cases. Due to these findings the typical floor beams were determined to be stable and not fracture critical.

The locations of the various studies and a brief description of the outcome are summarized in Table 1 below.

5   CONCLUSIONS

Fracture considerations for cable-stayed bridges should be carefully considered and coordinated between the Owner and Engineer. The redundant nature of many cable-stayed bridges may merit an analytical approach that would demonstrate that the bridge meets its performance requirements after the rupture of certain elements.

REFERENCES

AASHTO LRFD *Bridge Design Specifications*, 2012 (AASHTO LRFD).
FHWA Memorandum of June 20, 2012, *Clarification of Requirements for Fracture Critical Members*.
Post-Tensioning Institute *Recommendations for Stay Cable Design, Testing and Installation*, Sixth Edition, 2006 (PTI Guide Specifications).

Chapter 4

Repair project on cable stayed bridge "Binh bridge" damaged by ship collision in Vietnam

T. Tokuchi & S. Kaifuku
IHI Infrastructure Systems Co., Ltd., Osaka, Japan

ABSTRACT: Binh bridge, located in Haiphong city, Vietnam, a 3-span cable-stayed bridge with composite girders and a center span of 260 m, was opened to traffic in 2005 (Construction Report for "Binh bridge" in Vietnam). In 2010, 3 ships swept upstream by a typhoon collided with the bridge, resulting in serious damage to the edge girder and several cables. Repair and rehabilitation works on the bridge were awarded to IHI Infrastructure Asia Co., Ltd.

The complexity of repairing a composite girder cable-stayed bridge required the utilization of innovative methods in the analysis, planning and execution of the repair works. The repair works included, cutting of the damaged steel girders and on-site welding of new plates and stay cable replacement utilizing the adjacent stay cables in temporary hanger system that reduced erection loads during repair works.

1 INTRODUCTION

Binh bridge, located in Haiphong city, Vietnam, is a 3-span cable-stayed bridge with composite girders and a center span of 260 m that was completed in 2005. Figure 1 shows a complete view of the Binh bridge at completion. In July 17th, 2010, three cargo ships that had been moored for repair at a shipyard near the port of Haiphong were carried approximately 600 m upstream by typhoon No. 1 and collided with the composite girder of the Binh bridge. Figure 2 shows a general view of one of the ship's bridge stuck onto the composite edge girder after collision.

In response to a request from the Binh Bridge Management Company, IHI Corporate Research & Development, IHI Infrastructure Systems Co., Ltd. (IIS) and IHI Infrastructure Asia Co., Ltd. (IIA) working together as the IHI Group, initiated investigations and evaluations of the degree of damage immediately after the accident. A damage report was submitted together with a proposal of the repair method. The works were classified as part of the ODA (Official Development Assistance) emergency assistance program and in March 2012, IHI Group was awarded the repair works of the bridge.

Figure 1. Binh bridge at completion.

Figure 2.   Ship collision at edge girder.

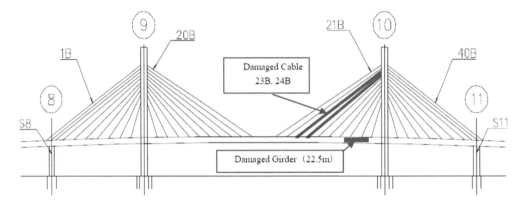

Figure 3.   Damaged location on bridge.

The Binh Bridge is designed as a cable-stayed bridge with a composite girder of 2 steel edge girders and a reinforced concrete deck slab, making the stress conditions in the edge girder to be influenced by the bridge's construction steps. Accordingly, to evaluate the stress levels in the damaged sections, it was necessary to faithfully simulate the construction stages in an analytical model. By conducting these analyses particularly carefully, the repair of the girder and replacement of the stay cable were performed without incident.

This may have been the first attempt in the world to repair the edge girder of a composite-girder cable-stayed bridge. In addition, replacing the cables of a cable-stayed bridge was a first experience for IHI, and only a small number of cases have been reported in the world. It is for these reasons that we are reporting on this valuable experience.

2   DAMAGE INVESTIGATION

2.1   *Damage outline*

The main areas of damage were; lower-flange of the edge girder (22.5 m), two stay cables, the guard railing and a navigation sign board in downstream side (see Figure 3, 4).

The edge girder lower-flange was deformed in the out-of-plane direction with the vertical stiffeners exhibiting complete buckling. Figure 5(a) and 5(b) show the damaged edge girder. Figure 5(c) shows the buckled stiffeners. However, the floor beams and concrete deck slabs were undamaged and sound.

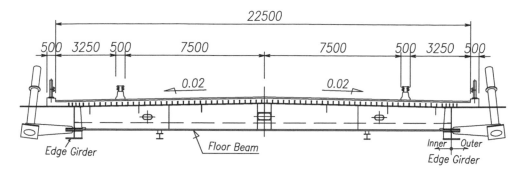

Figure 4.   Section profile.

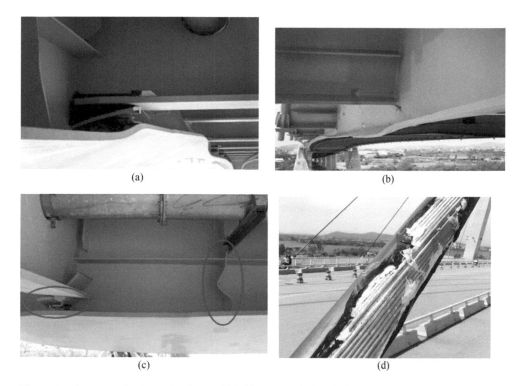

Figure 5.   a) Damaged edge girder (Inner side), b) Damaged edge girder (Outer side), c) Buckled stiffeners and d) Cable Damage.

The polyethylene cover (PE cover) of the two stay cables was peeled off completely exposing the wire strands which had partial damage to the galvanizing cover. Figure 5(d) shows the damaged cable. From this damage, it is predicted that so much salty water gets into the stay cable due to heavy rain for typhoon of this night and taking 4 days until temporary protection is covered.

2.2   *Traffic control*

After the accident, we restricted traffic until safety was confirmed. Only one of the lanes out of the four lanes was open to traffic (upstream side). This lane was opened as a two-ways lane for use by two-wheeled vehicles.

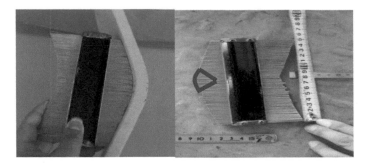

Figure 6.   Surface curvature of bent plate.

Subsequently, we conducted analysis on the assumption that the lower half of the cross section of the edge girder was damaged. It was thus confirmed that the edge girder would have a slight margin of stress, so we eased the traffic control to allow passenger cars with a mass of 20 kN or less to pass through one lane and the sidewalk on the upstream side.

Our repair work started in May 2012, and in order to use heavy machines such as construction cranes, passenger cars were restricted crossing the bridge all day to constrain the traffic of heavy objects other than construction machines and materials for the repair work. Automobiles crossing the river were asked to use a temporarily operated ferry or another bridge located 5 km upstream during the period of this traffic control.

2.3   *Evaluation of damage to girder*

The edge girder was significantly bent on the bottom flange side by the collision and had residual plastic deformation. A web was pulled by the bottom flange until it developed out-of-plane deformation. Webs had cracks in places near cable anchorages and were shifted by a distance nearly equal to the plate thickness (20 mm) in an out-of-plate direction. Stiffeners on the inner side of the edge girder were pushed upward by the bottom flange until they were buckled, and it was impossible to replace the stiffeners. Figure 5(c) shows the buckled stiffeners.

We conducted evaluation in terms of the strain of the plates with the aid of Expression (1) below. In the evaluation, we used cold bending radius $5t$ ($\varepsilon = 10\%$) that is stated in the Specifications for highway bridges (Japan Road Association, 2002) as a guide, in order to achieve Charpy absorbed energy higher than or equal to the required performance (Journal of JSCE) of steel materials from the viewpoint of ensuring the toughness of steel plates. Figure 6 shows the surface curvature of bent plate being measured.

$$\varepsilon = t/(2R) \qquad (1)$$

t: Plate thickness
R: Surface curvature

The simple strain evaluation by this method revealed that most portions had a strain of 3% (corresponding to a load of 15 t) or lower. The strains were at levels that would not cause problems. However, we determined to replace portions with cracks and those with apparent residual deformation anyway. The bridge portion that would be replaced is approximately 22.5 m long in the longitudinal direction and 1050 mm from the bottom of the edge girder, which is slightly higher than half of the girder height. Figure 7 shows the portion that would be replaced in the edge girder.

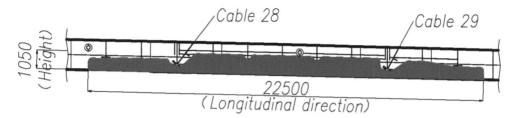

Figure 7.   Replaced edge girder range.

2.4   *Evaluation of damage to cables*

Since the wires of the damaged cables were not broken, we judged that their tensile strength was not degraded. The durability of a cable is affected by scratches and corrosion on the surface of its wires. This is because their fatigue strength varies depending on scratches and corrosion. For cable-stayed bridges, the durability of their cables means the durability of the bridges. Accordingly, measures to repair damage to cables need to be taken as quickly as possible.

In this investigation, damage to PE cover, scratches on cable wires, damage to galvanized paint on the wires, development of white rust were observed and so much salty water gets into the stay cable due to heavy rain for typhoon of this accident's night and taking 4 days until temporary protection is covered. These types of damage and the situation which galvanized wire exposed in the typhoon pose severe problems for cables. So we recommended to replace the damaged stay cable.

2.5   *Opening investigation of damaged stay cable*

After replacement the damaged stay cable, we performed the opening investigation for corrosion situation of wire of stay cable 24B. About 5.7 m position from cable socket was into the guide pipe of cable anchorage and stay cable cover of about 7.6 m position apart from there was damaged and wire of stay cable was exposed. Stay cable was exposed in the rain due to typhoon and 4 days later temporary protection of stay cable is covered.

In our investigation, about 1.7 m area from the socket including 1.5 m portion cowered with polybutadiene rubber was peel. Opening investigation result is shown in Figure 8.

As a result, regarding wire from the edge of cable to 1.2 m position some area is same as new product partially but the white rust could be found over the top of cable. Also in slight area the red rust could be found. As above mentioned, nevertheless it is difficult to evaluate the effect of corrosion; the replacement of damaged stay cable was decided in consideration to safety of long span. In our internal investigation, it was confirmed that wire was corroded two years later after damage cover of stay cable. So if the cover of stay cable is damaged, the water invader should be prevented as soon as possible.

2.6   *Influence of scratches of wire*

Scratches deep enough to catch a fingernail were observed on parts of the surfaces of element wires that were exposed to the atmosphere with PE covers removed. The scratches were formed in the axial direction, making it easy to see that a steel material slid axially along the surfaces of the element wires.

Since the scratches on the surfaces of the element wires were small, it was presumed that the static strength of the wires had not been degraded. The fatigue strength of the wires, however, varies depending on the shape of the scratches, and even a scratch as long as 100 mm may cause strength degradation. Accordingly, the fatigue strength of damaged element wires should be regarded as degraded. However, it was presumed that only the element wires on the outermost layer of the steel wire bundles were scratched and that the element wires in the inner layers were sound. Therefore,

Figure 8.   Opening investigation result of damaged stay cable.

we concluded that cable fatigue evaluation with the influence of corrosion ignored should be performed on the premise that loads acting on the cable will be supported only by inner-layer element wires.

2.7   *Influence of corrosion of wires*

White rust developed on the surfaces of element wires that were exposed to the atmosphere as their PE covers had been stripped away. This fact suggests that after the PE covers had been stripped away, the surfaces of the element wires were exposed to rainwater. Accordingly, it is presumed that rainwater permeated the steel wire bundles and accumulated in the bottoms of the cables.

The zinc layers (minimum: 300 g/m^2) of galvanized steel wires immersed in seawater will likely corrode until they completely disappear in as soon as one year. In addition, after the zinc has been lost, the corrosion of steel wires will progress which degrades the fatigue strength.

Binh bridge is located near an estuary that is exposed to sea breezes. Accordingly, the rainwater collecting in cables presumably contains salt. Therefore, there was the possibility that in the rainwater-holding portions of the third (No. 23) and fourth (No. 24) cables from the top, zinc on the surfaces of element wires would be lost in as little as one year and that the cable fatigue strength would start to degrade.

It was difficult to accurately evaluate the durability of the damaged cables, so it was impossible to guarantee their quality for the future. Therefore, we decided to replace the damaged cables with new cables.

2.8   *Cable socket portions*

In three cable socket portions of other cables, abrasion that seems to have been made due to contact with a steel material was found. In the main bodies of some of these sockets, indentations were found in the corner portions at their ends, but no abnormalities were found at the anchorages, which are the most important parts. Some of the anchor caps were partially deformed and had damaged bolts (See Figure 9).

No cracks were found in the painting at the member boundary portions between cable sockets and shim plates and between shim plates and washers. This indicates that the cable sockets did not rotate or move in the accident. Therefore, we judged that the three anchors that had scratches on their cable sockets would function soundly such that cable replacement would not be required.

Figure 9.   Damaged cable socket.

3   REPAIR AND REHABILITATION WORKS

After long detailed investigation by the Employer, IHI was awarded this repair and rehabilitation works on the Binh Bridge in March 2012. These works included the entire repair and rehabilitation works (Edge Girder, Stay Cable, Railing, concrete deck slab and so on).

3.1   *Repair of girder*

Before partially replacing the damaged portion of the edge girder, we first examined section forces by analysis with consideration given to continuous composition performed at the construction stage of the bridge and thereby investigated the stresses exerted in the present state of damage. Since the bridge consists of a composite girder, the stress check was performed by adding the stresses exerted before the composition and those exerted after the composition. This process allowed us to estimate the stress condition before the accident. However, it was difficult to estimate the stiffness degradation of the portion damaged by the accident and to examine stress redistribution due to the stiffness degradation. Consequently, we were not able to accurately ascertain the stress condition after the accident.

Therefore, as our reinforcement policy, we determined to provide additional reinforcement ensuring sectional performance higher than or equal to the section stiffness of the edge girder in a sound state, so that when the edge girder was undergoing partial cutting, the section force of the cut portion and fluctuating loads would be supported. We decided to cut and replace the damaged portion under these conditions.

In addition, we designed reinforcements for supporting the section force according to the following policy. We assumed that after the accident, the bottom flange and web of the damaged girder were also supporting stresses occurring in a sound state. We also assumed that stresses released by cutting the damaged portion on the above assumption would be redistributed to the reinforcements and the remaining portion of the girder. Based on these assumptions, we designed the required section of the reinforcements.

In order to partially cut the edge girder and remove the cut portion to replace it with a new member, it was necessary to first attach reinforcements for supporting the section force. A large crane was not able to be placed immediately above the damaged portion, so we were only able to place a hydraulic crane with a lifting capacity of 1000 kN. Moreover, in order to perform the repair work, we needed to pass the boom of the crane through the space between existing stay cables. It would be impossible to transfer very large members through the space. Since we could only use limited equipment to reinforce the bridge and with limited space, we determined to perform the reinforcement work by using a truss structure (Temporary Bypass Truss). Figure 10 illustrates the structure of the truss.

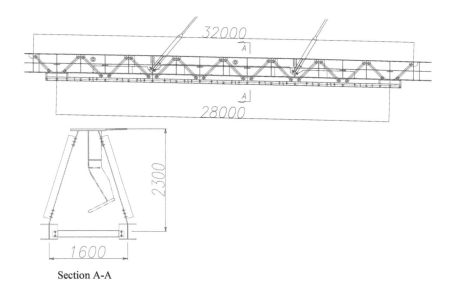

Section A-A

Figure 10.   Temporary Bypass Truss (DWG).

Figure 11.   Temporary bypass truss.

3.2   *Repair of edge girder*

In the process of installing scaffolds and the temporary bypass truss, the edge girder was first marked with girder cutting lines and reference lines. We arranged rails consisting of channel steels with rollers on the scaffolds and the temporary bypass truss in order to allow members to be transferred horizontally (refer to Figure 11).

By cutting a web of the damaged portion, the remaining side of the web would form a free end, and stress would be released. There was a possibility that the released stress might be redistributed near the free end causing the web to develop local buckling. Accordingly, before cutting the damaged member, we added a horizontal stiffener near the portion that would become the free end of the web in order to reinforce the portion.

In order to compare the actual stresses with those calculated for design, we attached a uniaxial strain gauges to the upper and bottom flanges of the edge girder and the temporary bypass truss and measured the stresses and strains at each construction stage to confirm safety. We also measured

Figure 12. Cutting of Damaged part.

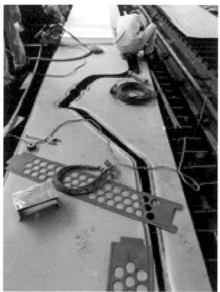

Figure 13. Fabrication of new part.   Figure 14. Installation of new parts at site.

the elevation of the edge girder at each major construction stage to confirm that no abnormal values were observed.

Cutting girder and groove making were performed on site. Dismantled girder was cut to approximately 2m pieces for easier handling when transferring the pieces from under bridge to over the deck slab easily. (Refer to Figure 12).

After exact measurement of the remaining girder piece and exact tracing using tracing paper and corrugated cardboard , the new girder member was fabricated by the factory of IHI Infrastructure Asia (refer to Figure 13).

The new girder pieces were delivered on site within one week after passing the cutting data to factory and installed piece by piece. After all pieces were assembled, welding work was performed. Due to the exact measurement and tracing, the misfit of member was not occurred for edge girder member's installation (refer to Figure 14).

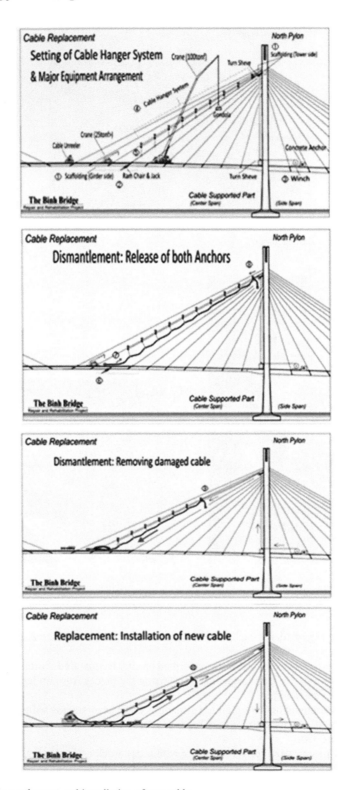

Figure 15.   Dismantlement and installation of stay cable.

Figure 16.   Dismantlement of stay cable.

3.3   *Replacement of damaged stay cable*

The cable adjacent and above the damaged cable was used as a Temporary Hanger System. This system reduced the sag of the cable being removed and the one being installed, resulting in efficient installation works. This replacement work's procedure is shown in the following Figure 15.

The installation step details are as follows:

1. Install the scaffolding around cable anchorage of pylon side and girder side.
2. Center-hole Jack and tension rod are installed on the anchorage of girder side.
3. Winch for lifting up or down device is installed.
4. Pulleys to be able to move on the cable are installed on the Stay Cable adjacent and above the damaged cable. Temporary hanger System is hanged from pulleys.
5. Band of Temporary Hanger System is fixed to replacement cable to support the self-weight of replacement cable with tensioning by Temporary Hanger System.
6. At the anchorage of girder side, the tension of existing stay cable is un-loaded by Center-Hole jack.
7. Remove the socket from the anchorage of girder side.
8. Remove the socket form the anchorage of pylon side.
9. Replacement cable with pulleys is lifting down by Winch (refer to Figure 16).

In case of installation of new stay cable, opposite works of abovementioned procedure is performed.

4   DESIGN AND ANALYSIS

Before performing repair works, it was confirmed that all effects to girder and cable due to repairing girder and due to replacement of stay cable by global frame analysis. In addition, regarding repair

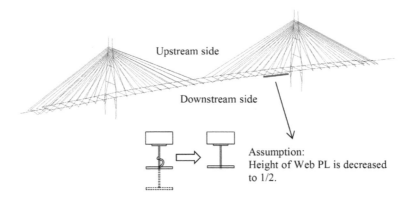

Figure 17.   Global analysis model which assumes damaged girder.

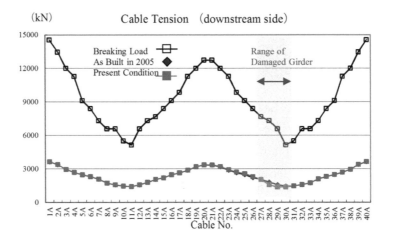

Figure 18.   Cable force after girder damage.

to girder, because stress concentrate is considered, local FEM analysis was performed to check the local stress around cutting area of web for edge girder.

4.1   *Global frame analysis*

4.1.1   *Repair the girder*

We made the model incorporated the effect of erection step during construction of this bridge by global frame model of Midas. After that, we decreased the section modulus of damaged girder for assumption that height of web PL is decreased to 1/2 due to effect of this accident. From this analysis, we confirmed that due to deformed condition, emergency countermeasure is not needed and before starting repair and rehabilitation works, limited traffic like motor bicycle for only 2 lanes can be opened. Figure 17 shows the global analysis model and assumption of damaged girder for model. Figure 18 shows the cable force's differences between damaged girder condition and as built condition without traffic. Figure 19 shows the girder stress's differences between damaged girder condition and as built condition without traffic.

Secondly, we made a model during repair girder work to add the bypass truss to the as-built model. To consider repairing the girder, we reduced the section modulus of damaged area form bypass truss model. This result shows the Figure 20–22.

Stress of Critical Sec. for Main Girder (downstream side)

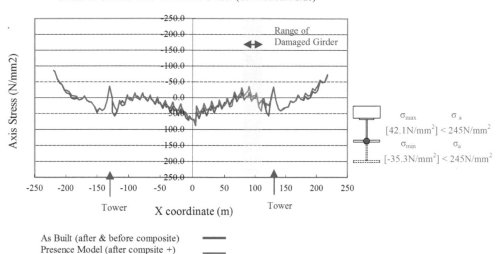

As Built (after & before composite)
Presence Model (after compsite +)

Figure 19.   Girder stress after girder damage.

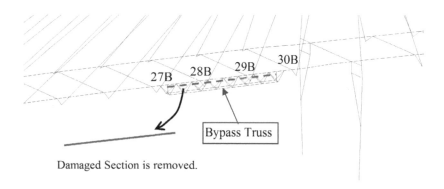

Damaged Section is removed.

Figure 20.   Global analysis model after removing damaged girder.

   In addition, we made an analysis model by incorporating the Temporary Bypass truss into a three-dimensional frame model of the entire cable-stayed bridge. We used it to conduct a sequential analysis for examining fluctuating loads in each construction step to sum the section forces and evaluated the total section force. Figure 23 illustrates the three-dimensional frame model.

   The local influence of factors such as the shape of the cut portion of the girder, and the structural influence of factors such as the effective width of the slab, cannot be accurately taken into consideration by performing only a stress check based on section forces obtained by the frame analysis. Accordingly, for the replacement portion of the girder, we also designed a Finite Element Method (FEM) model with consideration given to cable anchorages and the Temporary Bypass truss and applied the section forces obtained from the above frame model to the FEM model to evaluate local stress intensity. In this process of using the FEM model, we also simulated the sequential steps to confirm safety.

4.1.2   *Replace the stay cable*

After repairing the girder, we started to replace the stay cable. So we use the repaired girder model as above mentioned. And then until replacement of the cable, construction step analysis is performed.

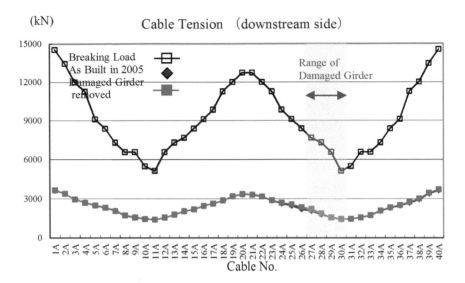

Figure 21.   Cable force after removing the damaged girder.

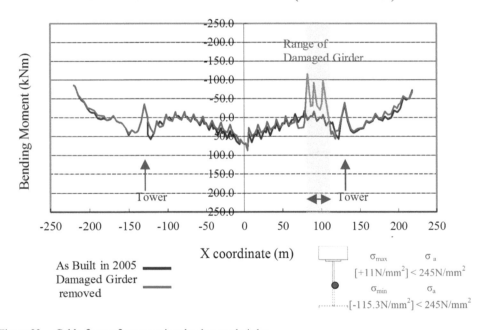

Figure 22.   Cable force after removing the damaged girder.

In our analysis result, it is confirmed that without transfer its load to temporary member and with one lane's traffic open, the deformation and cable tension force due to replacement of stay cable is not so big (refer to Figure 24 and Figure 25). Also the ratio of total stress on lower flange and strength in construction is small than 55% (refer to Figure 26). It can be concluded that girder withstand without the presence of one stay cable. Therefore, it is concluded that the stay cable is

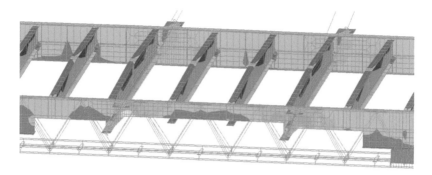

Figure 23.   FEM Model for checking local stress.

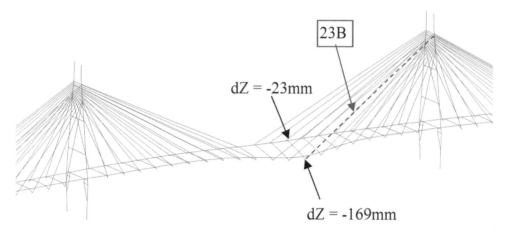

Figure 24.   Deflection after removing the damaged stay cable.

replaced without another temporary member to take replaced stay cable load and with one lane's traffic open.

4.2   *Stress monitoring*

To confirm the validity of the analysis result and the safety of construction, we also monitored stress by using strain gauges attached to the upper and lower flanges and the Temporary Bypass truss. The monitoring was performed at the following four stages with respect to stress values observed.

In Figure 27, load flow transferring to lower flange and Temporary Bypass Truss during repair is shown. In Figure 28, the result of monitoring measuring point (MP) 1 and 2 is shown by graph. For the vertical axis of this graph acting force is shown and for the horizontal axis construction step is shown. Regarding vertical bar of this graph, the changing force of lower flange and Temporary Bypass Truss per each construction step is shown, and regarding polyline of this graph the sum total of that is shown.

When cutting the lower flange deformed in the out-plain direction, the residual stress is released and the behavior which be back to non-deformed condition was confirmed. In the measuring result of monitoring, stress difference between edge girder of upstream side and that of downstream side due to out-plain deformation of edge girder was confirmed.

In the measuring result, the average value between the stress on upstream side and stress on downstream side of damaged girder is plotted to remove the influence of out-plain deformation.

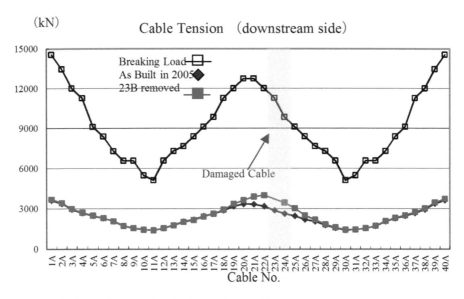

Figure 25.   Cable force after removing the damaged stay cable.

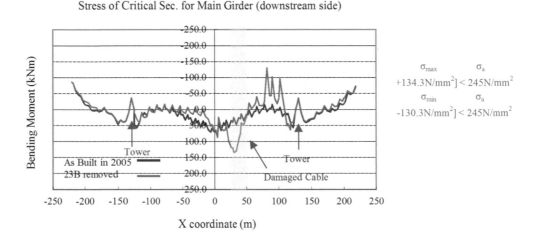

Figure 26.   Girder stress after removing the damaged stay cable.

When Temporary Bypass Truss was installed on the girder, there is not stress in this Temporary Bypass Truss but compression force is acted on the edge girder (Step 1). When cutting damaged edge girder, Temporary Bypass Truss is compressed due to re-distribution of compression force of cut edge girder (Step 2). When welding of newly parts of edge girder, Temporary Bypass Truss is compressed because of welding shrinkage. Welded newly part of edge girder is also tensioned (Step 3). When dismantling the Temporary Bypass Truss, due to release of compression force acted on Temporary Bypass Truss, this force is distributed to edge girder and concrete deck slab (Step 4). Finally distributed acting force on the Lower Flange is approximately 1,000 kN (for the stress, approximately 22 N/mm² (Section 900 mm × 60 mm)) including residual stress due to welding.

From the above result, we concluded that because Temporary Bypass Truss was transferred from the compression force acted on damaged edge girder, it is very efficient for this repair work.

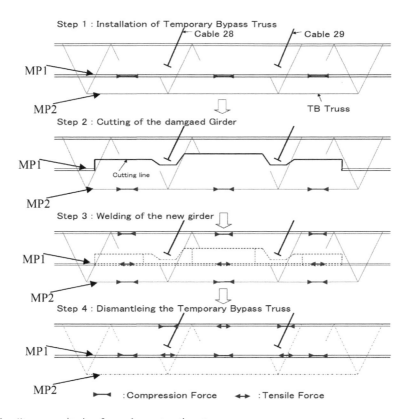

Figure 27.   Stress monitoring for each construction step.

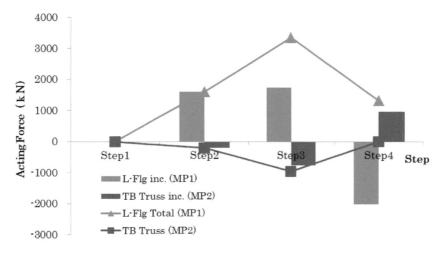

Figure 28.   Changing of force for each member during construction.

5   CONCLUSIONS

Large cargo ships, which do not normally pass under the Binh Bridge, were pushed along by a typhoon until they collided with the bridge damaging the edge girder and stay cables of the bridge. No one ever expected such a severe accident resulting in damage to the edge girder and part of the stay cables to happen to this two-main-composite-girder cable-stayed bridge. Fortunately, however, its slabs and floor beams were sound, and consequently, the redundancy of the bridge allowed it to deliver performance far beyond expectations and not collapse.

This repair work is characterized by the following aspects.

1. The repair of the girder was performed by the temporary bypass truss method using trusses with a triangular cross section. This method allowed us to safely complete cutting of the old girder and installation and welding of a new girder.
2. The stress behavior and displacement behavior of the complex structure of the bridge were examined by performing assembly calculations faithfully simulating each construction step with the aid of both frame analysis and FEM analysis, which allowed us to safely complete the repair work.
3. During the repair work of the girder, we performed computerized construction by conducting stress monitoring for safety confirmation.
4. In order to replace some cables contained in the cable plane, we used a temporary hanger system using pulleys able to move along a cable and were thus able to replace cables in a safe and accurate manner. This method will also be able to be used for various types of construction, such as cable replacement for aging cable-stayed bridges.

REFERENCES

Homma, Miki, Soya, Sasao, Okumura & Hara. 1997. A study on strain aging of cold worked structural steel and allowable cold working radius. *Journal of JSCE (No. 570/I-40)*: 53–162.
Japan Road Association. 2002. *Specifications for Highway Bridge*. I. Common, II. Steel Bridge.
Matsuno, Yamamoto, Nakayama, Oyama, Ueshima & Yabuno. 2006. Construction Report for "Binh Bridge" in Vietnam, *Bridge and Foundation Engineering*: Vol.40 2006/9: 18–23.

Bridge construction

Chapter 5

Reconstructing a bridge in ten days: New Jersey Route 46 over Musconetcong River accelerated replacement

R.J. Adams
Greenman-Pedersen, Inc, Lebanon, NJ, USA

S.J. Deeck
New Jersey Department of Transportation, Trenton, NJ, USA

ABSTRACT: This paper discusses the challenges faced during the design and construction of the Route 46 over Musconetcong River Bridge. The two-span through girder crossing was reconstructed over a ten-day road closure using Prefabricated Bridge Elements (PBEs) and the State of New Jersey's first application of Ultra-High Performance Concrete (UHPC).

1   INTRODUCTION

The Route 46 over Musconetcong River bridge (Structure No. 1208-162) connects the town of Hackettstown with Mount Olive and Washington Townships between New Jersey's Morris and Warren counties. The 127-foot long bridge, which carries 13,200 vehicles per day via one 15-foot lane in the east and west directions, is of significant importance to commuters and the surrounding residents and business communities. Built in 1924 and spanning a Class 1 waterway, the structure lies immediately adjacent to the remains of a historically significant grist mill in the northwest quadrant, and borders local businesses in the other three quadrants. In addition to these challenging site constraints, the bridge also carries ten underbridge conduits that were found to have insufficient slack for temporary relocation and several overhead utilities spanning across and around the structure. A plan view of the project site is shown in Figure 1.

1.1   *Existing structure*

The existing structure consists to two 61'-4 1/2" long simple spans composed of two concrete encased riveted steel through girders supporting concrete encased rolled steel floorbeams and a reinforced concrete deck. A concrete sidewalk is supported by cantilevered brackets off the south girder. A cross-section of the existing bridge is shown in Figure 2. Due to the two girder system, the structure is considered fracture critical. The substructure consists of concrete gravity abutments and wingwalls and a concrete pier, all supported on spread footings.

2   PROJECT NEED AND EVOLUTION

Initially identified to be in need of deck replacement through the New Jersey Department of Transportation's (NJDOT) Deck Replacement Program, Greenman-Pedersen, Inc. (GPI) performed an in-depth inspection to identify all conditions that warranted repair. As expected, the deck was noted to be in poor condition due to several shallow spalls, areas of fine pattern cracking and efflorescence, and evidence of seepage and chloride contamination. The superstructure, composed of concrete-encased built-up steel girders and floorbeams, was found to have extensive encasement

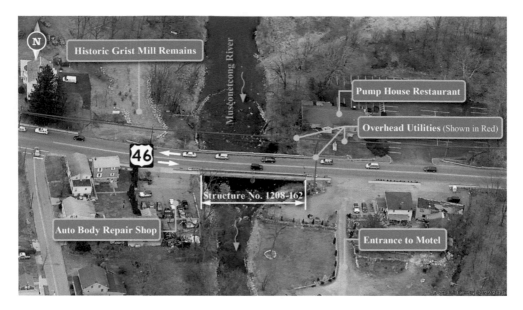

Figure 1.   Plan view of the Route 46 over Musconetcong River Bridge.

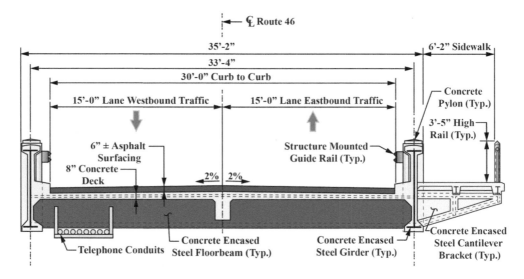

Figure 2.   Typical existing section.

deterioration, heavy corrosion with lamination, and section loss on exposed portions of the primary members. The substructure was found to be satisfactory, with localized areas of cracking and spalls, and cores indicating a compressive strength of over 11,000 psi.

Taking into account the condition of the existing structure, an alternatives analysis was performed to determine the best solution to meet the NJDOT's original project goal, and provide the most cost effective rehabilitation plan to extend the crossing's service life while minimizing traffic impacts to the surrounding communities. Several alternatives were considered:

a) Isolated deck repairs and rehabilitation;
b) Deck replacement using cast-in-place concrete;

c) Deck replacement using precast panels;
d) Deck/superstructure replacement using accelerated construction techniques; and
e) Complete bridge replacement.

Isolated deck repairs was the cheapest alternative with minimal traffic impacts as the work could be completed overnight under alternating single lane closures; however the repair process could induce further damage to the concrete encasement and sound deck areas. After taking into account the life span of the remaining bridge surfacing, this alternative was determined to not be cost effective. Deck replacement using traditional cast-in-place concrete was not considered feasible due to the severe traffic impacts that would arise from the concrete cure time. Additionally, the process of removing the concrete deck would have likely caused further damage to the concrete encasement, exposing more uncoated steel requiring repair and potential strengthening. Using precast panels in lieu of cast-in-place concrete would eliminate the cure time and reduce the construction duration, however, damage to the existing encasement was still a concern. Longitudinal and transverse post-tensioning would also complicate construction and future repairs, and the numerous joints between deck panels could be a long term performance issue with respect to leakage. A full bridge replacement would address all the deficiencies on the bridge, however the economic and right of way impacts needed for an alignment shift or temporary structure far exceeded the benefit of this alternative.

Taking into account the unique nature of the project location, the preferred alternative was determined to be a complete deck and superstructure replacement using accelerated construction techniques in concert with a short-term closure of Route 46. Utilizing accelerated construction would reduce both environmental and socio-economic impacts to the local community. While still subject to severe traffic impacts during the construction period(s), this solution provided for an extended service life as compared to the deck replacement option since the superstructure would also be replaced. The risk of complications arising from unforeseen repairs to the existing steel would also be eliminated. GPI initially proposed to design the reconstruction to occur over a series of modified weekend closings with a short term detour to handle the traffic during these off-peak cycles. Following numerous discussions with local public officials and stakeholders, it was agreed to perform the construction during a ten-day (240 hour) continuous road closure. This approach would minimize confusion for the traveling public stemming from the staggered weekend closings, and reduce the risk of unanticipated construction or weather problems causing a delay in the narrow closing windows.

The design team also took a proactive approach to inform the public of the short-term closure. Most often, in highly commercialized areas like the Route 46 project site, tenants and property owners are sometimes separate entities. Therefore, providing notice directly to property owners can result in information not getting to the people most affected by local developments. GPI and NJDOT staff made several trips to the surrounding project area and detour route, where they "pounded the pavement" and hand-delivered notices to local businesses and residents about the upcoming construction activities. Information was also broadcast through local cable television outlets and regional commuter traffic was informed via placement of multiple variable message signs five weeks prior to the road closure.

3   DESIGN CHALLENGES

3.1   *Selection of superstructure*

Upon receiving notice to proceed to begin the final design, GPI investigated several possible superstructure alternatives that could be constructed during the ten-day road closure:

a) *Precast Concrete Steel Composite Units* – The former proprietary system known as Inverset is a precast concrete and steel composite bridge superstructure system that utilizes an "upside-down" casting method using the force of gravity to prestress steel beams. The inverted casting process pre-compresses the concrete, yielding a crack resistant deck with high durability, and reduced superstructure depths. The units are of high quality due to a controlled environment construction and can be installed rapidly with a "traffic-ready" surface. The weight of the units would also be less than the existing superstructure, indicating that the substructure would be adequate for this superstructure type. Despite these advantages, the system is composed of steel beams, which would require future maintenance for cleaning and painting. Additionally, the cost of the system is approximately 50% more than that of a typical superstructure.

b) *Prefabricated Through Girder System* – A potential replacement alternative was to construct a new through girder system, similar to the existing superstructure, in a staging area and move it into place during the closure. The weight was estimated to be less than the existing superstructure, and the system would require limited reconstruction of the abutment and pier due to the similar bearing locations. However, this system would still be fracture critical, and does not offer as many advantages as the other types considered.

c) *Northeast Extreme Tee (NEXT) Beam* – The NEXT beam is a relatively new section on the bridge market developed by the Precast/Prestressed Concrete Institute Northeast (PCINE) and several Departments of Transportation. The NEXT Beam is a double-tee beam similar to those used in parking structures except the tee leg is wider to provide greater capacity needed when carrying bridge loads. Ranging in depths from 24"-40" and widths of 8'-0" to 12'-0", the NEXT beams were developed for spans ranging from 30 to 90 feet. Two types of NEXT Beams were developed for bridge use:

a) F (Form) Beam – a partial four-inch thick flange serves as a form for a cast-in-place deck.
b) D (Deck) Beam – full flange serves as a "traffic-ready" surface, which can be micro-milled to meet final grade elevations or covered with an overlay.

A closure pour is required longitudinally to ensure composite action between the beams. Due to the short closure duration, the Type F beam would not be feasible with a cast-in-place deck. The Type D beam would provide for a traffic-ready surface, and would ultimately weigh less than the existing superstructure.

After reviewing the possible alternatives, it was decided to move forward with designing the reconstruction using NEXT Type D beams. The ability to facilitate rapid construction along with prestressed concrete's advantages of an increased service life and limited future maintenance gave the NEXT beams an edge. Additionally, this superstructure type was found to be the least expensive option, resulting in it being the most cost effective alternative.

3.2   *Prefabricated superstructure connections*

When designing the connection system of the NEXT beams, it was important to specify a material that would provide a robust connection and eliminate any concerns about its short and long term performance. Additionally, since time was critical, a material with rapid strength gain was also important to deliver the project within the closure window. Working within these constraints, ultra-high performance concrete (UHPC), a steel fiber-reinforced cementitious composite with exceptionally high mechanical strength and durability properties, was chosen to connect the units along the longitudinal joints (Russell & Graybeal 2013). This material has the ability to obtain a compressive strength of 12,000 psi in 12 hours, and has superior durability and bonding properties when compared to other joint fill materials.

Due to the need to establish connections between structures composed of PBEs in the field during potentially unfavorable weather conditions and high intensity closure windows, connections between PBEs are naturally "weak links" in the structure and susceptible to future damage and deterioration. Having only been used in a handful of other states, New Jersey's first application of UHPC allowed GPI to minimize the joint widths between the NEXT beams. By taking advantage of UHPC's high strength, the space between adjacent units was set at six inches. This distance was sufficient to fully develop one mat of No. 16 headed rebar, staggered at six inch spacing as shown

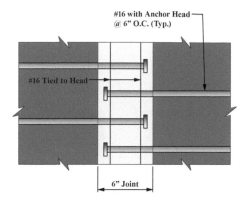

Figure 3.   Joint plan view.

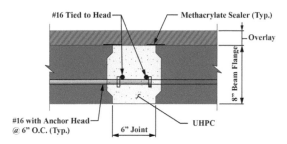

Figure 4.   Joint section.

in Figures 3–4. Additionally, the edges of the beams were notched and given an exposed aggregate finish to further enhance the bond and shear transfer across the joint.

The use of an exposed aggregate finish has been shown to greatly improve field-cast connections, however, the increased surface area will absorb water from the UHPC during the curing period, resulting in a weak bond. This issue was eliminated by including requirements in the project documents to pre-wet the joints for a period of 24 hours prior to casting the UHPC. The top surface interface between the precast beams and UHPC was then treated with a methacrylate crack sealer to prevent the introduction of water and ensure long-lasting performance.

3.3 *Continuity connection*

In order to limit future maintenance costs, GPI decided to eliminate the pier joint and make the spans continuous. Although AASHTO allows the effects of continuity to be taken into consideration when computing design forces for loads applied after continuity is established, the beams were designed as simple spans per AASHTO Section C5.14.1.4.1 (AASHTO 2012). A continuity diaphragm over the pier was then designed and detailed by placing additional deck reinforcement to resist the negative moments induced from the continuous action of the structure and positive moment reinforcement to resist restraint moments developing from time-dependent or other deformations. This approach provided a conservative design while still improving the structural integrity of the bridge, increasing its ability to resist extreme event and other unanticipated loadings.

The additional negative moment reinforcement was designed using force effects computed from the fully continuous structure. The top mat was connected by mechanical couplers and the bottom mat was extended down into the diaphragm with 90° hooks as shown in Figures 5–6. This required extensive detailing and careful placement in the field as the top bars between adjacent units needed to align within tight tolerances to make the mechanical connection and the bottom bars staggered

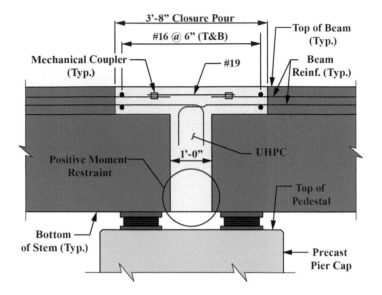

Figure 5. Continuity connection at beam webs.

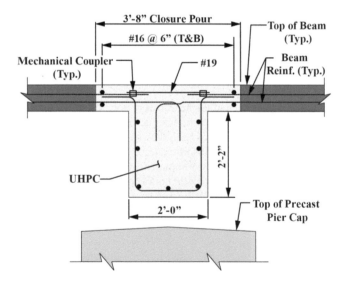

Figure 6. Continuity connection between beam webs.

to avoid hitting each other. Bending the bottom bars down into the diaphragm also meant that a full-depth diaphragm would need to be provided across the structure width in order to properly develop the bars.

Recent recommendations in the Transportation Research Board's SHRP 2 Report S2-R04-RR-2: ABC Toolkit (2013), show that continuity can be provided by connecting the deck reinforcement within the deck thickness and only extending the diaphragm to full-depth between adjacent beam webs. However, due to the proposed eight-inch thick deck and reinforcement geometry, this detail would not have been feasible. Future designs should incorporate a haunched deck near interior supports to facilitate this type of connection and reduce the depth of the diaphragm.

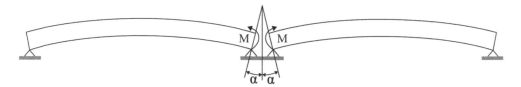

Figure 7.   Fixed end moments resulting from end restraints.

Due to the compressive stress caused by the prestressing force, the NEXT beams will creep. This creep will cause the beam camber to increase. In a simple span structure, this increase in camber is not accompanied by any stress as the beam ends are free to rotate. However, in continuous structures, the rotation from creep effects after the connection is established is restrained by the continuity connection (Hastak et al. 2003). This results in fixed end moments (FEM) that maintain the ends of the beams as flat, shown schematically in Figure 7.

If a Type-F beam was used, the differential shrinkage from the precast beam and cast-in-place deck would also cause similar forces. However, because the concrete deck is cast integral with the beam in a Type-D member, this additional force is eliminated.

The age of the beam at the time of application of the continuity connection has a great effect on the final creep moments. As the age of the beam increases before establishing continuity, the amount of creep and the resulting creep load effects will decrease. AASHTO recommends that the contract documents require the beams to be at least 90 days old before continuity is established, and that the positive moment connection be designed for a strength of $1.2M_{cr}$. This is because by 90 days, 60% of the creep and 70% of the shrinkage in a prestressed beam is gone, and a simplified design is sufficient (Miller et al. 2004). Overall, AASHTO gives designers four possible options to design a positive moment connection (Miller 2014):

a) Specify that the beams are at least 90 days old at the time continuity is established and design the connection with a strength of 1.2 Mcr.
b) Provide a positive moment connection with a strength of 1.2 Mcr and use the provisions of AASHTO Article 5.4.2.3, with ktd = 0.7 to establish the minimum age at which continuity can be established.
c) Consider the bridge continuous if the net stress at the bottom of the diaphragm from superimposed permanent loads, settlement, creep, shrinkage, temperature gradient, and 50% of the live load is compressive.
d) Calculate the actual restraint moments and design the connection for such.

Based on the very tight construction schedule for this project, GPI recognized that requiring a 90-day cure period was not feasible. The actual duration from the time the last beam was cast to establishing the continuity connection was conservatively estimated at 14 days. With this information, the connection was designed by computing the actual restraint moments arising from a 14-day cure. To compute the effects of creep, GPI used a method developed by the Portland Cement Association (PCA) (Freyermuth 1969). The fixed end moments required to restrain the ends of the simple span beam after continuity is established are computed and used to design the positive moment restraint connection. The method outlined in AASHTO Article 5.4.2.3.2 was used to compute the creep coefficient at infinite time and at the time of establishing continuity. The difference was then used to compute the creep correction factor C_{cr} per the PCA Method:

$$C_{cr} = 1 - e^{-\Phi} \qquad (1)$$

where Φ = the restrained creep coefficient in the beam after continuity is established.

Fixed end moments were then multiplied by this factor to obtain the positive moment restraint forces. Per NJDOT requirements, two details were designed: one using extended prestressing strands and the other using embedded mild reinforcing steel. The contractor ultimately decided to

Figure 8.   Continuity connection with extended strands.

use the extended strand detail where the strands are bent at a 90° angle upon exiting the web and lapped with the strands from the adjacent unit, as shown in Figure 8.

3.4   *Precast substructure units*

Due to the deteriorated condition of the existing substructure seats, it was decided to remove the top portion of the abutments and pier and replace them with monolithic precast concrete units. This facilitated easy placement of the NEXT beams and accommodated the shallower superstructure depth while providing increased service life. The existing substructure cores indicated that the concrete below the deteriorated seats was sound, and could easily support the precast units. Connection to the remaining substructure was achieved through a series of 18" diameter circular voids with vertical bars drilled and grouted into the existing concrete. A four-inch gap was detailed between the precast units and existing substructure to facilitate placing the units at the correct elevation. The gap and circular voids were then filled with a high-flow, low exotherm, non-shrink epoxy grout. A section through the abutment is shown in Figure 9.

The decision to use an epoxy grout over a typical cementitious grout arose from the need for a material that would provide rapid strength gain so the superstructure units could be set as quickly as possible. A performance specification was developed that called for an epoxy grout that could achieve 3,000 psi compressive strength in six hours. In addition to high early strength gain, epoxy grouts offer further benefits over cement grouts:

a) Ultimate compressive strengths reach 12,000–15,000 psi for epoxy grouts at seven days, compared to 5,000–7,000 psi for cement grouts.
b) Epoxy grouts exhibit a better bond strength to steel and concrete than cement grouts.
c) Epoxy grouts are far more resistant to chemical attack than cement grouts.
d) Most epoxy grouts have small amounts of shrinkage, whereas the product used for this project advertised zero shrinkage, typical design values assume 95% of the bearing contact area is effective. Cement grouts are also non-shrink but can exhibit expansion, which could have caused misalignment of the precast units.

Overall, epoxy grouts exhibit greater durability, strength, and toughness than cement grouts. The only major concern that needs to be investigated when specifying an epoxy grout is the difference

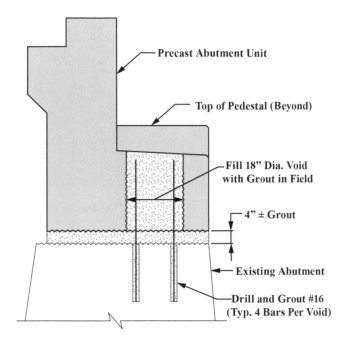

Figure 9.   Precast abutment unit connection detail (pier similar).

in the thermal coefficient of expansion (COE) between the grout and concrete. Most epoxy grouts have a COE on the order of two to ten times that of steel and concrete. When pouring an epoxy grout with large heat generation (high exotherm) against existing concrete, the rapid temperature changes could potentially cause cracking as the concrete will contract and expand at a different rate than the grout. This issue was mitigated by using a low exotherm product and extending the material with a 3/8" pea gravel in the large volume voids to further reduce heat generation. Additionally, the product used had a COE of approximately three times that of concrete, significantly lower than most commercially available epoxy grout materials.

4   CONSTRUCTION CHALLENGES

4.1   *PBE pre-assembly*

When performing accelerated bridge construction within a limited closure window it is important to minimize the chance of any surprises emerging during the shutdown. Prefabricated bridge elements must be fabricated to exact dimensions determined from accurate survey of the existing structure as any fit-up issues could prove disastrous. GPI added requirements in the contract documents that mandated all the prefabricated units be pre-assembled in their final configuration before being shipped to the jobsite. During the pre-assembly, the dimensions of each unit were verified and the overall geometry of the connected units checked against the requirements in the contract plans.

The Contractor used two separate fabricators to construct the precast substructure units and the NEXT beams. Located over 200 miles away from each other, assembling the substructure and superstructure units in one location would not have been feasible and therefore, separate operations were necessary. The substructure units were staged on a level surface and the pedestal elevations verified, shown in Figure 11. The width of the units were also checked and compared against the existing layout to ensure full bearing on the remaining portions of the substructure.

Figure 10.   Superstructure pre-assembly.

Figure 11.   Substructure pre-assembly.

At a separate location, the NEXT beams were assembled and blocked using the pedestal elevations obtained from the substructure assembly, shown in Figure 10. A major concern when connecting prestressed concrete units is differential camber, which could lead to fit-up and riding surface issues. Although the structure was to receive an overlay after the closure period, it was important to provide a relatively smooth riding surface in the interim and limit differences in joint elevations to facilitate proper forming of the UHPC joint pours. The pre-assembly results showed that all beam joint elevations were within the allowable tolerances. This was due in part to the rapid fabrication schedule of the beams and the use of identical prestressing patterns for all the units. If the relative vertical distance between joint flanges had exceeded 1/4", the contractor was required to reduce the high beam camber by applying additional dead load or using jacks. The beam lengths were also measured and imported into a CAD drawing of the existing surveyed plan along with the substructure units. This verified that all the PBEs would fit correctly during installation. The pre-assembly operations gave the project team confidence that the units would be erected with little to no issue. The units were ultimately installed in their final configuration to within less than 1/16" difference from the design locations and elevations.

4.2   *Grout pad*

The connection of the precast substructure units to the existing substructure to remain was detailed to allow the units to bear on a four inch grout pad. This dimension was set to provide the contractor flexibility during the demolition operations when removing the existing substructure to the required elevations. After removal, the surface would be surveyed and steel shims used to temporarily support the precast units prior to grouting operations. Figure 12 shows the precast substructure unit resting on the steel shims.

The methods used by the contractor to demolish the existing substructure allowed very precise elevations to be obtained. A wire saw was used to cleanly and accurately remove the top portion of the pier to the exact elevation required. This resulted in a very smooth surface and the corresponding shim heights varied from four to five inches thick. At the abutments, the front and rear faces were sawcut four inches deep at the necessary elevations, and a hydraulic breaker was used to demolish the existing seats. Hand operated hammers and grinders were then used to finish the surface to a near level elevation. This resulted in abutment shim heights ranging from three to five inches thick.

Following the grouting operations, shallow voids near the exterior surface of the grout seam were discovered in localized areas between the grout and precast units. These voids were potentially caused from small cracks in the substructure slowly absorbing some of the grout during the cure time and from form leakage. The epoxy grout is self-consolidating and therefore very fluid. Any small gap in the formwork would have led to leakage and caused a void to form along the top edge. Additionally, after reviewing the contractor's methods, it was noticed that as detailed, the

Figure 12.   Precast unit on steel shims.

Table 1.   UHPC placement rates.

Pour #	Location	Duration	Quantity	Placement Rate
1	West Approach Slabs	1.0 hrs	1.5 CY	1.50 CY/hr
2	West Span Joints	2.0 hrs	2.5 CY	1.25 CY/hr
3	East Span Joints	2.0 hrs	2.5 CY	1.25 CY/hr
4	East Approach Slabs	1.0 hrs	1.5 CY	1.50 CY/hr
5	West Abutment Closure	2.0 hrs	4.0 CY	2.00 CY/hr
6	East Abutment Closure	2.0 hrs	4.0 CY	2.00 CY/hr
7	Pier Closure	3.5 hrs	7.5 CY	2.14 CY/hr

four inch pad thickness was excessive. Discussions on site yielded that a two inch gap would have been optimal, which would have cut down on material costs and allowed the grouting activities to proceed much faster. The voids were repaired by pressure injecting epoxy grout and the surface treated with an epoxy waterproofing compound.

4.3   *UHPC pour and curing*

The UHPC material was batched on site using a mobile mixing system. Power buggys then transported the material from the mixers to the joint locations where it was carefully poured to avoid any spillage or overflow. The fluid nature of the material allows it to self-consolidate, however it also requires that the formwork be in complete contact with the edges of the joint. Any gap greater than 1/8" will lead to leakage and loss of material through the form. Pieces of galvanized rebar were used with coil loops spaced at two-foot intervals to support plastic-wrapped plywood beneath the longitudinal beam joints. This allowed the formwork to be supported internally and eliminated any insert voids at the top of the joint.

A pour sequence was specified for the UHPC material, whereby the approach slabs would be poured first, followed by the longitudinal span joints, abutments, and finish with the pier closure diaphragm. It was important to pour the pier diaphragm last, as continuity should always be established after the structure has been tied together. The UHPC manufacturer called for two mobile mixers and two power buggys to handle the pour operations. The placement rates achieved with this equipment setup are shown in Table 1 for each pour location.

As expected, the longitudinal joint pours represented the slowest placement operations. This was due to the narrow six-inch opening through which the UHPC had to be carefully poured.

In contrast, the abutment and pier closure pours went much faster due to the significantly wider openings, achieving placement rates almost double that of the longitudinal joints.

Immediately after the joints were filled, a top form was applied with plastic chimneys placed at either end. A 1/4″ thick compressible backer rod was installed along each edge of the form, and additional UHPC material was placed in the low side chimney to top off and overfill the joint. This procedure ensures that the joints are completely filled and eliminates any low spots. Due to the rapid set time of UHPC, it was critical that the top of the joints be formed and overfilled as quickly as possible. Even with the addition of ice into the mix to increase working time, the material will stiffen up within 20–30 minutes in ambient temperatures around 70°–80°F. Overfilling a joint that has already begun to solidify will create a cold joint near the surface between the separate pours. The strength required to load the structure was achieved after 22 hours of curing. Thermocouples were installed at several locations in the UHPC joints and temperature readings were compared against pre-established maturity curves to determine when the material had reached sufficient strength. After the top forms were removed, a methacrylate sealer was applied along each edge to ensure the joints would be completely waterproof.

5   CONCLUSIONS

The Route 46 over Musconetcong River Bridge was successfully reconstructed over a ten-day (240 hour) complete road closure. Through the use of innovative materials, careful detailing, and a determined construction team, the project was delivered on schedule and within budget. The short-term traffic impacts were found to be acceptable and easily justifiable to the surrounding residential and business communities given the increase in service life of this critical structure.

ACKNOWLEDGEMENTS

The authors would like to extend their gratitude and appreciation to the entire project team:

Owner: New Jersey Department of Transportation
Design Engineer: Greenman-Pedersen, Inc., Lebanon, NJ
Prime Contractor: J.F. Creamer & Son Joint Venture with Joseph Sanzari, Inc., Carlstadt, NJ
NEXT Beam Precaster: Precast Systems Inc., Allentown, NJ
Substructure & Approach Slab Precaster: The Fort Miller Co., Inc., Schuylerville, NY
UHPC Supplier: Lafarge North America Inc., Calgary, Alberta, Canada

REFERENCES

AASHTO LRFD Bridge Design Specifications. AASHTO, Washington, D.C., 2012.
Hastak, M., Mirmiran, A., Miller, R., Shah, R., Castrodale, R. *State of Practice for Positive Moment Connections in Prestressed Concrete Girders Made Continuous.* ASCE Journal of Bridge Engineering. September/October 2003.
Freyermuth, C.L. *Design of Continuous Highway Bridges with Precast, Prestressed Concrete Girders.* PCI Journal. pp 14–39. April 1969.
Innovative Bridge Designs for Rapid Renewal: ABC Toolkit. SHRP 2 Report S2-R04-RR-2. Transportation Research Board. Washington, D.C. 2013.
Miller, R.A., R. Castrodale, A. Mirmiran, and M. Hastak. *Connection of Simple-Span Precast Concrete Girders for Continuity*, National Cooperative Highway Research Program Report 519, Transportation Research Board, National Research Council, Washington, D.C., 2004.
Miller, R.A. *Special Note: Bridge Composed of Simple-Span, Precast Concrete Girders Made Continuous.* ASPIRE: The Concrete Bridge Magazine. Summer 2014.
Russell, H.G., Graybeal, B.A. *Ultra-High Performance Concrete: A State-of-the-Art Report for the Bridge Community.* FHWA Publication HRT-13-060. Federal Highway Administration. June 2013.

Chapter 6

Slip and creep performance for metallized connection faying surfaces used in steel bridge construction

M. Ampleman, C.-D. Annan & M. Fafard
Civil and Water Engineering Department, Université Laval, Quebec City, Quebec, Canada

J. Ocel
Federal Highway Administration, McLean, VA, USA

É. Lévesque
Canam-Bridges, Quebec City, Quebec, Canada

ABSTRACT: Steel bridge surfaces exposed to aggressive environment must be protected to preserve structural integrity and provide longevity. Metallization is a versatile thermal spray solution commonly used in steel bridge fabrication due to its long-term protection. Highway bridge design standards in North America specify values for slip coefficients to be used in slip-critical connections for various faying surfaces. Currently, these standards do not prescribe a slip coefficient value for metallized faying surfaces used with slip-critical bolted connections. Thus, bridge fabricators are compelled to mask off joint faying surfaces before metalizing, which is very time-consuming and expensive. This article presents results from two different research work carried by Université Laval, Quebec City, Canada, and the Federal Highway Administration, Virginia, United States, on the slip performance of metalized faying surfaces. Both short-duration slip tests under static load and long-term sustained creep tests, performed at the laboratories of the two partner institutions, are reported in this article. The slip resistance is then characterized based on the Canadian Highway bridge design code and the AASHTO LRFD bridge design code. Results of this study are likely to influence future code revisions in North America.

1 INTRODUCTION

Steel bridge elements are subjected to harsh environmental conditions and often require surface coating protection to preserve structural integrity. A surface coating method which has proved effective over the years and is fast becoming a commonly used surface protection solution for steel bridges in North American is metallization. Metallization is a term which describes the thermal spray of zinc, aluminum or both on metal surfaces, providing effective physical barrier as well as sacrificial protection against corrosion (Teruo 1999; Bayliss & Deacon 2002). Metallization is known to produce no volatile organic compound emission (Chang & al. 1999), and the metallized substrate is also known to be compatible with many different sealer types, recommended for use by a number of practical guidelines in the United States. There is no limit to the size of structural elements that can be metallized and the hardness and toughness of the metallized coating make it especially useful in high-wear applications (Bhatia 1999). It is noteworthy that the merit of a metallized coating protection depends on such factors as the coating thickness, steel surface preparation, coating type, and workmanship.

In order to be able to metallize an entire steel bridge element, including its connection areas, commonly referred to as faying surfaces, designers need to know the slip resistance of metallized

Figure 1.   Masked faying surfaces.

faying surfaces used with a slip-critical bolted connection. High strength bolted connections are designed as slip-critical if the joint would be subjected to load reversals, vibrations or heavy impact loads, where slip can be detrimental to the serviceability of the structure. The resistance of this type of connection is derived from the friction between the shear planes of the connected parts. The slip resistance is developed by the clamping force of the preloaded bolts on the connected parts. The faying surface condition is thus a critical parameter in the development of slip resistance.

Design standards such as the Canadian highway bridge design code, CAN/CSA S6-06 (CSA 2006) and the AASHTO LRFD bridge design code (AASHTO 2014) specify the conditions to which connection faying surfaces are to be prepared and their associated slip coefficient values. Both codes provide 3 classes of faying surface condition, A, B, and C. For both standards, Class A surfaces have a minimum slip coefficient of 0.33, representing unpainted clean mill scale steel surface or a blast-cleaned surface with a Class A certified coating. Class B faying surfaces have a minimum slip coefficient of 0.50, and represent unpainted blast-cleaned surfaces or surfaces with a Class B certified coating on blast-cleaned surface. The only variation between the two design standards lies in the slip coefficient value for Class C faying surfaces, representing hot-dip galvanized and wire brushed surfaces. The minimum slip coefficient value for this Class is 0.40 in the Canadian standard and 0.33 in the AASHTO LRFD design code. It is worth mentioning that the minimum slip coefficient value for Class B in the 2014 edition of the Canadian standard is revised to 0.52.

It is evident from the above-mentioned code specifications that there is no direct classification for metallized connection faying surfaces. From a steel bridge fabricator's perspective, there are advantages to metallize the entire steel bridge element, including the connection faying surfaces, to eliminate the extra labor for masking off all connection faying surfaces before metallizing (Figure 1) and doing touch-ups after the masks are taken off. In order to eliminate this time-consuming practice of masking off connection faying surfaces, it is important to characterize the slip resistance of metallized faying surfaces used with slip-critical connections in light of the prevailing design standards.

The Research Council on Structural Connections (RCSC) *Specification for Structural Joints Using High-Strength Bolts* (RCSC 2014) provides, in Appendix A, a widely used methodology to evaluate the slip resistance of a coated faying surface. Two sets of test are prescribed to characterize the slip resistance of a coating: the first is a short-duration static test to evaluate the mean slip coefficient. If the mean slip coefficient is found to be satisfactory, a long-term sustained tension creep test is carried out to ensure that the coating will not undergo significant creep and that creep will not adversely affect the long term slip resistance in the connection.

This paper presents results from research work carried by Université Laval, Canada, and the Federal Highway Administration (FHWA), United States, on the slip performance of 99.9% zinc-metallized faying surfaces (typically said as 100% zinc-metallized). FHWA also studied the slip performance for a 85/15 zinc/aluminum (Zn/Al) thermal-spray coated faying surfaces. Both short-duration slip tests under static load, and long-term tension creep tests performed independently at the laboratories of the two collaborating institutions are reported in this article. Based on the collective results from these experimental works, the slip resistance is characterized in light of the two North American bridge design codes.

2 TEST COMPONENTS FOR SLIP RESISTANCE

2.1 *Short-term compression slip tests*

Slip tests were designed according to RCSC specification (RCSC 2014) to establish the mean slip coefficient of the metallized coated surface. For this test, assemblies were made from three identical 5/8 inch thick steel plates, each measuring 4 in. by 4 in. The three test plates were mechanically clamped together and the middle plate was loaded in compression while the two exterior plates sat solidly on their flat edges. The clamping load of the bolt and the slip in the joint are monitored and recorded throughout the test. The slip coefficient (k_s) is calculated as follow:

$$k_s = \frac{\text{Slip load}}{\text{Clamping force x Number of slip planes (=2)}} \qquad (1)$$

where the slip load is determined as either the peak load on the load-slip curve or the load corresponding to 0.5 mm (0.02 in.) of slip, according to the RCSC (RCSC 2014).

For each set of test parameters, five replicate specimens are tested, and the mean slip coefficient is recorded. The RCSC specification (RCSC 2014) provides additional guidance on loading configuration, instrumentation, and loading protocols.

2.2 *Long-term tension creep tests*

Tension creep tests have been conducted to ensure the coating will not undergo significant creep deformation under service load (Yura & Frank 1985). The test consists of a set of three specimens assembled in series by ASTM A490 bolts. The bolts connecting different specimens are left loose, and plates making up individual specimens are clamped by fully pretensioned A490 bolts. The test chain is loaded in tension for 1000 hours. The sustained tension load applied during the test corresponds to the serviceability load level calculated as follow:

$$\text{Service Load} = k_s \times D \times T_m \times N_b \times m \qquad (2)$$

where k_s is the design slip coefficient, D is a slip probability factor (equal to 0.80), T_m is the minimum bolt pretension specified (equal to 218 kN for 7/8 inch diameter ASTM A490 bolts according to (Kulak et al. 2001), Nb is the number of bolts (equal to 1 in the present tests), and m is the number of slip planes (equal to 2).

The relative displacement between the middle plate and the two lap plates was recorded on both side of each assembly. Creep deformation measurements were recorded after the first 30 minutes of loading and measurements continued under the sustained tension loading for 1000 hours. The creep deformation for each specimen was obtained as the average of the two measurements, and this deformation measurement is deemed satisfactory if it is found to be less or equal to 0.127 mm

Table 1.   Parameters tested – Université Laval.

#	Parameters	Variables
1	Composition of coating	Zn – 100-percent zinc
2	Target thickness of coating	6 m – 6 mils
		12 m – 12 mils
3	Presence of burrs	s – burrs cleaned
		a – burrs left in place

(0.005 inch). Subsequently, the assembly is loaded in tension up to the design slip load, equal to the real average clamping load times the design slip coefficient times the number of slip planes (=2). This post-creep slip test is carried out in order to ensure that the loss of clamping force in the bolt does not reduce the slip load below that associated with the design slip coefficient. If the average post-creep slip deformation that occurs at this load level is less than 0.381 mm (0.015 inch) for the three identical specimens, the coated faying surface tested is considered to meet the requirements for the design slip coefficient. If any of the above-mentioned creep and post-creep slip requirements is not respected, the coating is considered to have failed for the design slip coefficient and a new test is required with a lesser design slip resistance. The RCSC specification (RCSC 2014) provides additional guidance on loading configuration, instrumentation, and loading protocols.

3 TESTS FROM UNIVERSITÉ LAVAL

3.1 *Plate preparation and test matrix*

Test plates were machined from steel grade 350AT and were zinc-metallized in the shop under controlled environmental conditions. Thermal spray coating was applied from a 99.9% zinc (hereafter said as 100% for simplicity) wire through an electric arc in accordance with SSPC-CS 23.00/AWS C2.23M/NACE No. 12 (SSPC/AWS/NACE 2003). The steel substrate for each plate was prepared to white metal finish SSPC-SP 5 and the angular profile depth was measured in shop to certify the standard requirement. All plates were coated on both sides. Thus, there are six layers of coating between the bolt head and the nut in each assembly. The thickness of the coating was measured using a Positector magnetic gage on each test plate in order to mate plates with similar average coating thickness. In accordance with the Society for Protective Coatings SSPC-PA 2 (SSPC 2012) standard, five different readings were taken on each plate faying surface and the average thickness was determined.

In some of the cases tested, small burrs produced by bolt hole drilling in plates were left in place and not cleaned. However, all burrs were within the limit imposed by RCSC bolt specification (RCSC 2014), since their depths were far less than 1/16 in. Table 1 contains the parameters of the specimens tested at the laboratory of Université Laval. Each specimen has been identified following the variables used in Table 1.

For example, specimen Zn-6 m-s refers to a faying surface with a 6 mils thick 100-percent zinc-metallized coating, where small burrs produced by drilling are removed.

3.2 *Short-term compression slip tests: instrumentation and results*

The compression slip tests were performed on a 1500 kN MTS hydraulic Universal Testing machine. The applied loading rate was 100 kN/minute. The clamping load was applied using a 7/8 inch diameter ASTM A325 bolt preloaded to about 70% of its tension capacity. ASTM A325 bolts were preferred for use in the slip tests since it is the most commonly used bolt in Canada. The pre-tensioned force was monitored using a washer-type load cell of 500 kN Omega installed in

Table 2. Slip results – Université Laval.

Specimen	k_1	k_2	k_3	k_4	k_5	$k_{average}$	Standard deviation
Zn-6m-s	0.88	0.81	0.77	0.80	0.84	0.82	0.04
Zn-12m-s	0.80	0.76	0.91	0.92	0.86	0.85	0.07
Zn-6m-a	0.90	0.82	0.96	1.00	-	0.92	0.08
Zn-12m-a	0.71	0.88	0.92	0.89	0.89	0.86	0.08

Extracted from Annan & Chiza (2013).

series with the clamped test plate assembly from time of assembly to the end of the test. The relative displacement between the loaded middle plate and the two side plates is found to be the slip displacement and was measured using LVDT displacement transducers. A data acquisition system was used to monitor and record the applied loading and the associated slip. It also served to monitor the clamping force during the test.

Table 2 reproduce individual specimen results of the short-term slip tests conducted by Annan & Chiza (2013). The mean slip coefficient and the standards deviation are also shown for each set of parameters. The mean slip coefficient of 0.85 was obtained for the 12 mils zinc-metallized coating thickness with burrs removed. When small burrs within the limit imposed by RCSC specification (RCSC 2014) are left in place, the mean slip coefficient observed remained almost unchanged at 0.86 for the 12 mils zinc-metallized coating thickness. For the 6 mils zinc-metallized coating thickness with cleaned burrs and with burrs left in place, the slip coefficients were obtained as 0.82 and 0.86, respectively. These results are significantly greater than the specified slip coefficient value of 0.50 for Class B faying surfaces by both the Canadian and American standards.

For the 6 mils targeted thickness, average coating thicknesses ranged between 5 and 8 mils, and between 11 and 14 mils for the 12 mils nominal coating thickness.

3.3 *Long-term tension creep tests: instrumentation and results*

Tension creep tests were performed on a 500 kN MTS hydraulic Universal Testing machine. For each specimen, the relative displacement between the middle plate and the two lap plates was measured using two MTS extensometers, on each side of the assembly. The displacement recorded for each assembly is the average of the two measurements. The bolt preload was manually applied by using a hand-held ratchet. This was continuously monitored from the time of assembly through to the end of testing by using a washer-type load cell of 500 kN Omega installed in series with the clamped test plate assembly. ASTM A490 high strength bolts were used as prescribed by the RCSC specification (RCSC 2014). A data acquisition system was used to monitor and record the applied loading, the extensometers measurements and the load cells measurements. For creep tests, a design slip coefficient must be chosen in order to calculate the tension service load and to verify the creep performance under that load associated to that slip coefficient. The creep performance at two different slip coefficient values (0.5 and 0.55) was assessed for the metallized faying surface. A parameter was thus added to each specimen notation to represent the assessed design slip coefficients of 0.50 and 0.55, as shown in Table 3.

The individual specimen dry film thicknesses and results of the long-term tension creep tests are shown in Table 3. All of the creep displacements were less than 0.127 mm (0.005 in.) after 1000-h for each parameters tested. Each chain slipped less than 0.381 mm (0.015 in.) after 1000-h when the load was increased to the design slip load. So, each parameter tested respected the requirement of the RCSC specification (RCSC 2014) for their respective design slip coefficient. Table 3 also shows that all 12 mils thick metallized coating present a creep deformation greater than 6 mils thick coating. The presence of small burrs does not seem to have any effect on the creep performance, since the final displacement after the post-creep slip test was 0.1012 mm for 12 mils zinc-metallized coating thickness with burrs removed and 0.1013 mm for the same parameters, but with small burrs

Table 3.   Specimen dry film thicknesses and creep results – Université Laval.

Specimen	Assembly	Left Outer Panel mils	Middle Plate, Left Face mils	Middle Plate, Right Face mils	Right Outer Panel mils	1000-h Creep displacement mm
Zn-6m-s-0.50	1	7.5	7.6	7.0	7.5	0.0449[a]
	2	7.5	7.6	6.4	7.6	0.0570[a]
	3	7.6	7.3	6.4	7.7	0.0556[a]
Zn-12m-s-0.50	1	12.9	13.4	12.3	12.8	0.0702[b]
	2	12.5	12.5	12.5	12.6	0.0798[b]
	3	12.1	12.3	12.3	12.1	0.0991[b]
	1	6.7	7.4	6.4	7.2	0.0417[c]
Zn-6m-a-0.55	2	7.9	6.5	6.8	7.8	0.0492[c]
	3	7.5	7.4	6.3	7.6	0.0375[c]
	1	12.9	13.6	12.0	13.2	0.0861[d]
Zn-12m-a-0.55	2	13.0	13.0	12.9	12.8	0.0711[d]
	3	12.4	12.2	12.7	12.6	0.0869[d]

[a]Chain of three specimens slipped 0.0682 mm when loading to the design load.
[b]Chain of three specimens slipped 0.1012 mm when loading to the design load.
[c]Chain of three specimens slipped 0.0548 mm when loading to the design load.
[d]Chain of three specimens slipped 0.1013 mm when loading to the design load.

left in place. Also, testing with a slip coefficient of 0.55 instead of 0.50 did not produce a greater creep deformation even if the applied tension load was higher.

4   TESTS FROM FEDERAL HIGHWAY ADMINISTRATION

4.1   *Test matrix and plate preparation*

Similar to the plate substrate preparation in the study by Université Laval, each plate was first blast cleaned to white metal finish SSPC-SP 5 condition. This was achieved with G40 steel grit to target 3.5- to 4.5-mil surface profile. The surface profile was verified with replica tape according to ASTM D4417-11. The measured profile on the faying surfaces ranged from 3.2 to 3.9 mils. Metallization was applied on both side in accordance with SSPC-CS 23.00/AWS C2.23M/NACE No. 12 (SSPC/AWS/NACE 2003) to all specimens on the same day. The target metallized-coating thickness was 14-mil thickness in each case (i.e. a nominal 12 mil thickness plus an additional 2 mils per RCSC for accidental overspray). A separate spray unit was used for zinc and zinc/aluminum alloy wires. The spray distance varied between 4 and 6 inches. The thickness of the thermal spray coating was measured on both sides of each specimen with a calibrated type 2 (electronic) non-destructive dry film thickness (DFT) gauge. Because of the relatively small panel size, only two spot measurements were obtained from each panel face.

Table 4 contains the parameters of the specimens tested. Each specimen has been identified following the variables used in Table 4.

4.2   *Short-term compression slip tests: instrumentation and results*

The slip tests were performed using a 450 kN (100 kips) jack. The applied compression loading rate did not exceed either 111 kN/minute (25 kips/min) or 0.076 mm/min (0.003 inches/min). A 7/8 inch diameter threaded rod is inserted through the holes and tensioned and maintained to 218 kN to represent the clamping force from an ASTM A490 bolt. The compression load applied and the displacement is recorded.

Table 4.   Parameters tested – FHWA.

#	Parameters	Variables
1	Composition of coating	Zn – 100-percent zinc
		Zn/Al – 85/15-percent zinc/aluminum
2	Target thickness of coating	14 m – 14 mils
3	Presence of burrs	s – burrs cleaned

Table 5.   Slip results – FHWA.

Specimen	k_1	k_2	k_3	k_4	k_5	$k_{average}$	Standard deviation
Zn-14m-s	>0.857	0.743	0.641	>0.824	>0.833	>0.78	0.08
Zn/Al-14m-s	>0.837	>0.857	0.702	>0.865	0.596	>0.77	0.11

Extracted from Ocel (2014)

Table 5 reports the individual specimen results of the short-term slip tests conducted by FHWA (2014). Slip coefficient is reported for each assembly tested. Many of these results are reported with a greater-than symbol because the vertical load was nearing the limits of the load frame and the test was terminated for safety reasons. The 14 mils zinc metallized-coating thickness yielded a mean slip coefficient greater than 0.78 with a standard deviation of 0.08. The 14 mils thick zinc/aluminum metallized-coating yield a mean slip coefficient greater than 0.77 with a standard deviation of 0.11. These results far exceed the 0.50 requirement for Class B by North American standards and are consistent with results from Université Laval. For the 14 mils targeted thickness, average coating thicknesses ranged between 13.1 and 15.6 mils.

4.3   *Long-term tension creep tests: instrumentation and results*

Tension creep tests were performed using a 450 kN (100 kips) jack. ASTM F2280 (twist-off equivalent to ASTM A490) high strength bolts were used and tensioned with an electric wrench. Real bolt tension was the average of three bolts from the lot verified with a bolt-tension calibration device. Creep displacement for each specimen was monitored manually with a dial gauge magnetically affixed to each specimen on both sides. The displacement recorded for each assembly has been the average of the two measurements. Design slip coefficient for both creep test was selected as 0.50 for qualification as Class B.

The individual specimen dry film thicknesses and results of the long-term tension creep tests from FHWA (2014) are shown in Table 6. All of the creep displacements were less than 0.127 mm (0.005 in.) after 1000 hours for each of the alloys. Each of the two chains slipped less than 0.381 mm (0.015 in.) after 1000 hours when the load was increased to the design slip load. In fact, zinc and zinc-aluminum alloy metallized-coating yielded a slip displacement of 0.1016 mm and 0.1473 mm in the post-creep slip test, respectively. Therefore, both zinc and zinc/aluminum alloy metallized-coating passed the creep test, according to RCSC specification (RCSC 2014).

5   COMPARISON OF RESULTS

Figure 2 shows the mean slip coefficient obtained for zinc-metallized coating for 6, 12 and 14 nominal coating thickness. Figure 3 presents the creep displacement obtained for the same specimens. By observing the results in Figure 2, it appears that the coating thickness doesn't have any effects on the slip coefficient. However, Figure 3 shows that the coating thickness influences the creep behavior. In fact, as the thickness of the coating increases, the creep displacement increases.

Table 6.   Specimen dry film thicknesses and creep results – FHWA.

Specimen	Assembly	Left Outer Panel mils	Middle Plate, Left Face mils	Middle Plate, Right Face mils	Right Outer Panel mils	1000 h Creep displacement mm
Zn-14m-s-0.50	1	14.6	14.0	13.8	14.6	0.0889[a]
	2	14.2	14.1	14.7	14.1	0.0914[a]
	3	13.8	15.1	14.9	14.2	0.1016[a]
Zn/Al-14m-s-0.50	1	13.5	14.0	14.4	13.4	0.1524[b,c]
	2	14.8	14.5	13.9	13.8	0.0737[c]
	3	14.2	14.3	14.7	14.5	0.0838[c]

Extracted from Ocel (2014)
[a] Chain of three specimens slipped 0.1016 mm when loading to the design load.
[b] Testing on neighboring chain failed, causing movement. Chain passed final loading, and slip in excess of 0.1270 mm is thought to be result of shock loading from neighboring chain.
[c] Chain of three specimens slipped 0.1473 mm when loading to the design load.

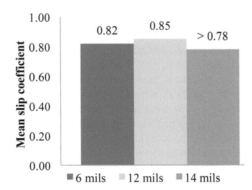

Figure 2.   Mean slip coefficient of zinc-metallized coating.

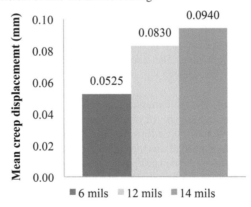

Figure 3.   Mean creep displacement of zinc-metallized coating for a design slip coefficient of 0.50.

Figure 4 presents the mean creep displacement obtained for a design slip coefficient of 0.50 and 0.55 for both 6 and 12 nominal zinc-metallized coating thickness. Results show that the creep behavior is not affected by increasing the sustained tension service load from a design slip coefficient of 0.50 to 0.55.

Figure 5 shows the mean creep displacement for 14 mils thick zinc and 85/15 zinc/aluminum-metallized coating for a design slip coefficient of 0.50. Zinc/aluminum-metallized coating yields a

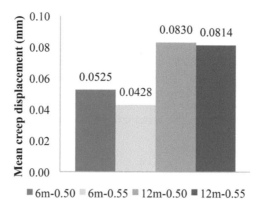

Figure 4.   Mean creep displacement of zinc-metallized coating for a design slip coefficient of 0.50 and 0.55.

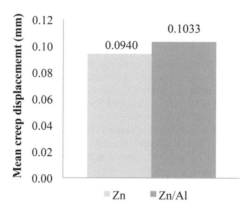

Figure 5.   Mean creep displacement of 14 mils thick zinc and zinc/aluminum-metallized coating for a design slip coefficient of 0.50.

greater creep displacement than 100-percent zinc. However, both composition of metallized coating presents a creep deformation lesser than the limit permitted by RCSC specification (RCSC 2014).

6   CONCLUSIONS

In order to be able to metallize an entire steel bridge element, including its faying surfaces, the slip resistance of metallized faying surfaces must be known. In this paper, results for short and long-term slip performance from two independent laboratories are reported. Conclusions from the studies are summarized as follows:

1. In the tests conducted by Université Laval, the minimum mean slip coefficient obtained was 0.82 representing the 6 mils thick zinc-metallized coating surfaces. This slip coefficient is significantly greater than that specified for Class B faying surface by the North American standards.
2. In the tests conducted by FHWA, the mean slip coefficient was reported as greater than 0.78 and greater than 0.77 for the 14 mils thick zinc-metallized coating surfaces and the 14 mils thick zinc/aluminum thermal spray coated surfaces, respectively. These slip coefficients are significantly greater than that for Class B faying surfaces by North American standards.

3. For the creep tests conducted by the two institutions, the maximum creep deformation obtained for the zinc-metallized coating was 0.1016 mm for the 14 mils thick coating tested at a design slip coefficient of 0.50. For the zinc/aluminum thermal spray coating, the maximum creep displacement was obtained as 0.0838 mm for the 14 mils thick coating tested at a design slip coefficient of 0.50. These deformations are lesser than 0.127 mm, the requirement for qualification prescribed by RCSC specification (RCSC 2014).

4. Additional tests are in progress after which recommendation will be made to the Canadian and American codes committees to include metallized faying surfaces in Class B.

ACKNOWLEDGEMENT

The authors from Université Laval would like to acknowledge the financial support of the Natural Sciences and Engineering Research Council of Canada (NSERC), the Fonds de recherche du Québec – Nature et technologie (FRQNT) and Canam-Bridges, a division of the Canam group.

REFERENCES

AASHTO. 2014. AASHTO LRFD Bridge Design Specifications, 7th Edition, Washington, DC.

Annan, C-D. & Chiza, A. 2013. Characterization of Slip Resistance of High Strength Bolted Connections with Zinc-Based Metallized Faying Surfaces, Elsevier, Engineering Structures 56.

ASTM International 2011. ASTM D4417-11— Standard Test Methods for Field Measurement of Surface Profile of Blast Cleaned Steel, West Conshohocken, PA.

ASTM International 2011. ASTM D4541-09e1—Standard Test Method for Pull-Off Strength of Coatings Using Portable Adhesion Testers, West Conshohocken, PA.

Bayliss, D. A. & Deacon, D. H. 2002, Steelwork Corrosion Control, Second Edition, Spon Press, London, England.

Bhatia, A. 1999. Thermal Spraying Technology and Applications, Course No: T04-002, Continuing Education and Development, Inc., Stony Point, New-York.

CAN/CSA S6-06. 2006. Canadian Highway Bridge Design Code, Canadian Standards Association, Mississauga, Canada.

Chang, L. M., Zayed, T. & Fricker, J. D. 1999, Steel Bridge Protection Policy: Metalization of Steel Bridges: Research and Practice, Purdue e-Pubs, Purdue University.

Kulak, G. L., Fisher, J. W. & Struik, J. H. A. 2001. Guide to Design Criteria for Bolted and Riveted Joints, 2nd Edition, Research Council on Structural Connections.

Ocel, J. 2014, "Techbrief: Slip and Creep of Thermal Spray Coatings," FHWA-HRT-14-083, Federal Highway Administration, McLean, Virginia, United States.

Research Council on Structural Connections (RCSC). 2014. Specification for Structural Joints Using High-Strength Bolts, American Institute of Steel Construction, Chicago, Illinois.

SSPC/AWS/NACE. 2003. Specification for the Application of Thermal Spray Coatings (Metallizing) of Aluminum, Zinc, and Their Alloys and Composites for the Corrosion Protection of Steel, Joint International Standard SSPC-CS 23.00/AWS C.2.23M/NACE No.12.

SSPC 2012. Procedure for Determining Conformance to Dry Coating Thickness Requirements, Paint Application Specification No. 2, SSPC: The Society for Protective Coatings.

Teruo, K. 1999. The Status of Metal Spray Corrosion Resistance Method under Normal Temperature, Japan Society of Corrosion Engineering, Japan.

Yura, J. A. & Frank, K. H. 1985. Testing Method to Determine the Slip Coefficient for Coatings Used in Bolted Joints, Engineering Journal, American Institute of Steel Construction, Third Quarter, Pg. 151–155.

Chapter 7

Design and construction guidelines for skewed/curved steel I-girder bridges

V.L. Liang, W.S. Johnsen & B.P. McFadden
Greenman-Pedersen, Inc., Lebanon, NJ, USA

C. Titze
Cambridge Systematics, New York, NY, USA

G. Venkiteela
New Jersey Department of Transportation Research Bureau, Trenton, NJ, USA

ABSTRACT: The New Jersey Department of Transportation (NJDOT) Research Bureau retained Cambridge Systematics (CS) and Greenman-Pedersen Inc. (GPI) to develop design and construction engineering guidelines and checklists to instruct designers on how to properly address out-of-plumb issues for skewed and curved steel I-Girder bridges during the design phase of the project. The guidelines were developed based on current AASHTO design specifications, available research papers and reports, and GPI's past project experience. This paper focuses on the design portion of the research project which covers the following aspects: theory and analysis, out-of-plumb tolerances, techniques to minimize girder differential deflection and associated out-of-plumb issues, cross-frames detailing methods, design considerations for cross-frames, understanding thermal behavior and determining bearing fixity layout and guided bearing orientation, design consideration for deck joints, and connection detailing. Additionally, construction engineering guidelines and notes to be included in the contract plans are discussed.

1 INTRODUCTION

Skewed and/or horizontally curved steel I-girder bridges make up a significant portion of the steel bridge population in the United States. The structural behavior of such bridges is more complicated than their non-skewed, tangent counterparts due to additional effects from skew/curvature. Therefore, supplemental guidance is recommended for both the design and construction phases. Figures 1–2 are examples of a severely skewed and a highly curved bridge.

Prior to the research project completed by CS and GPI, NJDOT did not have specific guidance on how designers and contractors should address issues associated with skewed and/or horizontally curved steel I-girder bridges. Other state DOTs offer guidance on design, detailing, and fabrication policies but guidance varies from state to state and may even be contradictory on certain issues. The objective of the research project completed by CS and GPI was to generate and compile guidelines for both design and construction engineering in order to fill this void. The objective of this paper is to provide the reader with practical design and analysis guidelines that can be used on the general level.

2 THEORY AND ANALYSIS

2.1 *Knowing and understanding the challenges of designing skewed and/or curved girder bridges*

Girders deflect under vertical load and this deflection varies along the length of the girders. The deflection of a girder is greatest near mid-span, varies along the length of the span, and is equal to

Figure 1.   A severely skewed bridge in plan and elevation.

Figure 2.   A highly curved bridge in plan and elevation.

zero at the supports. For a non-skewed tangent bridge, the girder deflections at any cross-section of the bridge due to the deck weight are roughly the same assuming relatively equal girder sections and spacing. In contrast, the girder deflections at a cross-section of a skewed bridge are not equal, because the girders are longitudinally offset from each other due to the skew. Therefore, differential deflections will occur between the girders across any section of the bridge as shown in Figure 3. However, the girders cannot realize these differential deflections without twisting because they are tied together by relatively rigid cross-frames (Beckmann et al, 2008). As the dead load of the deck is applied, the change in the shape of the cross-frames is relatively minor as compared to the deflection of the girders. Prior to connection of the cross-frames, steel I-girders are torsionally

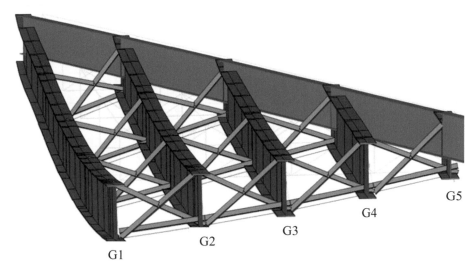

Figure 3.   3D Model of the Garden State Parkway (GSP) Bridge over Route 9/166 showing girder twisting and differential deflection, with 3.66″ max deflection at G1 and 0″ deflection at G5 near support.

flexible, however, once the cross-frames are connected the girders will twist (out-of-plumb) due to the differential deflection.

For straight skewed bridges, it is recommended that the girders should be plumb (within a reasonable tolerance) when the deck construction is complete (Beckmann et al, 2008). Therefore, the girders must be out-of-plumb before the deck pour so that as the deck is placed, the girders will deflect and untwist into the plumb condition. This out-of-plumb condition prior to deck placement is commonly referred to as web layover.

As with skewed bridges, deflections complicate girder behavior on curved bridges. On curved bridges with radial supports, deflection differences occur because the girders in a curved span have different lengths. Also, load applied at mid-span of a curved girder is at an offset from a chord line joining the supports, which induces eccentric load on the girder and framing system causing differential deflection between girders. If the substructures are skewed instead of radial, the skew induces additional deflection differences. Long spans and continuous spans further complicate the issue. As long spans tend to deflect more, the differential deflection between adjacent girders also increases. On continuous bridges, girder deflections are influenced by adjacent spans. In combination with skewed supports and/or curved geometry, the differential deflection between adjacent girders in continuous bridges is further amplified, especially for bridges with an unbalanced span arrangement.

2.2   *Comparison of Three Analysis Methods – 1D Line Girder Analysis, 2D Grillage or Plate and Eccentric Beam Analysis, and 3D Finite Element Analysis*

During the design process, the designer is faced with a decision on how to accurately model bridge complexities (skew, curve, etc.) to ensure an adequate design. NCHRP Report 725 developed a system based on bridge geometry, span continuity, and number of cross-frames that can be used by designers to select an appropriate level of analysis for skewed and curved I-girder bridges. The system evaluates the accuracy of traditional 1D and 2D analysis techniques against 3D analysis benchmark solutions and provides a scoring rubric for easy comparison of the different techniques.

The report developed a connectivity index (I_C) and skew index (I_S), which essentially rate the severity of the curve/skew and allow the scoring rubric to be further subcategorized (highly skewed, slightly skewed, skewed and curved, etc.).

Table 1.   Simplified matrix for recommended level of analysis: I-girder bridges.

Simplified Matrix for Recommended Level of Analysis - I-girder bridges										
Geometry	Major-Axis Bending Stress		Vertical Displacement		Cross-Frame Forces		Flange Lateral Bending Stress		Girder Layover at Bearings	
	Traditional 2D	1D-Line Girder	Traditional 2D	1D-Line Girder	Traditional 2D	1D-Line Girder	Traditional 2D	1D-Line Girder	Traditional 2D	1D-Line Girder
C (I_C <=1)	A	B	A	B	B	B	B	B	NA	NA
C (I_C>1)	B	C	F	C	C	C	C	C	NA	NA
S (I_S<0.30)	A	A	A	A	NA	NA	NA	NA	A	A
S (0.30<I_S<0.65)	B	B	A	B	F	F	F	F	A	B
S (I_S >= 0.65)	C	C	C	C	F	F	F	F	C	C
C & S (I_C>0.5 & I_S>0.1)	B	C	F	C	F	F	F	F	F	C

NCHRP Report 725 Table 3-1 is a summary of this detailed evaluation system and the authors urge the reader to review this table and Chapter 3.1.2 of the report in order to determine the applicability and accuracy of 1D, 2D and 3D analysis methods.

As part of the research project completed for the NJDOT, GPI compiled a simplified evaluation matrix, which uses the "Mode of Scores" portion of Table 3-1 in the NCHRP report but ignores the "Worst-Case Scores", as shown in Table 1 below.

Please note that the scores for Traditional 2D Analysis presented in the matrix above and the matrix in Report 725 Table 3-1 are based on software evaluation performed in 2011 under Report 725. As 2D analysis software is continuously being updated, the actual scores may differ from the scores shown above. The "C" under Geometry represents curved bridges, and "S" represents skewed bridges. "C&S" represents curved and skewed bridges. Table 1 above uses the same letter grade scoring system developed under NCHRP Report 725. For additional discussions on the letter grading system, and the calculation of the connectivity/skew index, see NCHRP Report 725.

3   DESIGN GUIDELINES

3.1   *Out-of-plumb tolerance and web layover*

In review of available literature (AASHTO/NSBA 2007; Linzell et al, 2010; White et al, 2012) the out-of-plumb tolerance under total deck placement for skewed and/or curved I-girders should be within D/96 or 0.01 radians, where D is the girder web depth in inches. Note that some agencies do require stricter tolerances (particularly New York State Department of Transportation) however; it is the authors' opinion that stricter requirements than those mentioned above are unnecessary and may result in higher construction/fabrication costs.

Based on the above, an out-of-plumb tolerance of D/96 under total deck placement is recommended, as the superimposed dead load will not be able to further twist the girder after the steel girders were made composite with the deck. If the computed out-of-plumb rotation is within this tolerance under total deck placement, the designer may still opt to detail the bridge as plumb under total dead load (which will induce a layover prior to deck placement), but is not required to since the tolerance criteria will be satisfied. If the designer decides not to induce web layover the girder shall be truly plumb under steel dead load to ensure the girder web will be within D/96 under total dead load.

3.2   *Minimizing girder differential deflections*

There are some options the designer can explore to try and minimize girder differential deflections. If vertical clearance and/or profile is not a concern, the designer can make the girder deeper which will increase stiffness and reduce overall deflections, and subsequently, differential deflections

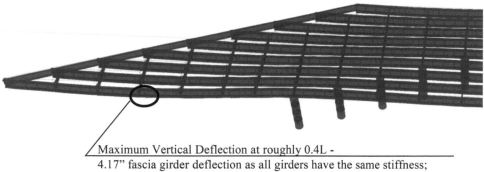

Maximum Vertical Deflection at roughly 0.4L -
4.17" fascia girder deflection as all girders have the same stiffness;
3.66" fascia girder deflection as fascia girder stiffened by 33%

Figure 4.   The deformed shape of a framing plan.

and out-of-plane rotation will be reduced. The designer can also consider minimizing differential deflections by increasing girder sections in strategic locations. For instance, fascia girders in skewed bridges and outer girders (with largest radius/length) for non-skewed curved bridges tend to deflect more even if all girders have similar lengths and stiffness. Stiffening only those girders subject to max deflection is more effective than stiffening all girders.

Figure 4 below shows the deformed shape of a severely skewed bridge under non-composite dead load. For a framing having the same section at all girders, it has a maximum vertical deflection of 4.17″, lateral deflection of 0.93″ and torsional rotation of 0.0149 radians at about 0.4L of this 2-span continuous girder. For the same framing but with stiffer fascia girders, it has a max vertical deflection of 3.66″, lateral deflection of 0.72″, and torsional rotation of 0.0124 radians at the same location. The fascia girder cross-sectional area was increased by approximately 25%, which represents a 33% increase to the moment of inertia, and reduces the structural response at the referenced location by 13%, 23%, and 13% for vertical deflection, lateral deflection, and rotation, respectively.

Further stiffening the fascia girder can reduce the out-of-plane rotation. However, as the fascia girder is stiffened, it will attract additional load and make this process less effective. It is the designer's responsibility to seek a balance between an increase in steel fabrication cost versus the reduction in out-of-plane rotation and improved constructability in the field.

3.3   *Cross-frame detailing methods*

As shown in Figure 3, differential deflection will cause racking of the cross section, which will try to distort the cross-frame. However, because the stiffness of the cross-frame is quite large in comparison to the lateral stiffness of the girders, the girders will rotate out-of-plane instead. Excessive out-of-plane rotation of girders in their final constructed position is undesirable as it may cause unforeseen bending and undermine the capacity, stability, and visual appearance of the structure. In an effort to alleviate out-of-plane rotation in skewed and curved structures, the designer should specify the erected position of the girders and the loading condition under which that position should be theoretically achieved, which then must be clearly conveyed on contract documents in order to achieve a plumb or near plumb position after the deck is poured.

Erected girder positions can be specified as either web plumb or web out-of-plumb under no load, steel dead load, or total dead load loading conditions. Note that theoretically, girder webs can be plumb only under one of the loading conditions. Web plumbness in different loading conditions is achieved by cross-frame and/or diaphragm detailing. Cross-frame members are detailed with the intent of forcing the girders to be web-plumb at the specified loading condition chosen by the designer.

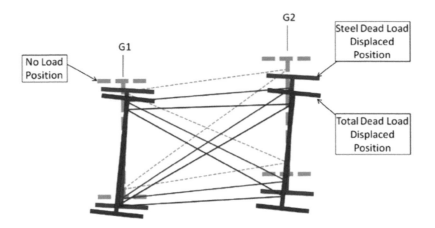

Figure 5.   Girder and cross frame positions when detailed for NLF.
(Source: AASHTO/NSBA G13 Guidelines for Steel Girder Bridge Analysis.)

In No Load Fit (NLF), cross-frames are detailed to fit girders with the webs vertically plumb as though no dead load deflections have taken place. And once the self-weight of the steel and deck weight have been realized, the girder webs will rotate out of plumb as shown in Figure 5.

Similarly, for Steel Dead Load Fit (SDLF), cross-frames are detailed to force the girder webs to be theoretically plumb once the girders have displaced due to steel dead load. In Total Dead Load Fit (TDLF), cross-frames are detailed to force the girder webs to be theoretically plumb once the girders have displaced due to total non-composite dead load. In all cases, the diaphragm or cross-frame members are detailed and fabricated so the girders are plumb at the intended load condition. This is achieved in fabrication by changing the lengths and angles between cross-frame members so they hold the girders plumb in the intended final position. In doing so, the girders are often twisted out-of-plane prior to their final position in order to install the cross-frames. The twist of the girders is achieved by an application of an external force or fit-up force.

For additional information, the reader is directed to the AASHTO/NSBA G13 Guidelines for Steel Girder Bridge Analysis which describes the detailing of cross-frames and girders for the intended position at great length. Additionally, NCHRP Report 725 Chapter 3.5 provides a guideline for the selection of cross-frame detailing methods based on the skew, curvature, and span length which should be consulted during design. Note that Report 725 does not define short, moderate and long span lengths. However, based on correspondence with the lead author of Report 725, short span lengths can be considered as anything less than 150 feet, moderate spans as 150 to 200 feet, and long spans as longer than 200 feet.

3.4   *Design considerations for cross-frames from a force perspective*

The differential deflections associated with severely skewed and/or curved bridges induce forces on the cross-frames. Higher axial forces on cross-frames are anticipated at locations with larger differential deflection or larger differential shear force. These forces are usually computed by 2D or 3D structural analysis and should not be ignored during design of the cross-frames and connections (Figure 6 and 7).

In order to minimize the forces in the cross-frames and optimize the layout, it is recommended that the designer try several different cross-frame layouts using a 2D or 3D model. In skewed bridges, NCHRP Report 725 recommends the first intermediate cross-frame be positioned at an offset distance "a" from the support, which is the greater of 1.5D and 0.4b, where D is the girder depth and b is the second unbraced length within the span adjacent to the offset from the bearing

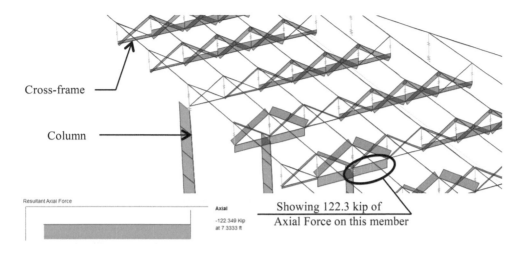

Figure 6.   Force contour in a 3D model showing axial forces in cross-frames and columns.

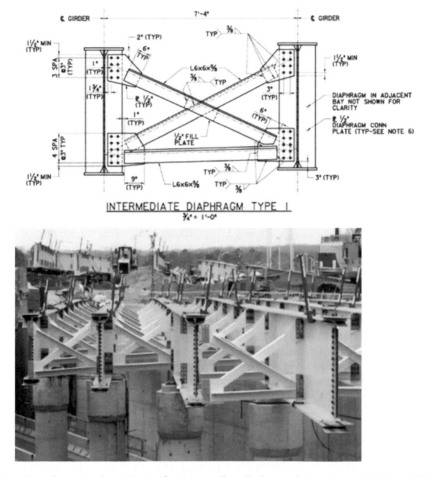

Figure 7.   Cross-frame detail and photo of an intermediate diaphragm designed to resist high axial forces in a severely skewed bridge.

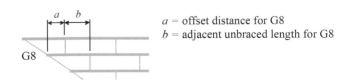

Figure 8.   Illustration of cross-frame offset.
            (Source: NCHRP Report 725).

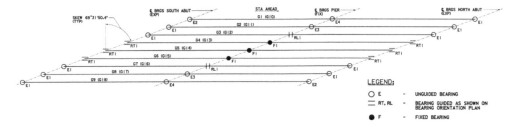

Figure 9.   Typical bearing orientation plan for a skewed bridge.

line (Figure 8). This is intended to alleviate local spikes in the cross-frame forces and reduces the potential for fit-up difficulty at these locations.

3.5   *Thermal behavior and bearing orientation*

A bridge framing plan subject to thermal load will expand and contract both longitudinally and transversely, unless restrained. For severely skewed or curved bridges, thermal movements can be difficult to predict without performing 2D or 3D thermal analysis. In general, longitudinal and transverse thermal movements at the acute corners can be quite different from those seen at the obtuse corners, and a proposed bearing fixity and guided bearing orientation layout should allow such thermal displacement instead of preventing them from moving.

Figure 9 shows the proposed bearing fixity and orientation layout of a skewed bridge. Note the center three fixed bearings over the pier and the location and orientation of the guided bearings at the piers and abutments. For this particularly bridge, fixing and guiding the bearings at the center girders minimized the forces due to restraint at the bearings, while allowing the outermost girders, which exhibited the most movement, to move freely with unguided bearings.

While the fascia girders near the acute corners may exhibit less lateral movement than the obtuse corners, placing longitudinally guided bearings at locations having significant out-of-plane rotations is not recommended due to potential for binding of the guide bars stemming from twisting and untwisting of the girders during erection and deck placement. Instead, longitudinally guided bearings are placed along the bridge centerline.

For curved bridges, the deformed shape under thermal load varies depending on curvature and the extended angle of the curve. Figure 10 shows the deformed shape and bearing layout of a 9-span bridge comprised of one 6-span continuous unit and a one 3-span continuous unit.

As shown, longitudinally guided bearings (restrained transversely) are placed at each end to allow longitudinal thermal movement, while restraining the transverse movement due to other loads. The second support from the right is a frame-in pier rigidly connected to the steel framing. For the rest of the supports, seismic isolation bearings are used to restrain braking forces, seismic and wind loads, while allowing thermal movement in both directions.

In general, 2D or 3D thermal analysis is recommended to understand and visualize the thermal behavior in order to come up with a bearing fixity and guided bearing orientation layout that minimizes thermal load on the bearings, while restraining lateral movements from braking, seismic,

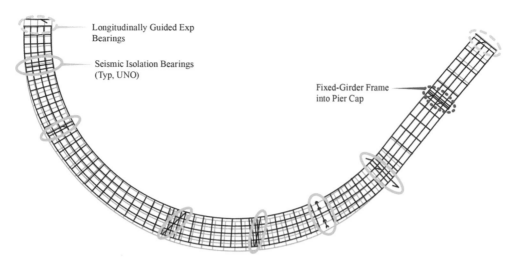

Figure 10.   Thermal deformed shape and bearing layout of a severely curved framing – where lighter lines show the original shape, as darker lines show the deformed shape under a 100°F temperature drop.

and wind loads. For long span severely curved bridges, isolation bearings are a viable alternate to conventional HLMR bearings because isolators become flexible under slowly applied force (thermal load) but maintain stiffness under sudden non-seismic lateral loads. Although the cost difference between isolation and HLMR bearings is minimal, there is additional engineering effort to develop a set of seismic isolation bearing design parameters.

Lastly, bearing reactions and rotations in skewed or curved bridges are uneven. The designer is advised to check for any bearing uplift and/or excessive rotations during different phases/stages of construction or under the final condition, and provide adequate measures (such as tie-downs) to counteract uplift and/or specify bearings with higher rotational capacity.

3.6   *Design considerations for deck joints*

As discussed previously, longitudinal and transverse thermal movements at the acute corners of a skewed bridge can be quite different from those at the obtuse corners. The deck joints are therefore subject to racking movements and shall be designed accordingly.

Additionally, as girders twist due to differential deflection, the entire framing plan including end diaphragms will twist during deck pour. If a TDLF method is required to pre-twist the girders (so that the girder web will be twisted back to a plumb position after deck pour), it may be prudent to block out the last few feet of the deck at the deck joint for a separate final pour to avoid unnecessary additional stress on the deck joint support assembly and end diaphragm.

3.7   *Oversized holes in connections*

In general, the use of oversized holes should be avoided. In NCHRP Report 725, White et al (2012) stated, "The use of oversized or slotted holes in gusset and connection plates can significantly decrease the stability bracing efficiency of cross-frames. In addition, the control of the deformed bridge geometry can also be affected since cross-frames are necessary to integrate the girders and make them deform as a unit rather than as independent components. Therefore, it is not recommended to use this scheme as a solution to erecting cross-frames at stiff locations such as the regions near skewed supports." A review of other literature yielded a similar message.

However, past project experience has shown that skewed girders can be successfully constructed using oversized holes locally near supports, finger-tight bolts prior to deck placement, or even cross-frames installed after deck placement at locations near acute corners.

In the authors' opinion, the designer should absolutely avoid using oversized or slotted holes at all diaphragm and cross-frame connection plates, especially for long and/or wide bridges with severe skews or curvature. However, using oversized or slotted holes or finger-tight bolts locally in an area with high potential fit-up difficulty may be considered but shall be kept to a minimum. The effects of using these methods should be studied diligently before their use and should be considered a last resort.

4   CONTRACT PLAN NOTES

The following notes, which were developed for bridges that TDLF is specified for detailing, should be considered for inclusion in the contract plans for such bridges. It is imperative to include notes on the plans that not only state the responsibilities of the Contractor, but also alert the Contractor that web layover will occur at a given position. The following notes are only examples and should be reviewed and modified by designers to be project specific.

- The Contractor is responsible for detailing the steel beams, cross-frames and diaphragms using "Total (or Full) Dead Load Fit", accounting for the manner in which the bridge was designed (i.e. diaphragms and cross-frames are connected to the girders with fully tightened bolts prior to placing the deck slab, unless noted otherwise).
- The girders are required to be erected with web layover (webs out of plumb) such that, after the deck slab is poured the webs will become plumb. The Contractor may need to use come-alongs or other similar equipment to erect the girders with layover. At erection, it is anticipated that webs will range from being out of plumb at some locations* to vertical at other locations*. (Note: Designer need to specify these locations.)
- Under full dead load, girder ends and all bearing stiffeners shall be plumb and girder webs shall be vertical within fabrication and construction tolerances of $1/8''$ per foot (equal to $D/96$ in inches) (AASHTO/NSBA, 2007). The Contractor shall account for deflections and rotations of the girders such that, under full dead load, connection plates and bearing stiffeners are plumb and the girder webs are vertical.
- The Contractor is responsible for stability of the girders during erection.

5   GUIDELINES FOR CONSTRUCTION ENGINEERING

Until this point, this paper has focused primarily on the engineer and the efforts associated with the design of skewed and curved bridges and potential issues to address and avoid. However, the final stage in successfully completing any bridge comes during construction. It is important to identify and alert the Contractor to potential difficulties during the preparation of the contract documents, but it is equally important to make sure the Contractor takes the necessary precautions to ensure the safe construction of the bridge. To achieve this we rely on the contractor's shop drawings, and more specifically, the erection plan and accompanying calculations.

5.1   *Erection plan and accompanying calculations*

Typically, agencies have standard language in their specifications that discuss the erection plan and minimum requirements to be met. For highly curved and severely skewed bridges, it is recommended that these sections be significantly strengthened with additional and more specific requirements. By doing so, the contractor, fabricator, and erector must realize and address potential erection issues during the shop drawing stage. Additionally, by requiring erection procedures and calculations the potential for unforeseen obstacles can be minimized, which will avoid costly delays to the construction schedule. For a detailed checklist of additional requirements and language to consider, please see the "Guidelines for Construction Engineering Provisions and Checklists" section of NJDOT Research Report NJ-2014-003 (Louch et al, 2013) completed by CS and GPI titled, "Steel Erection Out-of-Plumb."

5.2 *Additional Erector Qualifications*

In addition to broadening the requirements for the erection plan and calculations, it is also recommended that the erector qualifications be refined based on the severity of the skewed/curved bridge to be constructed. For bridges with sharp skew angles (measured from bridge centerline to centerline of substructure) and/or with a tight radius of curvature, the Contractor and/or Erector shall meet the following additional qualification requirements:

– For bridges with skew angles ranging from 30 to 60 degrees, and/or with a radius of curvature between 600 and 1200 feet, the Contractor shall provide evidence to demonstrate that the Contractor and/or Erector have successfully erected similar bridges within this geometric range, or a more severe geometric range within the past five years.
– For bridges with skew angles less than 30 degrees, and/or with a radius of curvature less than 600 feet, the Contractor shall provide evidence to demonstrate that the Contractor or Erector have successfully erected similar bridges within this geometric range in the past five years.
– Skewed bridge experience cannot substitute for curved bridge experience and vice versa.

6 CONCLUSIONS

Structural behavior of severely skewed or curved steel I-girder bridges can often times be complicated and difficult to predict. This paper discusses only a few important issues in an ever expanding field of knowledge. Performing a detailed 3D analysis allows the designer to better understand the behavior of these complicated structures, which leads to better design and detailing practices. While skewed and curved girder bridges are not a new concept, research on the behavior and effects of these bridges is ongoing. With this ongoing effort more streamlined procedures are emerging to aid the engineer in successfully designing a complex structure by avoiding common mistakes.

It is of paramount importance that the engineer and contractor are aware of the potential issues, and that they work together to minimize or eliminate any issues related to skew and/or curvature that could cause costly delays or claims during construction. As construction is the final stage of a successful bridge design, it is important to identify and alert the Contractor to potential difficulties during the preparation of the contract documents. It is equally important to make sure the Contractor takes the necessary precautions to ensure the safe construction of the bridge. To achieve this we rely on the contractor's shop drawings, and more specifically, the erection plan and accompanying calculations.

ACKNOWLEDGEMENTS

The authors would like to thank Sean Sheehy of NJDOT Construction, Camille Crichton-Sumners of NJDOT Research Bureau, and Hugh Louch of Cambridge Systematics for giving us the opportunity to work on the research project and develop the "Steel Erection Out-of-Plumb" report which is the basis of this paper. We would also like to thank Ronald Medlock and Bob Cisneros of High Steel Structures, Inc.; Professor Donald White of Georgia Institute of Technology; and Ralph Csogi, Han Lee, Lenny Lembersky, Peter Mahally and David Wagner of GPI for their input on this paper and the research project.

REFERENCES

AASHTO/NSBA Steel Bridge Collaboration. *G12.1 Guidelines for Design for Constructability*. Washington, DC, 2003.

AASHTO/NSBA Steel Bridge Collaboration. *S10.1 Steel Bridge Erection Guide Specification*. Washington, DC, 2007.

AASHTO/NSBA Steel Bridge Collaboration. *G13.1 Guidelines for Steel Girder Bridge Analysis*, 1st edition. Washington, DC, 2011.

AASHTO. *AASHTO LRFD Bridge Design Specifications*, 6th edition. Washington, DC, 2012.

Beckmann, F., Cisneros, B., Nedlock, R., White, D. *A Skewed Perspective*. Modern Steel Construction. December 2008.

Linzell, D., Chen, A., Sharafbayani, M., Seo, J., Nevling, D., Jaissa-Ard, T., Ashour, O. *Guidelines for Analyzing Curved and Skewed Bridges and Designing Them for Construction*. Pennsylvania Department of Transportation (PennDOT). Harrisburg, PA, 2010.

Louch, H., Johnsen, S., Liang, V. *Steel Erection Out-of-Plumb*. Report No. NJ-2014-003. New Jersey Department of Transportation (NJDOT). Trenton, NJ, 2014.

New York State Department of Transportation (NYSDOT). *NYSDOT Steel Construction Manual*, 3rd edition. Albany, NY, 2008.

White, D. et al. *Guidelines for Analysis Methods and Construction Engineering of Curved and Skewed Steel Girder Bridges – NCHRP Report 725*. Transportation Research Board. Washington, DC, 2012.

Chapter 8

Construction of the Nhat Tan Bridge superstructure

K. Matsuno & N. Taki

IHI Infrastructure Systems Co., Ltd., Osaka, Japan

ABSTRACT: The main bridge of Nhat Tan bridge is a 1500 m long, 6-span cable stayed bridge with 8 traffic lanes. The bridge is the "edge girder" type that is a continuous composite girder consisting of a 35.6 m precast reinforced concrete deck on 2 I-beam steel edge girders. Balanced cantilever erection was adopted for the whole span of bridge, resulting in a 4 times closure works were required. Additionally, according to specific regulation of Southeast Asia, occurrence of crack on concrete deck and pylon was impermissible during erection even a negligibly small level of crack that would not affect structural soundness. Therefore, the cantilever step must be determined to minimize occurrence of crack on any concrete structure.

This paper describes the challenges encountered in the development and application of construction methods, solutions applied to overcome these challenges and the results that are also presented.

1   INTRODUCTION

The main bridge of Nhat Tan bridge (hereinafter referred to as main bridge) is one part of new route from Noi Bai new international airport to downtown in Hanoi, Vietnam which was opened to traffic in January 2015.

The main bridge is a 1500 m long, 6-span cable stayed bridge with 8 traffic lanes. This scale of multiple span cable stayed bridge is the first application in Southeast Asia and also very rare type of bridge in the world (Kensetsu Tosho. 2014). A general view of the main bridge is shown in Figure 1. Steel pipe and sheet pile (SPSP) is adopted for the foundation of pylons, with a total weight of steel pipe equal to 14,200 ton. Height of A-shaped reinforced concrete pylons is around 110 m and steel anchorage boxes are embedded to support stay cables on top part of pylon as shown in Figure 2. The bridge has 2 I-beam steel edge girders and reinforced concrete deck, which is called as the "edge girder" type, and designed as composite girder as shown in Figure 2. Total weight of anchorage box and steel girder is around 15,000 ton. New prefabricated parallel wire strands (New-PWS) having 1,770 Mpa tensile strength is used for stay cable and maximum diameter of cable is 155 mm (313 numbers of $\phi7$ mm diameter galvanized steel wires). Total numbers of stay cable is 220 and total weight of stay cables is approximately 1800 ton.

2   CONSTRUCTION

2.1   *General construction method*

The main bridge across and over the 1 km wide river plays an important role as navigation route, so that any construction method that occupies the river (such as construction bent on river) was not allowed. Therefore, inclined bent method is adopted around each pylon and balanced cantilever construction method was adopted in the remaining parts including side span of bridge as shown in Figure 3. Side span usually is constructed in advance with use of bent, so application of cantilever

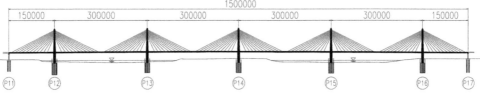

Figure 1.   General view of the Main Bridge.

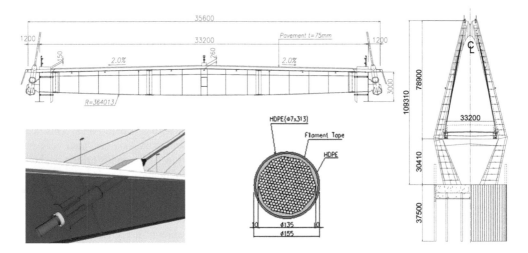

Figure 2.   Main girder, stay cable and pylon.

erection into whole span in large scale of multiple span stay cable bridge would be the first challenge of its kind in the world (Japan Society of Civil Engineers. 2010).

Construction methods are categorized into three parts (construction by inclined bent, cantilever erection and closure erection) as shown in the Figure 4.

2.2   *Cantilever erection*

Balanced cantilever method is applied in most part of the bridge, by adjusting the elevation of girder at each construction stage predicting bridge camber at bridge completion. Girder and floor beam were erected by 150 ton crane and 50 ton crane which are arranged on deck as shown in Figure 5.

Figure 3. View of construction by inclined bent and cantilever erection.

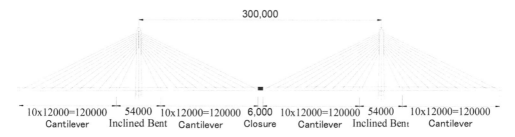

Figure 4. General construction method.

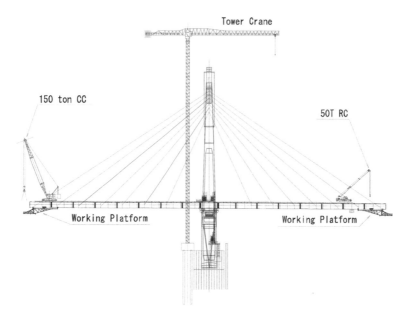

Figure 5. Cantilever erection.

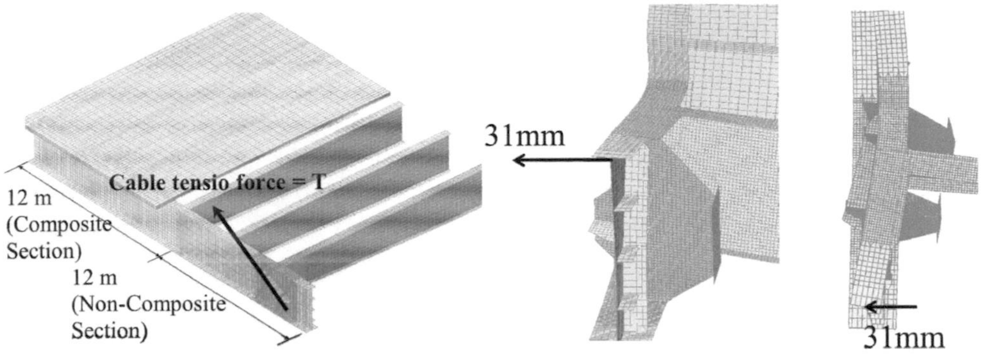

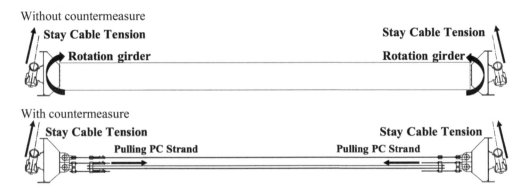

Figure 6.   FEM model and results.

Table 1.   Stress and deformation with and without countermeasure

	Without Countermeasure	With Countermeasure	Allowable Value
Stress	172 MPa	96 MPa	140 MPa
Deformation	31 mm	2 mm	5 mm

Without countermeasure

With countermeasure

Figure 7.   Girder deformation prevention device.

2.3   *Countermeasure for out of plane deformation during tensioning of stay cable*

In order to avoid crack appearance on deck during stay cable tensioning work, tensioning work needs to be performed under girders while being a non-composite section. This bridge is edge girder type and distance between girders is 33.2 m, meaning very flexible structure especially in out of plane. Additionally, cable anchorage is located at outside web of girder, so high level of stress and deformation were the concerns during stay cable tensioning.

In order to evaluate level of stress and deformation on girders during stay cable tensioning work, finite element (hereinafter referred to as FEM) analysis was performed. Shell element for girders and solid element for deck slab were used in the analysis model and half model was applied to reduce calculation processing time as shown in Figure 6.

Analysis result on outermost stay cable where maximum tension appears is shown in Figure 6 and Table 1. High level of deformation and stress can be confirmed from analysis result. As countermeasure for this, girder deformation prevention device is installed to pull girders around 150 ton toward inside by PC strands during stay cable tensioning work as shown in Figure 7.

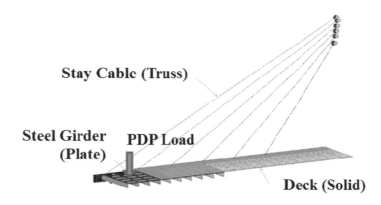

Figure 8.   Analysis model.

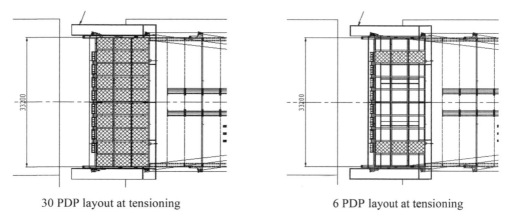

30 PDP layout at tensioning   6 PDP layout at tensioning

Figure 9.   PDP layout at tensioning.

As shown in Table 1, deformation and stress level are significantly improved after countermeasure is applied. From the further analysis and evaluation, this countermeasure was applied on 5 stay cables from outermost stay cable.

2.4   *Crack prevention during cantilever erection*

Maximum allowable crack width of 0.2 mm during construction was specified in the project technical specification. However according to a specific custom in Vietnam, even a negligibly small level of crack, which is less than 0.2 mm, must also be minimized. Therefore, cantilever step must be determined to minimize occurrence of crack on any concrete structure.

Precast deck panels (hereinafter referred to as PDP) were used on most part of cantilever erection and 30 pieces of PDP were needed to be arranged per a cycle step. However, if all PDP were arranged before stay cable tensioning was made; there may be occurrence of crack on deck slab due to high level of negative moment.

In order to evaluate level of tensile stress on deck slab, FEM analysis was performed. FEM model is shown in Figure 8.

According to analysis, the number of PDP prior to stay cable tensioning is reduced from 30 pieces to 6 pieces as shown in Figure 9 which reduced negative flexural moments in the deck by 40%, resulting in maximum tensile stress which was under allowable value of 7.5 MPa as shown in Figure 10.

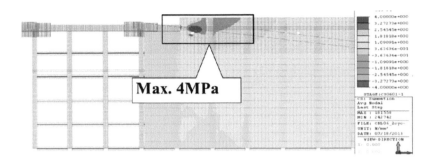

Figure 10.   Maximum stress.

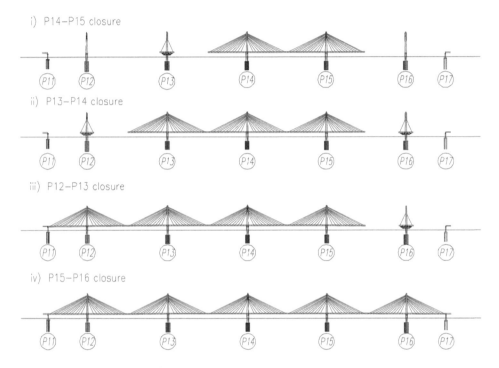

Figure 11.   Bridge closing procedure.

2.5   *Closure work*

Stay cable bridges around each pylon are constructed respectively, resulting in 4 times of bridge closing and 2 times bridge end closing required as shown in Figure 11. In the latter part of closing work on multiple span cable stayed bridge, both very stiff structures needs to be closed together so severe geometry control till just before closure and adequate preparation to adjust in each directions are required.

In order to match both elevations of girders, counterweight concretes were arranged on deck according to analysis result as shown in Figure 12. In addition, 150 ton crawler crane and 50 ton rafter crane were moved on bridge for minor adjustment. On adjustment in longitudinal direction, jacks are arranged on all bearings to set backward and/or set forward bearings.

To adjust alignment of girders, a special device was used. Edge girder is a very flexible structure before it becomes a composite section, so not only local deformation of girders but also global

Figure 12.   Closing work and counter weight concrete for closing work.

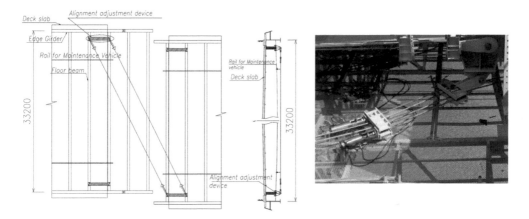

Figure 13.   Alignment adjustment device.

deformation shall be regarded. Additionally, from the point of view to avoid crack appearance on deck, device shall be installed at location as low as possible to minimize out of plane moment on girders; however, adjustment device must be installed under girder so that it may not interfere with girders and floor beams. Therefore, FEM was performed to confirm level of tensile stress on deck and global and local deformation of girders, resulting in the installation of alignment adjustment device under floor beam. This device is installing piece and drawing each girders by 2 center hole jack 50 ton and ϕ 28.6 PC strands, having 80 mm alignment adjustment capability as shown in Figure 13.

3   STEP ANALYSIS AND GEOMETRY CONTROL

3.1   *Step analysis*

In order to confirm the structural integrity of the bridge during erection, forward step analysis was performed. Modelling and analysis were performed using proprietary software, MIDAS CIVIL 2013. It is a noticeable point on this model that fishbone model with two separate beam elements representing the steel girders and concrete deck slab was applied as shown in Figure 14, thereby, it realizes effectively modeling of creep and shrinkage on concrete deck that is much closer to actual than non-separation model usually used in other project as well as realizing easy modeling of composite and non-composite section effect on girders.

Figure 14.　Analysis Model.

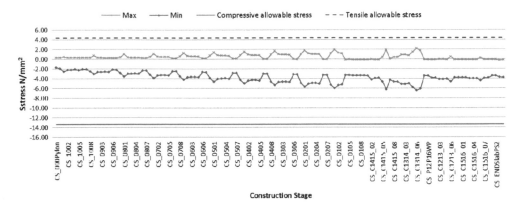

Figure 15.　Stress envelope curves of pylon leg.

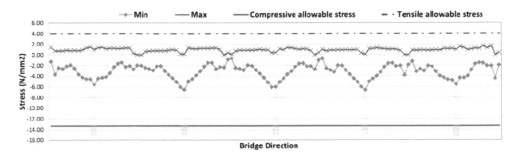

Figure 16.　Stress envelope curves of upper surface of concrete deck.

3.1.1　*Pylon*

Sectional forces from the pylon legs were used to evaluate the stress levels during the construction process. The stresses were confirmed to be below allowable compression values. For the tensile stresses, the stress at which a 0.2 mm crack appears was used as the maximum allowable value. Figure 15 below shows the envelope curve of stresses during the construction process.

3.1.2　*Composite Girder*

The flexural stress and axial stresses were evaluated for each construction stage in the analysis.

The maximum and minimum stress values were confirmed to be within allowable values as shown in the following Figures 16 to 18.

3.1.3　*Stay Cables*

The stay cable forces from the analysis were compared to the ultimate tensile strength. It was confirmed that the forces were below the allowable value of 56% of the ultimate strength. Figure 19 below shows the actual values after completion of erection works in comparison to analysis.

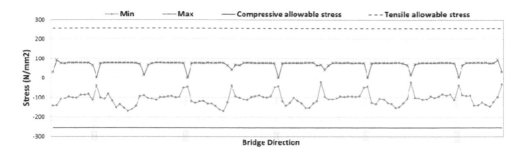

Figure 17. Stress envelope curves of upper flange of steel girders.

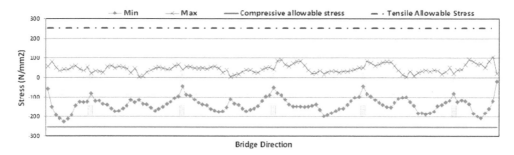

Figure 18. Stress envelope curves of lower flange of steel girders.

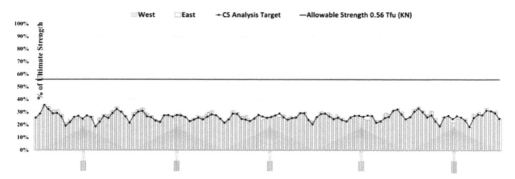

Figure 19. Stay cable tension forces at completion of erection works.

3.2 *Geometry Control*

During cantilever erection, the bridge has a high degree of freedom thereby showing a highly sensitive displacement response. Therefore, to achieve the required geometry at bridge completion, it was necessary to predict the effects of any adjustments made at each erection stage on the geometry at bridge completion. Construction stage results were used as target values for comparison during stay cable tensioning works.

Fabrication errors on edge girder, anchor box and stay cable were considered and the targets were adjusted accordingly. Additionally, influence matrixes of temperature on the geometry at each erection stage were evaluated in advance and used to calibrate geometry measurements performed at site depending on actual temperature conditions.

A system that incorporated all of the above was developed and applied for the installation and adjustments of all stay cables. Comparison of the target displacement (analysis) and calibrated

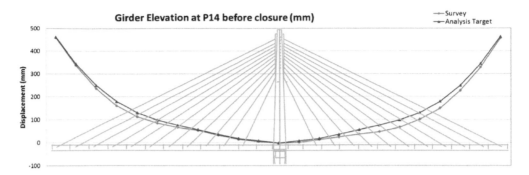

Figure 20.   Comparison of target and actual girder displacement during cantilever erection.

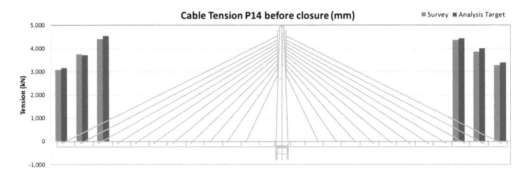

Figure 21.   Comparison of target and actual stay cable tension forces during cantilever erection.

Table 2.   Allowable tolerance specified in technical specification.

Pylon Inclination	$-70\,\text{mm} \sim +70\,\text{mm}$
Girder Elevation (Side Span)	$-60\,\text{mm} \sim +100\,\text{mm}$
Girder Elevation (Center Span)	$-100\,\text{mm} \sim +170\,\text{mm}$

measurement values is shown in Figure 20. Comparison of the stay cable tensions is also shown in Figure 21.

Allowable tolerances specified in technical specification are shown in Table 2.

According to error factor analysis, to achieve being within these allowable values, it is clarified that elevation control is the most important factor, so elevation was controlled during cantilever erection by using equation (1) and (2), which are defined by parameter L (L: Length of cantilever).

$$\text{Center span: } 0.5 \times (25 + L/2) \tag{1}$$

$$\text{Side span: } 0.3 \times (25 + L/2) \tag{2}$$

Following the application of the geometry control, the elevation after closure of all spans satisfied all allowable deviations without requiring any additional stay cable force adjustments.

4 CONCLUSIONS

The Nhat Tan bridge is a 6-span cable stayed bridge with a composite girder and was erected using the cantilever method. As a conclusion, knowledge and experiences acquired through evaluation of this project can be summarized as follows:

1. To prevent deformation and reduce stress on girder during cable tensioning, deformation prevention device was installed and good workability was confirmed throughout the work.
2. Arrangement work of PDP was done separately 2 times before and after cable tensioning work to reduce negative flexural moments and tensile stress in the deck, resulting in the non-observance of any crack over 0.2 mm width on the deck slab. Crack less than 0.2 mm was successfully minimized.
3. Alignment device was installed under girders in order to adjust alignment at closure of both girders.
4. In the evaluation of the structural integrity of the bridge during erection and to provide target values for geometry control, construction stage analysis was applied.
5. By using geometry control system which we developed and set of elevation criteria acquired from error factor analysis, after closure of all spans, all criteria required in technical specification satisfied all allowable deviations without requiring any additional stay cable force adjustments.

The above features could contribute to the successful completion with shortening construction period by 4 months, and satisfy the client's requirement on minimizing crack on deck at the same time. We believe that these can contribute further development on construction technology of multiple stay cabled bridges and stay cabled bridges in Southeast Asia.

REFERENCES

Japan Society of Civil Engineers. 2010. *Steel cable-stayed bridges.*
Kensetsu Tosho. 2014. Construction work of Nhat Tan bridge (Japan Vietnam Friendship bridge) Superstructure. *Bridge and Foundation Engineering* Vol.48: 02–12.

Bridge analysis & design

Chapter 9

Limit analysis for steel beams connection nodes

M. Arquier & X. Cespedes
Strains Civil Engineering, Paris, France

ABSTRACT: The analysis of Ultimate Limit State (ULS) of a structure requires a stability study until failure. This mechanical behavior is complex to compute with standard tools. Cracking, damage, elastic-plastic law, etc., are phenomena which often lead to numerical problem of convergence and interpretation of results. It is therefore often advised to use codes instead (Eurocodes, AASHTO, etc.), but this solution comes at the expense of accurate analysis of the physical behavior of failure. An alternate solution is limit analysis, which combines two parallel and complementary methods. Used on a finite element mesh for rigid-plastic calculations, these two methods lead to a full determination of the physical failure: mechanism, stresses distribution and safety factor. Strains presents a software program using limit analysis for steel beam connections nodes, taking into account such phenomena as contact, separation, friction, welding, plasticity and pre-stressed bolts.

1 INTRODUCTION

In order to check the stability of a structure at failure, engineers typically use the concept of ultimate limit state as defined in various codes (Eurocodes, AASHTO, SNIP, etc.). This state is often studied under an elastic hypothesis, and doesn't take into account all nonlinear phenomena linked to failure: plasticity, cracking, damage, etc.

When possible, a simplified approach used by engineers is to perform elastic studies mainly by using simple software programs or doing manual calculations. This seldom takes into account nonlinear aspects. Instead, safety is generally built in the computations by increasing loads and by curbing the limit strength of the materials, following rules provided in the codes.

The drawback of this method is that it does not take into account the physical behavior of the structure. As said previously, failure is typically nonlinear and an elastic analysis, even when safety factors are included, does not account for the real physical behavior (displacements and stresses).

Therefore, engineers have to perform elasto-plastic analysis. Not only can this take time in order to create the full 3D model, especially in the field of metallic beam connections, but this often leads to numerical problem of convergence, and complex interpretation of results.

Another way to assess the failure state is by using the limit analysis theory.

2 FROM LIMIT ANALYSIS

By definition, limit analysis aims at studying a structure at its limits, meaning at failure, by assuming all materials have reached (and withstood) their limit strength criterion. The elastic behavior is therefore not included in the analysis, therefore no elasto-plastic iterations need to be performed. But the underlying assumption is that the materials allow high ductility deformations.

The plasticity is defined thanks to a criterion which limits the stresses. It's usually defined by a function f over an admissible stress value domain G such as: $\sigma \in G \Leftrightarrow f(\sigma) \leq 0$.

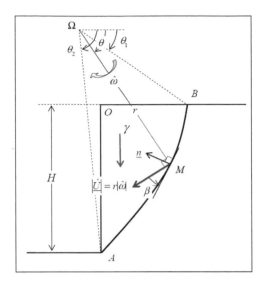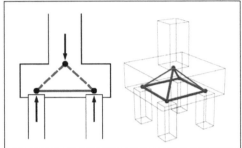

Figure 1.　Limit analysis examples. Left: collapse of the rigid block OAB under its weight with AB a logarithmic spiral. Refer to de Buhan (2007). Right: strut-and-tie patter of a massive reinforced concrete
piece. Refer to Gao (2012).

As material criterion (or law), here is a list of few examples, which vary according to the chosen
model:

- $N^c \leq N \leq N^t$ for a 1D beam model under normal force
- $M^1 \leq M \leq M^2$ for a 1D beam model under bending moment
- Von Mises for multidimensional steel model
- Drucker-Prager for multidimensional soil model

For a more detailed theory, and more different criteria, please refer to Salençon (1983) or de
Buhan (2007).

This theory, developed in the 80s, is currently used in very specific fields: in soil mechanic when
the engineer makes a 'guess' on a failure mechanism by blocs separated by logarithmic spirals, or
in reinforced concrete when the engineer needs to find a strut-and-tie pattern (Figure 1).

This theory is not widely used in civil engineering because of its main drawback: the engineer
(or the software program) has to make a 'guess' on what the solution is likely to be: either a failure
mechanism (by blocs for instance), or a stress distribution (as in strut-and-tie method).

Most of the time engineers don't know in advance the solution, so they need to try different
solutions until finding the optimal one. As this can be very tricky, especially in case of complex
structures, this theory has limited use (2D analysis or examples above), and will need careful review
by senior engineers.

In short, limit analysis is the research of an optimal solution either in stress distribution or in
failure displacements. That is why it's typically split in two approaches: a static one and a kinematic
one.

2.1 *Static approach*

In a static approach and for a given external load F, one wants to find the safety factor λ which
leads to failure. To do so, a first stress distribution σ^1, in equilibrium with F, is guessed. Then,
the maximal multiplier α^1, such that $\alpha^1 \sigma^1$ withstands the given material criterion $f(\alpha^1 \sigma^1) \leq 0$, is
found.

As only one stress distribution has been analyzed, one does not know if it is the optimal distribution. Hence, $\alpha^1 \leq \lambda$. By implementing this method with another distribution σ^2, still in equilibrium with F, one gets a α^2 which can be higher or lower than α^1.

By trying all admissible distributions in equilibrium with F, the static approach leads to the calculation of the maximum of all the multiplier α^i. And as all mathematical possible distributions (infinite dimensional vector space) can't be tried, $\alpha^{opt} = \max(\alpha^i)$ is computed as a lower bound of λ.

The output of this approach is α^{opt}, lower bound of the load multiplier, and the corresponding stress distribution. The drawback is the lack of information regarding the displacement field.

2.2 *Kinematic approach*

The kinematic approach is based on the research of an optimal displacements field which minimizes the deformation energy.

For a given external load F, the Virtual Work Principle (1) over the structure gives:

$$\forall \hat{U} \qquad F.\hat{U} = \iiint_V \sigma : \varepsilon(\hat{U}) \, dV \qquad (1)$$

As, by hypothesis, all material have reached their limit strength criterion (2):

$$\sigma : \varepsilon(\hat{U}) \leq \max[\sigma : \varepsilon(\hat{U}), \forall \sigma \in G] \overset{\text{def}}{=} \pi(\hat{U}) \qquad (2)$$

where π is named the support function of the criterion. See referenced books for few examples.

Equations (1) and (2) combined leads to this inequality:

$$\forall \hat{U} \qquad F.\hat{U} \leq \iiint_V \pi(\hat{U}) \, dV \qquad (3)$$

If it's verified for all virtual displacements $\hat{U}$, the structure withstands the load F. On the contrary, if at least one displacement $\hat{U}^o$ that does not verify this inequality is found, that means the structure will collapse under F, with $\hat{U}^o$ as failure mechanism.

Let's define β as follow:

$$\beta(\hat{U}) = \frac{\iiint_V \pi(\hat{U}) dV}{F.\hat{U}} \implies VWP : \forall \hat{U}, \beta(\hat{U}) \geq 1 \qquad (4)$$

So, $\beta^{opt} = \min \beta(\hat{U})$ is computed, for all possible $\hat{U}$. Two possibilities appear:

- $\beta^{opt} \leq 1 \Longrightarrow$ collapse of the structure under F
- $\beta^{opt} > 1 \Longrightarrow$ the structure withstands F

But, (4) can also be written:

$$\beta^{opt} = \min \frac{\iiint_V \pi(\hat{U}) dV}{F.\hat{U}} \implies \min \frac{\iiint_V \pi(\hat{U}) dV}{(\beta^{opt}F).\hat{U}} = 1 \qquad (5)$$

Hence, β^{opt} can be seen in (5) as the load multiplier to the failure. As in the static approach, all possible $\hat{U}$ cannot be tried, and β^{opt} is calculated as an upper bound of the safety factor: $\lambda \leq \beta^{opt}$.

The output of the kinematic approach is an upper bound of the safety factor and the corresponding failure mechanism. The drawback is the lack of information regarding the stress distribution.

2.3 *Optimal safety factor*

The combination of these two approaches, provides the bounds of the safety factor λ:

$$\alpha^{opt} \leq \lambda \leq \beta^{opt} \qquad (6)$$

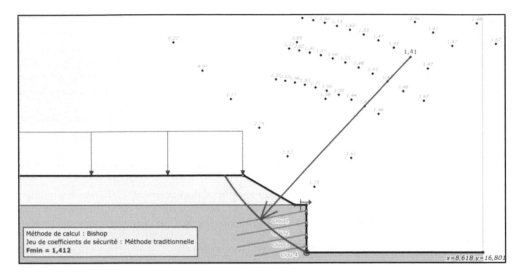

Figure 2.   Caption of TALREN©: a 2D soil stability software program. Refer to Terrasol firm web-site.

An optimal analysis would give the same value for the upper and lower bounds. But, as all possible solutions cannot be tried, the equality of the bounds is seldom found.

Hence, the precision on the safety factor depends on the values of both bounds: the closer they are, the more precise the safety factor is.

Therefore, the limit analysis provides the load multiplier to the failure, its displacement field and stress distribution.

3   STEEL FRAME CONNECTIONS SOFTWARE PROGRAM

Few software programs use limit analysis. The classic field is in soil mechanics, to check the stability of the ground. But, not only do they usually stay in 2D, but they also focus on the kinematic approach. This is due to the numerical method which often uses strict displacement field hypothesis (Figure 2).

This issue is at stake in several disciplines of civil engineering. Strains is therefore seeking to expand the scope of use of limit analysis to a new field: steel beam connection nodes.

3.1   *Objectives*

This software solution aims at analyzing the stability of different kinds of steel beam connections nodes (bridges trusses or buildings frameworks for instance) in a full 3D model.

Thanks to a meshed geometry, with given boundary conditions (in displacements and forces), the static and kinematic approaches are computed via a very efficient external optimizer tool.

As output, the failure behavior is obtained.

3.2   *Phenomena taken into account*

In order to be realistic, the real behavior of all components of the structure has to be taken into account:

– Rigid-plastic behavior of steel (Von Mises 3D law)
– Welded profiled beams, sometimes thanks to plates
– Pre-stressed or not bolts, to force contact between pieces
– Friction coefficient of a contact area (non-associated law, see below)

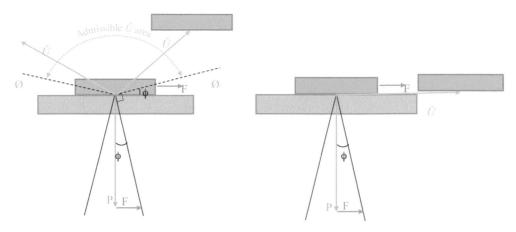

Figure 3.   Frictional behavior: associated law (left), non-associated (right). P is the contact normal force, F the friction force and ϕ the friction angle.

A major difficulty resides in the friction behavior, because of the non-associated rule. In theory, the plastic flow must be normal to the criterion surface ("admissible $\hat{U}$ area"), which leads to the loss of contact in case of friction. However, a volume can slide tangentially to another volume, without losing contact (Figure 3).

As limit analysis is based on normal plastic flow rule, a modified law is used to represent the real behavior of friction.

3.3 *Calculations*

To perform an efficient analysis without constraining the admissible guessed fields, solid finite elements are created to assemble all the data necessary to calculate both extrema of the static and kinematic approaches.

The mathematical problem can also be written in terms of conic quadratic optimization, which can be solved very quickly by an external software solution used for this purpose.

The static approach is done first to assess the stresses distribution and to calculate the normal force on all contact areas. Then, according to this first result, the kinematic problem is solved to get the displacements field, and so, the failure mechanism.

3.4 *Mesh refinement*

A key feature of the finite elements used is the ability to refine the mesh according to a first analysis. This is done automatically, by focusing on the more stressed/deformed area to be refined, and by letting loose the rigid parts.

Moreover, thanks to our partnership with Distene firm, new kind of mesh, with orthotropic elements, is implemented. These elements are distorted on purpose, along one direction of no deformation. Hence it's possible to have a refined mesh over a beam cross section, but a coarse mesh along the beam axis.

This saves a lot of engineer-time and computer-time.

3.5 *Output*

Both approaches combined provides full knowledge mechanism behavior: stress distribution, displacements field and the bounds of the safety factor (Figure 4).

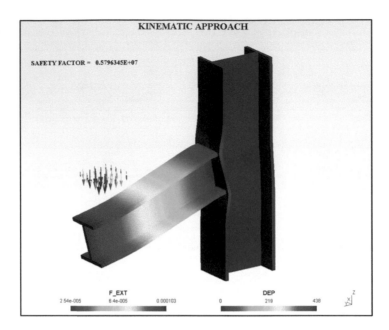

Figure 4.   Kinematic approach output: failure mechanism and safety factor upper bound.

4   CONCLUSIONS AND PERSPECTIVES

The combination of limit analysis theory, supported by a user-friendly CAD, a powerful mesher and a very efficient conic quadratic optimizer leads to a modern tool for civil engineers that allows them to check the resistance of all their steel beam connection nodes.

Strains aims at expanding the scope of use of limit analysis by creating more tools based on the limit analysis theory. For instance, 3d soil stability or 3d reinforced concrete massive pieces.

REFERENCES

de Buhan, P. (ed.) 2007. *Plasticité et calcul à la rupture.* Paris: Presse de l'école nationale des ponts et chaussées.
Gao, N. 2012. Strut-and-tie A to Z. *https://studystructural.wordpress.com/2012/12/03/stm-a-to-z/*
Salençon, J. (ed.) 1983. *Calcul à la rupture et analyse limite.* Paris: Presse de l'école nationale des ponts et chaussées.
Terrasol. TALREN V5. *http://www.terrasol.fr/fr/logiciels/logiciels-terrasol/talren-v5*

Chapter 10

Integral abutment bridges and the modeling of soil-structure interaction

S. Rhodes
LUSAS, Kingston upon Thames, UK

T. Cakebread
LUSAS, New York, NY, USA

ABSTRACT: No standard approach for the analysis of integral abutment bridges appears in the AASHTO LRFD Bridge Design Specifications or other international codes. This paper considers the approaches most suitable for modeling common integral abutment bridge forms, expanding upon recent UK guidance regarding soil-structure interaction approaches. Issues including material properties, initial stress state and the incorporation of the effects of soil ratcheting are discussed and both continuum and spring-type ('subgrade modulus') finite element models are explored.

1 INTRODUCTION

Owing to durability problems associated with movement joints, it is widely accepted that short and medium length bridges are best designed without such joints. This has led to a rise in popularity of integral abutment bridges (no movement joints, no bearings) and semi-integral bridges (no movement joints) for new construction internationally. Both integral and semi-integral bridges accommodate the thermal expansion and contraction of the superstructure by movement of the abutments or end-screens, which are retaining structures.

Often retaining structures are analyzed representing the soil as merely a load – the stiffness of the soil is not modeled. The design proceeds considering only limiting active and passive lateral earth pressures. However, if movements/ deflections of the structure are insufficient to mobilize the limiting values, intermediate values of earth pressure occur, as illustrated in Figure 1 below.

The lateral earth pressure depends on the strain in the soil, which in turn depends on movements in the structure. Structural movements depend on the stiffness of both structure and soil, and on lateral earth pressures. Unless the assumption of limiting earth pressures can be deemed conservative and acceptable, an analysis which somehow reflects this loop is required.

The lateral earth pressure is illustrated in Figure 1 by reference to an earth pressure coefficient (K) which links lateral pressures to vertical stress in the soil without specific reference to cohesion, for simplicity. The principle stands for various soil types. Also see AASHTO (2012) C3.11.1.

In some retaining structures, use of limiting earth pressures can be demonstrably either over-conservative or unconservative. During the summer expansion of the superstructure in an integral abutment bridge, lateral earth pressures on the abutments can approach the theoretical passive state, especially in the upper portion, where horizontal displacements are largest – pressures may be an order of magnitude greater than those experienced by the abutments of a non-integral abutment bridge. For some integral abutment bridge arrangements, it is sufficient to carry out the design on the basis of some assumed lateral earth pressure distribution, i.e. a limiting equilibrium approach. For others, the soil stiffness plays a more significant part in the behavior of the system and an

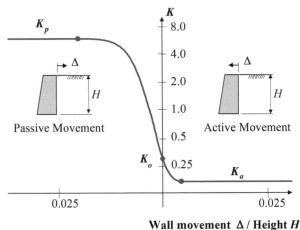

Wall movement Δ / Height *H*

Figure 1.   Pressure/ deflection curve, after NCHRP Report 343 (Barker et al. 1991).

analysis which models the behavior of both soil and structure – a soil-structure interaction (SSI) analysis – is required. Both approaches are described below.

2   SOIL RATCHETING AND LIMITING EQUILIBRIUM APPROACHES FOR INTEGRAL ABUTMENT BRIDGES

The repeated thermal movements of integral bridge abutments cause particle realignment in granular backfill materials. Year-on-year, the lateral earth pressures in summer increase, as the backfill becomes stiffer; a phenomenon known as soil 'ratcheting' (England and Tsang 2001). The effect of ratcheting is that soil stiffness and therefore the maximum lateral pressure can be significantly greater than the 'intermediate' value (K) that would be obtained from a pressure/ deflection curve such as Figure 1 for the expected movement (Δ). By the same token, the relative movements required to reach active or passive conditions suggested by AASHTO (2012) Table C3.11.1-1 or EN1997-1 (CEN 2004) Figure C.3 are not applicable.

After 100–200 cycles, the increase in stiffness tails off, with maximum (summer) pressures tending to a value which has been empirically linked to backfill properties, geometry and the movement range. Similar peak values are reached even if the backfill was not very well compacted at placement – see BA42/96 (Highways Agency 2003) clause 3.2.

'Limiting equilibrium' approaches for the design of integral abutment bridges generally use an assumed lateral earth pressure distribution and earth pressure coefficient, commonly denoted K^*. Where the abutment retains granular material, the pressure distribution and value of K^* used should be based on a theory which takes ratcheting into account.

No standard approach for earth pressure distribution behind integral abutments, or for determination of K^* appears in the LRFD Bridge Design Specifications (AASHTO 2012) or the Eurocodes (CEN 2004) and practices vary (Kunin and Alampalli 1999), with some methods making no allowance for ratcheting.

Figure 2 shows the assumed pressure distribution given by PD6694-1 (BSI 2011) for a full height abutment on flexible foundations.

PD6694-1 (BSI 2011) clause 9.4.3 indicates that, for this case, K^* can be conservatively calculated using:

$$K^* = K_0 + \left(\frac{Cd'_d}{H}\right)^{0.6} K_{p;t} \tag{1}$$

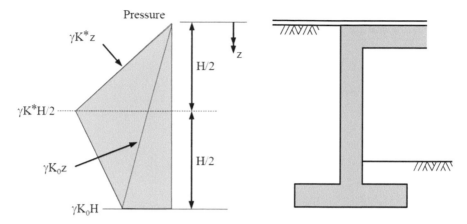

Figure 2.   Assumed earth pressure distribution for full height abutment on flexible foundations after PD6694-1 (BSI 2011) Fig 5.

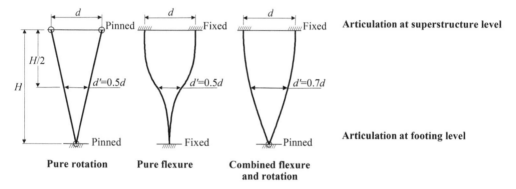

Figure 3.   Comparison of different types of rotational and flexural abutment movements after Stage 1 Report (Highways Agency/Arup 2009), Fig 9.

where
H height of the wall
d'_d wall movement range at $H/2$ below ground level, taken as 0.5 to 0.7 times the design value of the movement range at the top of the wall, based on an assessment of the rotation and flexure in the system. See Figure 3 below.
K_o coefficient of at-rest earth pressure
$K_{p;t}$ coefficient of passive earth pressure determined using the design value of the triaxial ϕ'
C coefficient dependent upon the elastic modulus of the subgrade (E_s):

$$C = 0.35E_s + 14.9 \text{ for } E_s \text{ in ksi; } 20 \leq C \leq 66 \tag{2}$$

For the shorter height bank pad abutments that accommodate thermal movements by translation without rotation, a simple triangular pressure distribution may be assumed and PD6694-1 (BSI 2011) clause 9.4.4 gives the following expression for K^*:

$$K^* = K_0 + \left(\frac{40d'_d}{H}\right)^{0.4} K_{p;t} \tag{3}$$

These expressions and the recommendations of PD6694-1 (BSI 2011) in general reflect a thorough review of research in the field (Highways Agency/Arup 2009), much improved by comparison

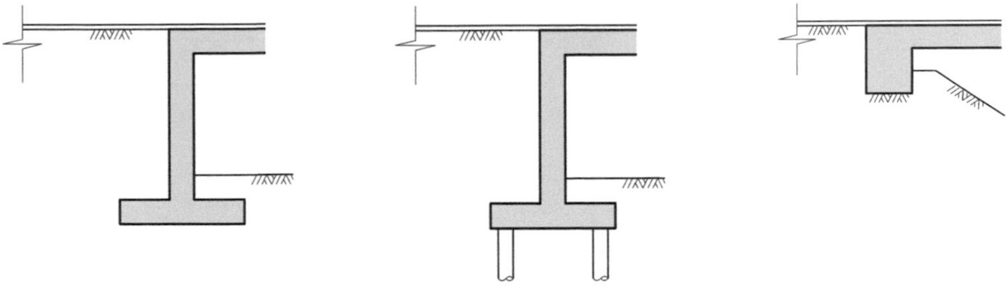

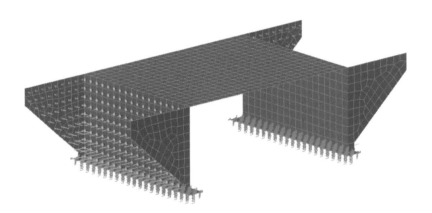

(i) Full height integral
abutment on pad footing

(ii) Full height integral
abutment on piles

(iii) Bank pad

Figure 4.   Integral abutment types which can typically be designed with K^* approaches.

Figure 5.   Typical 3D (shell) model of a full-height integral abutment bridge.

to the previous UK guidance in BA42/96 (Highways Agency 2003), as described by Denton et al (2010).

In the review (Highways Agency/Arup 2009), limiting equilibrium approaches based on K^* have been found to be appropriate for some common integral bridge abutment types such as those illustrated in Figure 4 below. It is usual for these abutment types to be constructed with a free-draining granular material used for backfill – it should be noted that the effects of soil ratcheting may be ignored when the material behind the abutment is a cohesive soil.

In the limiting equilibrium approach, the assumed earth pressures are applied to an analysis model of the structure in question. A 2D frame (beam element) model or 3D model may be used. The grid (grillage) analogy is routinely used for bridge superstructure analysis and some texts advocate extending such models into three dimensions to include abutment walls (Nicholson 1998). However, in-plane effects would be expected to be significant and these can create misleading local in-plane distortions of grid members in a 3D analysis (O'Brien and Keogh 1999). Therefore 3D shell element models or mixed element models would probably be more appropriate (as in Figure 5 below). But, whatever type of model is used, the soil is represented only as a load – this is not a soil-structure interaction analysis; it is a way of *avoiding* a soil-structure analysis.

From such a structure-only model, load effects (bending moments, shear forces etc) are obtained for design purposes. Care must be taken to ensure that all components – superstructure, abutment

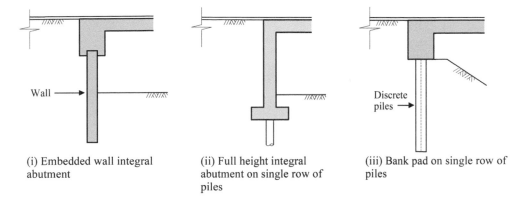

(i) Embedded wall integral
abutment

(ii) Full height integral
abutment on single row of
piles

(iii) Bank pad on single row of
piles

Figure 6.   Integral abutment types which typically require SSI analysis.

walls or end screens, foundation members/ piles – are designed considering suitable maximum or minimum earth pressures in combination with all other applicable loads, including the corresponding bridge temperature.

The pressures which should be considered in conjunction with bridge expansion and contraction are illustrated in PD6694-1 (BSI 2011) Figure 6 and the Finnish Transport Agency Guideline (Liikenneviraston 2012) Figure 12. Notably the latter identifies the minimum pressure as being close to zero (rather than K_a) with not only contraction of the bridge but also freezing of the soil at the lowest temperatures. As the bridge expands but temperatures remain below zero, soil pressures increase rapidly as the ground is still frozen, reducing somewhat on thawing.

PD6694-1 (BSI 2011) suggests that the lateral earth pressure on wingwalls of abutments which support K^* pressures should themselves be subject to a pressure distribution similar to that illustrated in Figure 2, but calculated using the greater of K_o or $K_a \times K^*$ (clause 9.9).

In skew integral abutment bridges, the earth pressures (normal to the walls) create a couple which, unresisted, would cause the bridge to rotate on plan. Furthermore, for bridges such as types (i) and (ii) of both Figures 5 and 6, thermal expansion results in twisting of the top of the abutment walls relative to their bases, along with any intermediate piers. PD6694-1 (BSI 2011) discusses the rotation and twisting (clause 9.8) but a simple calculation approach is not given. Where such effects are deemed to be significant, a 3D analysis and perhaps SSI analysis may be appropriate.

3   SOIL STRUCTURE INTERACTION APPROACHES

The limiting equilibrium approach is not appropriate where soil stiffness plays a more significant part in the behavior of the system. This is the situation for a number of common integral abutment bridge types such as those illustrated in Figure 6 below.

SSI can be handled by a number of calculation methods including closed form solutions, although these typically consider only simple cases, not extending to integral abutment bridge arrangements. Most real project cases require numerical integration for which the finite difference, boundary element or finite element (FE) methods may be employed. FE approaches to SSI tend to fall into those which represent the soil using continuum elements and those which represent the soil using springs.

Bank pad abutments, often supported on steel H-piles – type (iii) in Figure 6 – are perhaps the most popular integral abutment bridge type where space allows. For these, a 'subgrade modulus' model, where the soil stiffness is represented using springs, is probably most appropriate. When – as is common practice in the US – the upper portion of the piles is sleeved, SSI is largely eliminated, but the movement of the bank pads in and out of the backfill will be relatively greater, due to the

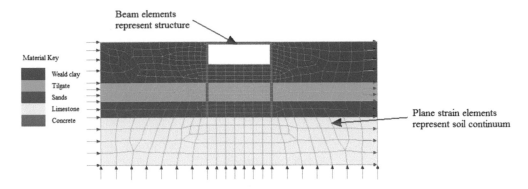

Figure 7.   Typical 2D model of a single-span underpass.

lower resistance from the foundations. Soil ratcheting will still occur, resulting in increased earth pressures on the bank pads, but the total lateral load arising will be quite limited due to the low height of the pads. Structural models with springs and K^* pressures may be adequate, as discussed in Section 5 below. The ratcheting effect may be further mitigated by the use of pea gravel or other specialized backfill.

Embedded walls (contiguous piled, secant or diaphragm) – type (i) in Figure 6 – are perhaps most popular for short-span underpasses in congested urban areas (Connal 2004). For such structures, a full continuum model may be more suitable.

4   FE CONTINUUM MODELS

4.1   *Element selection and considerations*

For the representation of soil masses, beam and shell elements (most familiar to bridge engineers) are inappropriate and continuum elements must be used. In some bridges, 3D effects are of concern; however, often 2D models are sufficient and are certainly recommended for preliminary studies of SSI issues – for example, to assess the sensitivity of results to possible variation in certain parameters. Figure 7 shows the analysis model for an embedded wall underpass (type (i) in Figure 6), where 2D plane strain elements are used to represent the soil and 2D beam elements used to represent the structural members.

As in all FE analyses it is important to ensure that the number of elements used is sufficiently large that any inaccuracy arising from the division strategy may be deemed negligible by comparison to other assumptions inherent in the analysis. For FE continuum analyses we must also consider that elements give best accuracy at an aspect ratio of 1:1 and equal internal angles, although pragmatically ratios up to 1:3 are usually acceptable in areas of interest, and ratios up to 1:10 may be acceptable in remote regions of the model.

When modeling a structure interacting with a soil mass, the extent of the model is not straight-forward to define: vertical and horizontal boundaries must be imposed on the soil mass at some distance from the structure. Where such boundaries cannot be reasonably defined to match physical boundaries (e.g. free soil face, bedrock) they need to be determined by comparing key results from several models which are identical except for the assumed width or depth. Where the stiffness of the soil (E') has been assumed constant with depth, the predicted deflection under vertical load at the surface will increase as the depth of soil below the structure is increased so other key results should be used for comparison. If E' increases with depth, this effect is less pronounced.

Where dynamic effects are required to be considered, non-reflective boundaries are typically required. Such considerations are outside the scope of this paper.

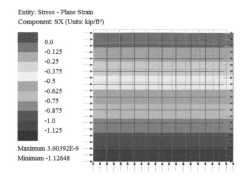

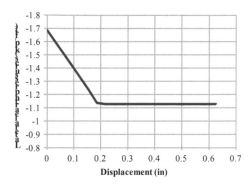

Figure 8.   Results from Mohr-Coulomb test model where $\phi' = 30$.

4.2   *Soil material properties and initial stresses*

The simplest SSI models can assume isotropic linear elastic material properties for the soil. These would require only two parameters in their definition: elastic modulus, (E', perhaps varying with depth) and Poisson's ratio (ν), or shear modulus and bulk modulus. However, most situations which demand SSI analysis also demand a more realistic mathematical representation of the soil material to be employed.

There are many nonlinear material models designed to represent soil behavior. Amongst these, the Mohr-Coulomb model is probably the most widely used (Muir Wood 2004) section 3.4.1. Yield is based on a critical shear stress which is dependent on the normal pressure, making it applicable for soils where strength increases with balanced confining stresses. When yield is exceeded, volumetric plastic straining (dilatancy) occurs, and isotropic hardening may be assumed.

To describe a granular material using a Mohr-Coulomb material model, elastic properties must be given (E', ν') along with the initial and final internal angles of friction (ϕ_1', ϕ_2'), cohesion (c') and dilation angle (ψ) (LUSAS 2013). Effective stress parameters are referenced and the concept of effective stress is important because the *stiffness* of a saturated soil is dependent on whether an increase in load may be carried by fluids, fluids and soil skeleton or skeleton only and the *shear strength* of a soil is dependent on the effective normal stress.

Simple test models can be used to show that the internal stresses in a block of Mohr-Coulomb material correspond to active and passive pressures as predicted by Rankine-Bell equations. Figure 8 illustrates this, showing a test case (plane strain soil block of 30ft × 30ft) with active pressures achieved by movement of the right boundary; movement in the opposite direction will similarly achieve the expected passive pressure.

The results of Figure 8 may be corroborated simply:

Vertical stress at 30ft depth $\quad$ $S_Y = \rho g H = 3.5\text{E-}3 \times -32.185 \times 30 = -3.379\text{ksf}$ $\hfill$ (4)

Maximum (initial) lateral $\quad$ $S_X = S_Z = K_o \times S_Y$ where K_o was taken to be 0.5 $\hfill$ (5)
stress at 30ft depth

Thus S_X (initial) $= 0.5 \times -3.379 = -1.70\text{ksf}$

Minimum (final) lateral $\quad$ $S_X = S_Z = K_a \times S_Y$ where $\hfill$ (6)
stress at 30ft depth

$$K_a = \frac{1 - \sin\phi'}{1 + \sin\phi'} = 0.333 \hfill (7)$$

Thus S_X (final) $= 0.333 \times -3.379 = -1.12\text{ksf}$

The displacement of the wall (or in this case, boundary) which is required to mobilize the limiting active or passive pressure must be understood to be dependent upon not only the elastic modulus

Table 1.   Effect of assumed value for K_o on movement required to mobilize limiting earth pressures.

K_o	Limiting pressure at 30ft depth mobilized at (in)	
	Active	Passive
0.1	0	$3^1/_8$
0.5	$3/_{16}$	$2^7/_8$
1.0	$1/_2$	$2^5/_{16}$
2.0	$1^5/_8$	$1^3/_{16}$

of the material (as might be immediately anticipated) but also upon the *initial stress* in the soil. By varying the assumed value of K_o, the displacements required for full active or passive pressure to be reached is altered, as shown in Table 1.

The values of Table 1 illustrate the importance of initial stresses in SSI analyses. The applied loads for the initial (equilibrium) state in an FE analysis must include an initial stress which varies with depth, usually based on an assumed K_o.

4.3   *Representing interfaces in continuum-based models*

The interface between soil and structure typically needs some special consideration in any SSI analysis. Comprehensive FE systems offer a range of options in this area such as those outlined by LUSAS (2013):

1. Joint elements & materials. Known as 'link', 'hook' or 'fuse' elements in some software, joint elements notionally have no length but instead provide a means of connecting two adjacent elements without full fixity, introducing options such as frictional or yielding behavior.
2. Contact slidelines. Contact algorithms in software such as LUSAS enable the proximity of elements to each other to be detected, allowing transfer of load between one 'component' and another without adjacent elements actually sharing nodes. In the context of SSI, the components would be the soil and the structure, and frictional slidelines would typically be of interest.
3. Elasto-plastic interface materials. A layer of elasto-plastic material (assigned to plane strain or 3D continuum elements) can represent the friction-contact relationship between the soil and the structure. The material reproduces the nonlinear response of a system containing planes of weakness governed by Mohr-Coulomb type laws.

Whichever of these options is utilized, for retaining structures and integral abutment bridges, the crucial consideration is usually back of wall friction, δ. The value of δ cannot be less than zero (a notionally smooth wall) nor exceed that of ϕ' for the material being retained. For many retaining structures, lower bound ϕ' and δ are deemed critical for design and $\delta = 0$ is used, as suggested in PD6694-1 (BSI 2011) clause 7.2.2 and CIRIA C580 (Gaba et al. 2003) section 4.1.4, whereas for integral abutment bridges upper bound values may also be critical. For the design of integral bridge abutments, BA42/96 (Highways Agency 2003) clause 3.3 states that wall friction should be taken as $\delta = \phi'/2$.

4.4   *Representing the structure*

It is important to remind ourselves that relative stiffness is crucial to the distribution of loads in an FE model. In the case of a wholly concrete structure, for example, the accuracy of the value used for elastic modulus affects deflections but generally has little effect on load distribution, since the *relative* stiffness is accurate. However, in the case of a concrete structure in contact with the ground, a reasonably accurate relative stiffness may demand more consideration of issues such as concrete cracking and creep deformation. While these are considerable topics in their own right, it should be underlined that reinforced concrete (RC) is generally cracked and therefore has a stiffness significantly less than that which would be assessed using the gross section and the elastic modulus from a code of practice. See Rombach (2011) section 2.4.2.

4.5 *When and how to incorporate the effects of soil ratcheting*

Often, the popular embedded wall integral abutment (Figure 7 and type (i) in Figure 6), is constructed as part of a top-down scheme in cohesive soil. As noted earlier, for such soils, the effects of strain ratcheting may be ignored (see PD6694-1 clauses 9.4.5.2 and A3.2) and so a suitable SSI analysis (perhaps utilizing a Cam Clay material model) may be used with no further special considerations. The software used must be capable of modeling the staged construction process in conjunction with the use of the preferred nonlinear soil material.

For integral abutment bridges with embedded walls (type (i) in Figure 6) or full height integral abutments on a single row of piles (type (ii) in Figure 6), retaining granular materials, guidance for a suitable SSI analysis incorporating ratcheting is given in PD6694-1 Annex A with further background given in Stage 2 report (Highways Agency/Arup 2009), Section 5. The recommended approach essentially entails the soil being modeled as a continuum, with an elastic modulus (E') which varies with depth according to an assumed lateral earth pressure profile which can be regarded as a 'quasi-passive' limit, similar in nature to the K^* profile described for the limiting equilibrium method above. Along with this, lateral earth pressures are restricted to lie between the active limit and the quasi-passive limit. Therefore a continuum model with Mohr-Coulomb material can be used, together with joint elements (option 1 in the 'Interfaces' list above) which yield at the quasi-passive limit. The software used must be capable of handling the variation of the material properties (E', K^*) with depth in the Mohr-Coulomb and joint materials.

For a bank pad abutment on piles (type (iii) in Figure 6), lateral movement at the pile head does not infer plane lateral movement of the soil in the way that lateral movement of a retaining wall does. There may be some arching of soil between piles but judgments are needed. This makes 2D plane strain models less appropriate for such bridges and the use of spring models more attractive.

5   USE OF SPRINGS TO REPRESENT SOIL

An alternative means of representing soil in an SSI analysis is to use springs, with such analyses sometimes referred to as 'Winkler spring' or 'subgrade modulus' models. The springs may be used to represent the vertical or horizontal resistance of the soil; in the context of retaining structures and integral abutment bridges it is the horizontal stiffness, characterized by a spring stiffness, k_h (force/length^3), which is of interest. Spring models are suggested in EN1997-1 (CEN 2004) clause 9.5.4.

Nonlinear springs or joints may be used within an FE model to generate lateral earth pressures for a retaining wall design based on a pressure/deflection relationship such as that in Figure 9 below:

Critically Figure 9 incorporates not only the modulus of subgrade reaction, k_h, but also at-rest earth pressures (σ_o, based on K_o). Neither quantities are considered when retaining walls are designed using limiting earth pressure methods, however they are essential components of SSI analyses (see CIRIA C580 (Gaba et al. 2003), section 5.1). Typically all the quantities represented in Figure 9 – active and passive 'yield points', the spring stiffness, k_h, and the at-rest pressure – vary with depth.

The yielding spring approach illustrated is advocated by Frank et al. (2004). It is also applied to integral abutment bridge analysis (Faraji et al. 2001) and, with suitable values of K_a and K^* (varying with depth) as the yield points, suitable stiffness, k_h (varying with depth) and initial stresses, might provide an alternative to continuum analyses for embedded wall integral abutments (type (i) in Figure 6). However, it is in the analysis of abutments on piles (types (ii) and (iii) of Figure 6) that the use of springs seems most appropriate.

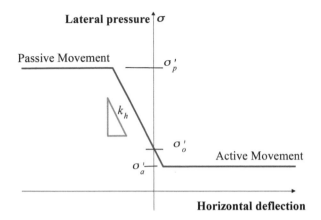

Figure 9.   Nonlinear 'soil joint' pressure/ deflection graph (LUSAS 2013).

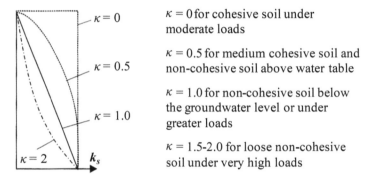

$\kappa = 0$ for cohesive soil under moderate loads

$\kappa = 0.5$ for medium cohesive soil and non-cohesive soil above water table

$\kappa = 1.0$ for non-cohesive soil below the groundwater level or under greater loads

$\kappa = 1.5\text{-}2.0$ for loose non-cohesive soil under very high loads

Figure 10.   Typical variations of modulus of subgrade reaction with depth for piles after Rombach (2011) Figure 2.39.

Where piles are installed in level ground, the at-rest pressures are in equilibrium and so are not of interest for most analysis/design purposes. When lateral strains are expected to be small, it may be reasonable to model piles using beam elements, supported by linear elastic lateral springs. For cohesive soils, it is generally considered that spring stiffness may be assumed constant with depth – as in Liikenneviraston (2012) section 4.3.5.1 and Rombach (2011). For granular soils, a linear variation with depth may be used. Indeed Finnish Transport Agency Guideline (Liikenneviraston 2012) suggests a linear variation up to a depth of $10d$ and thereafter a constant value for k_h. It may be helpful to consider the variation to be a polynomial of the form:

$$k_h = A + Bz^\kappa \tag{8}$$

In this, A, B and κ are empirical constants. Typical variations of k_h with depth for different soil types are suggested by Rombach (2011) and illustrated in Figure 10.

Values for k_h are notoriously difficult to obtain, since the spring stiffness is not a fundamental soil property. However some guidance may be found in the Finnish Guidelines (Liikenneviraston 2012) section 4.3.5.1 (including a correlation between ϕ' and k_h), while RP2A (API 2000) section 6.8 describes methods for defining pressure-deflection (p-y) curves for laterally loaded piles appropriate to various soil materials. Comprehensive FE software is capable of handling such curves within a nonlinear joint material as in Figure 11 below and the matter is covered in more detail by Reese and Van Impe (2000).

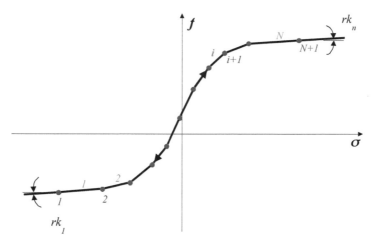

Figure 11.   Piecewise joint definition for nonlinear joint material, from LUSAS (2013).

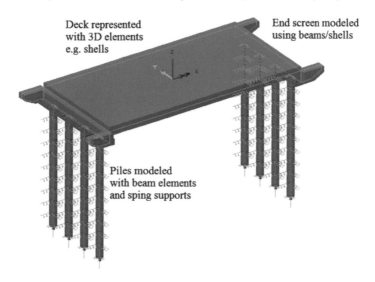

Figure 12.   3D FE model of integral abutment bridge with bank pads on piles.

PD6694-1 (BSI 2011) clause 6.4.9 indicates that SSI analysis is required for bank pad abutments on piles (type (iii) in Figure 6), and draws attention to a particular problem: forward of the piles, the ground slopes away and so the ground stiffness and limiting pressures would be different on each side of the line of piles (see CIRIA C580 (Gaba et al. 2003), clause 7.2).

The remolding of soil over many thermal cycles should also be taken into account as described by Wasserman (2007). This is reflected in RP2A (API 2000), which describes cyclic loads as causing a deterioration of lateral resistance as compared to that observed for static loads, and gives p–y curves for both static and cyclical loading.

For bank pad abutments on piles, then, suitable lateral earth pressures for the end screen may be determined by adopting a K^* approach (PD6694-1 (BSI 2011) clause 9.4.6), while the pile-soil interaction can be handled using linear or nonlinear springs (or 'joints') to represent the soil as illustrated in Figure 12 below. Such approaches to pile-soil modeling have been used for integral abutment bridges – and validated using field measurements and sub-models in the past – see Jayaraman et al. (2001), Krizek and Studnièka (2007), and Albhaisi et al. (2012).

Table 2.   Suggested analysis approach by integral abutment type.

Integral abutment type	Limiting equilibrium	SSI	Notes and reference within PD6694-1 (BSI 2011)
Full height wall on pad footing	Yes		Granular backfill. Assumed earth pressure distribution from Fig 5 and K^* from clause 9.4.3 incorporates ratcheting.
Full height wall on piled footing	Yes		
Bank pad	Yes		Granular backfill. Triangular earth pressure distribution and K^* from clause 9.4.4 incorporates ratcheting.
Embedded wall		Yes	Soil modeled using continuum. For granular soils or backfill, modify E' with depth to suit Annex A and restrict pressures to K^* from clause 9.4.3. Alternatively nonlinear spring model with similar considerations.
Full height wall on single row of piles		Yes	Soil modeled using nonlinear springs. For granular backfill, modify stiffnesses and limiting pressures to suit Annex A.
Bank pad on single row of piles		Yes	Soil modeled using nonlinear springs. Reduced stiffness and limiting pressures for front face of piles. Granular backfill to end screen. Triangular earth pressure distribution and K^* from clause 9.4.4 incorporates ratcheting.

6   CONCLUSIONS

Typically the abutment walls or endscreens of integral abutment bridges are backfilled with granular material, where the effects of soil ratcheting should be taken into account. Where cohesive soil lies behind an embedded wall, ratcheting can be neglected. Limiting earth pressure and SSI analysis options have been explored and suggested approaches are shown in Table 2 below.

REFERENCES

AASHTO LRFD Bridge Design Specifications, American Association of State Highway and Transportation Officials, Washington D.C., 2012

Albhaisi, S., Nassif, H., and Hwang, E. 2012. Effect of Substructure Stiffness on Performance of Steel Integral Abutment Bridges Under Thermal Loads, *Journal of the Transportation Research Board,* Transportation Research Board of the National Academies, 2313: 22–32

API Recommended Practice RP2A-WSD. Recommended Practice for Planning, Designing and Constructing Fixed Offshore Platforms – Working Stress Design, 21st Edition, Dec 2000. American Petroleum Institute, Washington D.C.

Barker, R.M., Duncan, J.M., Rojiani, K.B., Ooi, P.S.K., Tan, C.K. and Kim, S.G. 1991. *NCHRP Report 343: Manuals for the Design of Bridge Foundations,* Transport Research Board/ National Research Council, Washington D.C., USA

BSI. 2011. *PD 6694-1:2011 Recommendations for the Design of Structures Subject to Traffic Loading to BS EN 1997-1:2004*, British Standards Institute, London, UK

CEN (Comité Européen de Normalisation). 2004. *BS EN1997-1:2004 Eurocode 7: Geotechnical design — Part 1: General rules*, British Standards Institute, London, UK.

Connal, J. 2004. Integral Abutment Bridges – Australian and US Practice. *Austroads 5th Bridge Conference*, Austroads, Sydney, Australia.

Denton, S., Riches, O., Christie, T. and Kidd, A. 2010. Developments in Integral Bridge Design, *Bridge Design to Eurocodes: UK Implementation*, ICE Publishing, London, UK

England, G.L. and Tsang, N.C.M. 2001. *Technical Paper No 2: Towards the Design of Soil Loading for Integral Bridges*, Concrete Bridge Development Group, Camberley, Surrey UK

Faraji, S., Ting, J.M., Crovo, D.S., and Ernst., H. 2001. Nonlinear Analysis of Integral Bridges: Finite Element Model, *ASCE Journal of Geotechnical and Geoenvironmental Engineering*, 127 (5): 454–461

Frank, R., Bauduin, C., Driscoll, R., Kavvadas, M., Krebs Ovesen, N., Orr, T. and Schuppener, B. 2004. *Designers' Guide to EN 1997-1, Eurocode 7: Geotechnical design – General rules,* Thomas Telford Ltd, London, UK

Gaba, A.R., Simpson, B., Powrie, W. and Beadman, D. R. 2003. *CIRIA C580 Embedded Retaining Walls – Guidance for Economic Design*, Construction Industry Research and Information Association, London, UK

Highways Agency. 2003. *BA42/96 Amendment No 1, The Design of Integral Bridges*, The Stationery Office, London, UK

Highways Agency/Arup, 2009. Integral Bridges – Best Practice Review and New Research, Phase 2b – Review of Existing Data, Back-Analysis of Measured Performance and Recommendations (Stages 1, 2 and 3)

Jayaraman, R., Merz, P.B. and McLellan Pte Ltd. 2001. Integral Bridge Concept Applied to Rehabilitate an Existing Bridge and Construct a Dual-use Bridge, *26th Conference on Our World in Concrete & Structures*, Singapore.

Krizek, J. and Studnička, J. 2007. Integral Bridges and Soil–Structure Interaction, *Modern Building Materials, Structures and Techniques 9th International Conference*, Vilnius, Lithuania

Kunin, J. and Alampalli. 1999. *Special Report 132: Integral Abutment Bridges: Current Practice in the United States and Canada*, Transportation Research and Development Bureau, New York State Department of Transportation, New York, NY, US.

Liikenneviraston, 2012. *Sillan geotekninen suunnittelu – Sillat ja muut taitorakenteet*, Liikenneviraston ohjeita (Transport Agency Guidelines) 11/2012, Helsinki

LUSAS. 2013. *Theory Manual Volume 1*, LUSAS, Kingston Upon Thames, Surrey, UK

Muir Wood, D. 2004. *Geotechnical Modeling*, Spon Press, New York, NY, USA

Nicholson, B.A. 1998. *Integral Abutments for Prestressed Beam Bridges,* Prestressed Concrete Association, Leicester, UK

O'Brien, E.J. and Keogh, D.L. 1999. *Bridge Deck Analysis,* E&FN Spon, London, UK

Reese, L.C. and Van Impe, W. 2000. Single Piles and Pile Groups Under Lateral Loading, Taylor & Francis, London, UK.

Rombach, G.A. 2011. *Finite-element Design of Concrete Structures: Practical Problems and their Solutions (Second Edition)*, ICE Publishing, London, 2011

Wasserman, E.P. 2007. Integral Abutment Design (Practices in the United States), *First U.S.–Italy Seismic Bridge Workshop*, Pavia, Italy.

Chapter 11

Improving structural reliability using a post-tensioned concrete floor system for major non-redundant steel bridges

C. Chang & R.A. Lawrie

Hardesty & Hanover LLC, Alexandria, VA, USA

ABSTRACT: Many steel bridges built decades ago have redundancy issues since redundancy was not accommodated in the design. These major non-redundant steel bridges are in various forms, such as two-girder bridges, tied arch bridges with tension ties, and truss bridges. With the lack of redundancy, failure from one member of the bridge would lead to the failure of the entire bridge. Serious attention is necessary for this structural performance, structural reliability, and, most importantly, public safety issue. Instead of replacement of the entire bridge, rehabilitation might be favorable for some steel bridges due to their historical significance, materials viability for continuous service, cost effectiveness of rehabilitation over bridge replacement, etc. This paper provides a discussion on the structural reliability improvement using a post-tensioned concrete floor system for major non-redundant steel bridges rehabilitation. A non-redundant structure can be represented as a series system, in the reliability engineering aspect, in which when one of the system components fails, the entire system fails. A structure with redundancy, on the other hand, is considered as a combination system made of series and parallel configurations, where a parallel configuration is one that does not fail unless all the components fail. Illustrative examples are provided in this paper for further demonstration of the structural reliability improvement.

1 INTRODUCTION

It has been made clear in recent years that redundancy is an issue and a need. This paper discusses reliability improvement by using a post-tensioned concrete floor system combined with post-tensioned steel, based on a TY Lin Philosophy, (1955) in tension areas for rehabilitation on major non-redundant steel bridges. A two-girder steel bridge, a tied arch bridge with tension ties, and a major truss bridge are used in this paper to illustrate the reliability improvement and the strengthening obtained by using combined post-tensioning in concrete and structural steel. The issues of each of the three steel bridge systems are discussed in Section 2. Section 3 provides the concepts of reliability for different system configurations, while the post-tensioned combined concrete and steel structure is illustrated in Section 4. Section 5 provides examples illustrating reliability improvement by using post-tensioned structural steel, and Section 6 is the conclusion.

2 MAJOR NON-REDUNDANT STEEL BRIDGES

Many bridges built decades ago did not consider redundancy in the design. Several types of major non-redundant steel bridges such as a two-girder bridge, an arch bridge with tension ties, and a truss bridge are shown in this section to illustrate the issue of lack of redundancy

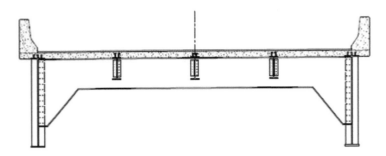

Figure 1.   A two-girder system cross section.

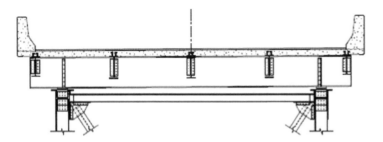

Figure 2.   A deck truss cross-section.

2.1 *Two-girder bridge*

Figure 1 shows a typical two-girder bridge cross section. This cross section utilizes two main girders with intermediate floorbeams and stringers. Failure of one tension flange in the two girders would result in the failure of the entire structure. Many of this type of bridge were built prior to 1960 because it showed exceptional economics.

2.2 *Tied arch bridge with tension ties*

In past years, the issue of tied arch bridges with tension ties showed itself in major cracks in the steel flanges of these ties. The cracking problem found on the tension ties put the entire bridge in danger. Redundancy could be added to such a structure by using a post-tensioning system in parallel with the structural steel tie.

2.3 *Truss bridge*

Figures 2 and 3 show the typical cross-sections of a deck truss and a through truss, respectively, based on JAL Waddell's layout for simple truss details (1925). Truss bridges designed decades ago are often with intermediate hinges. This lack of continuity is a separate issue to be addressed in rehabilitation. In addition, certain members of this type of bridge, especially the stringers and floorbeams of the floor system, suffer from severe corrosion, leaving a major loss of section.

3   RELIABILITY OF SYSTEMS

3.1 *Systems in series configuration*

A system in series configuration is represented in Figure 4. The connectivity demonstrates the contribution of component failures to the system failure and does not necessary represent the

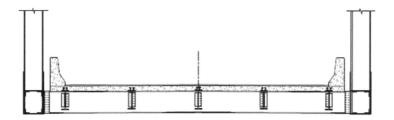

Figure 3.   A through truss cross-section

Figure 4.   A series system of N components.

physical connectivity of the system. As shown in Figure 4, any one of the N components fails, the entire system fails.

If the failure event of the components called E_i, where i = 1, 2, ..., N, the survival of the system can be denoted as $\bar{E}$ and modeled as following:

$$\bar{E} = \bar{E}_1 \cap \bar{E}_2 \cap \cdots \cap \bar{E}_N \tag{1}$$

where $\bar{E}_i$ is the survival event of component i. The probability of failure of the system can be presented as:

$$P(E) = 1 - P(\bar{E}) \tag{2}$$

For independent failure events of the components, Eq. (2) can be expressed as:

$$P(E) = 1 - [1 - P(E_1)][1 - P(E_2)] \cdots [1 - P(E_N)] \tag{3}$$

Considering a system in which the components are not fully dependent or independent, the failure probability of the system Pf can be presented as follows:

$$\max_i(Pf_i) < Pf < 1 - (1 - Pf_1)(1 - Pf_2) \cdots (1 - Pf_N) \tag{4}$$

where Pf_i is the failure probability of the ith component for $i = 1, 2, ..., N$. The lower bound indicates the failure probability where the components of the system are fully dependent while the upper bound represents the failure probability where the components of the system are fully independent.

3.2   *Systems in parallel configuration*

A system in parallel configuration is represented in Figure 5. Similar to the series configuration, the connectivity demonstrates the contribution of component failures to the system failure and does not necessary represent the physical connectivity of the system. The system fails only if all the components of the system fail.

If the failure event of the components called E_i, where $i = 1, 2, ..., N$, the failure of the system can be denoted as E and modeled as following:

$$E = E_1 \cap E_2 \cap \cdots \cap E_N \tag{5}$$

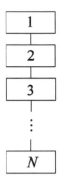

Figure 5.   A parallel system of N components.

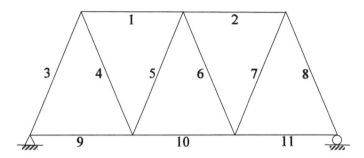

Figure 6.   A sloped truss.

The probability of failure of the system Pf can be presented as:

$$\prod_i Pf_i < Pf < \min(Pf_i) \qquad (6)$$

where Pf_i is the failure probability of the ith component for $i = 1, 2, \ldots, N$. The lower bound indicates the failure probability where the components of the system are fully independent while the upper bound represents the failure probability where the components of the system are fully dependent.

3.3   *Illustrative example using a simple truss bridge*

A steel through truss bridge shown in Figure 6 is utilized to illustrate the reliability improvement by replacing the steel floor system by a post-tensioned concrete floor system described in Section 4. The truss system shown in Figure 6 has 11 members. The connectivity of this system can be expressed by Figure 7, which shows that it is a system in series and if any one of the 11 members fails, the entire bridge would collapse. With a post-tensioned concrete floor system redundancy has been provided to the bottom chord and strengthening the system. Figure 8 shows the logical connectivity of the system with the post-tensioned concrete floor system, which indicates that the system becomes a mix system of series and parallel configuration and the reliability of the system has been improved.

3.4   *Reliability assessment*

Reliability needs to be integrated into serviceability and strength analysis so as to obtain the most efficient solution for the public's good and safety. It is suggested that the "series configuration"

Figure 7.   Logical connectivity of the truss shown in Figure 6.

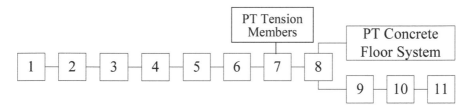

Figure 8.   Logical connectivity of the truss shown in Figure 6 with a post-tensioned concrete floor system.

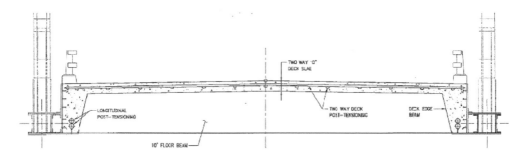

Figure 9.   A post-tensioned concrete floor system.

fails with the weakest element while the "parallel configuration" fails only as all components fail together.

4   POST-TENSIONED CONCRETE FLOOR SYSTEM

The post-tensioned concrete floor system shown in Figure 9, introduced by Chang and Lawrie (2015), has two concrete edge beams monolithic with the concrete deck and a 10" floorbeam transverse to the roadway adjacent to the panel point. The concrete deck has two-way deck post-tensioning. In addition, longitudinal post-tensioning is inside the two edge beams.

The concrete floor system can increase the dead load by as much as 50%; however, when compared to total load in a long truss span, the increase in total load is about 25%. This is based on several projects that have been studied as proposals.

For a 300-ft span, this dead load moment can be increased by approximately 17,000 kip-ft. A 2 × 4-0.6″ dia cable force of 100 kips on a tension chord can produce a resisting moment of 17,000 kip-ft with one cable on each side of tension members. This moment is based on chord deformation and sets up primary and secondary bending effects. Therefore, this seemingly large increase in bending can easily be offset by small post-tensioned cable resisting force.

Also to be noted is the ability to make the concrete edge beam composite with tension chord this strengthening the chord and the whole truss.

Shear must be checked since lateral loads do not produce direct vertical shear and all shear resistance must come from secondary action. To accommodate shear, the number of strands may need to be increased (increasing secondary action) or a portion deflected to create vertical components.

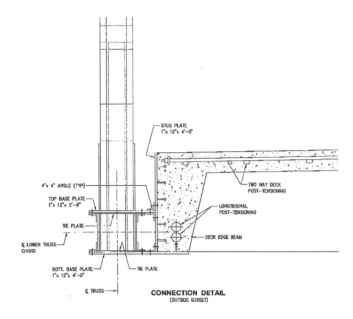

Figure 10.   Connection details.

Figure 11.   Berkley Bridge.

5   STRENGTHENING AND IMPROVED RELIABILITY WITH POST-TENSIONED STRUCTURAL SYSTEM

Experience has shown success in strengthening steel structures by post-tensioning:
    The Berkley Bridge in Norfolk, VA was strengthened in the positive moment region with post-tensioned cables (Figure 11).

Figure 12.   Bonners Ferry Bridge.

Figure 13.   High Bridge.

Bonner's Ferry Bridge in Idaho was strengthened by cables in the negative moment region with post-tensioned cables (Figure 12).

The High Bridge in Minneapolis, MN over the Mississippi River was post-tensioned in the tension tie with cables that provided redundancy. (Figure 13)

6   CONCLUSIONS

This paper discusses the reliability improvement by using a post-tensioned concrete floor system and also using post-tensioned structural steel. Because of the historical significance, cost effectiveness, etc., bridge rehabilitation may be favorable over replacement. In addition, the floor system is often the part of the structure that requires immediate attention for rehabilitation. The reasons for such could be the loss of steel section area due to corrosion, the need for widening the roadway, the inefficiency of tension ties due to cracking or other issues, etc. Examples have been provided to show the viability of post-tensioned steel. Additional dead load can be carried by adding post-tensioning cables to tension areas. Such can provide strengthening as well as providing redundancy, which enhances the reliability and safety of the structure.

REFERENCES

Chang, C. & Lawrie, R.A. 2015. Rehabilitation for Truss Bridge using Post-tensioned Concrete Floor System, *32nd Annual International Bridge Conference, June 2015.*
Lin, T.Y. 1955 *Prestressed Concrete Structures.* New York, NY: John Wiley & Sons.
Waddell, J.A.L. 1925 *Bridge Engineering.* New York, NY: John Wiley & Sons.

Chapter 12

AASHTO fatigue testing of modular expansion joints – setting new standards

G. Moor
Mageba USA, New York, NY, USA

S. Hoffmann & C. O'Suilleabhain
Mageba SA, Bulach, Switzerland

ABSTRACT: Fatigue performance of bridge expansion joints, a critical consideration in their selection and use, is discussed in relation to the most versatile type of expansion joint available: the modular joint. The recent fatigue testing, in the "infinite life regime", of a modular joint in accordance with AASHTO specifications – the most demanding by any major authority worldwide – is presented, with reference to the particularly demanding interpretation by leading American authorities of the provisions relating to the number of load cycles to be applied during testing.

1 INTRODUCTION

Laboratory testing of bridge components has an important role to play in verifying their long-term performance and thus minimizing their life-cycle costs. As noted by Spuler et al (2012), the life-cycle costs of a bridge's expansion joints are likely to be many times higher than the initial supply and installation costs. An expansion joint that offers better durability will, of course, need to be replaced fewer times during the bridge's life of 100 years or more, and it is during replacement works that the most significant costs of an expansion joint, to the bridge's owner and its users, arise.

The long-term performance of these critical bridge components, and their fatigue performance in particular, should thus be a key factor in their selection and design. While the long-term performance of a particular type of expansion joint, as manufactured by a particular supplier, can in many cases be evaluated on the basis of the performance to date of expansion joints that have been in service for many years, it is often desirable to require evidence in the form of standardized laboratory testing, as discussed below.

2 THE TENSA-MODULAR EXPANSION JOINT

Modular expansion joints (Figure 1) have a great deal to offer the designers and constructors of bridges everywhere, thanks to their ability to facilitate very large longitudinal movements and their great flexibility – no other type of joint can accommodate longitudinal movements of two meters or more while, where so designed, also facilitating movements in all directions and rotations about all axes. This has led to modular expansion joints being the preferred solution for many of the world's largest bridges in recent years, and to an increasing focus on performance standards and testing requirements for such joints by owners and engineers.

Modular expansion joints divide the total longitudinal movement requirement of the superstructure among individual, smaller gaps. The gaps are separated by centerbeams, which create the driving surface and which are supported at regular intervals by support bars underneath. The gaps are made watertight by means of rubber sealing profiles. Tensa-Modular is a modular expansion

Figure 1.   A modular expansion joint, viewed from above, showing the centerbeams and edgebeams that form its driving surface.

Figure 2.   Representation of an installed Tensa-Modular expansion joint (cross section at a support bar), showing stirrup connections to centerbeams.

Figure 3.   Installation of a Tensa-Modular expansion joint on a bridge with a concrete deck.

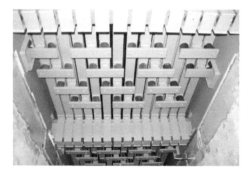

Figure 4.   An installed Tensa-Modular expansion joint, viewed from below.

joint of the single support bar type (with every support bar supporting all centerbeams), with pre-stressed, free-sliding, bolted stirrup connections between centerbeams and support bars (see Figures 2 to 4). The support bars themselves are supported by a similar system in the joist boxes at each end. Rubber control springs, positioned in sets below the centerbeams, coordinate the movements of the centerbeams. This elastic system avoids constraint forces and reduces the effects of loading on the joint and on the main structure, extending the life of the entire system.

3   FATIGUE TESTING OF MODULAR JOINTS – THE AMERICAN CONTEXT

In the American context, fatigue testing of modular expansion joints is specified, among many other aspects of bridge construction, by the American Association of State Highway and Transportation Officials (AASHTO) in its LRFD Bridge Construction Specifications (AASHTO, 2004). The section of these specifications that deals with testing of modular expansion joints, Appendix A19, was based on a detailed 1997 report by the Transportation Research Board of the National Research Council. This report, entitled "Fatigue Design of Modular Bridge Expansion Joints" (Dexter et al, 1997) was issued as Report No. 402 of the National Cooperative Highway Research Program (NCHRP), and was based on research which was sponsored by AASHTO in cooperation with the Federal Highway Administration, United States Department of Transportation. A subsequent report on modular expansion joints by the NCHRP, Report 467 from 2002 (Dexter et al, 2002), noted:

"When the root cause of an overall failure is a failure of the structural supports (i.e., the centerbeams and the support bars), it is usually the result of fatigue cracking. Research was previously conducted on this problem, and fatigue design and testing specifications were proposed in NCHRP Report 402. It is believed that implementing the design and testing specifications proposed in NCHRP Report 402 can substantially reduce the occurrence of fatigue cracking". NCHRP Report 402 can thus be recognized as having a great deal of legitimacy, and the testing it defines is the most comprehensive fatigue testing currently specified by any major authority with responsibility for bridge expansion joints.

NCHRP Report 402 presents a practical test procedure for the determination of the fatigue resistance of critical details in the joint's construction. The onerous testing required by this report, and consequently by AASHTO's LRFD Bridge Construction Specifications, simulates the fatigue-inducing movements and stresses of a service life on a full-scale section of a joint which contains all critical members and connections. It involves the subjecting of expansion joint specimens to an enormous number of load cycles, and its complexity increases with the complexity of the expansion joint itself. For a highly developed and particularly flexible type of modular joint such as Tensa-Modular, fatigue testing can be especially demanding.

Although fatigue testing is specified in great detail by NCHRP Report 402, one critical aspect is not clearly defined: the number of cycles to which each test specimen must be subjected. Although a lower bound of 200,000 cycles is indicated, this is far too low to be of any practical use today. In the past, a figure of two million load cycles was commonly applied in fatigue testing of expansion joint types and components. Although this figure appears to be very high, it can quickly appear entirely inadequate when the number of axle loads to which an expansion joint is subjected during a typical service life is considered. Supposing a bridge is crossed by 30,000 vehicles per day in each direction, this would result in approximately one billion axle loads during a service life of 40 years. But testing with just a few million cycles is indeed adequate, as explained below.

4   FATIGUE PERFORMANCE AND TESTING – AN INTRODUCTION

To understand fatigue performance of a device, such as an expansion joint, which is primarily made of steel, it is helpful to consider first the fatigue performance of steel in its simplest form – as a pure material. Fatigue performance is commonly represented by an S-N curve – a graph of the magnitude of a cyclic stress (S) against the number of cycles to failure (N), with N being on a logarithmic scale (see Figure 5). Typically, as might be expected, the higher the stress, the lower the number of cycles that will cause failure. As a consequence, the parameters (S and N) for testing fatigue performance can be selected anywhere along the S-N curve, in the knowledge that satisfying the requirements (i.e. achieving results above the curve on the graph) at any point on the curve is equivalent to satisfying requirements at any other point. Of course, for practical reasons, it is preferable to minimise the number of cycles required in testing by selecting a point as close to the left end of the curve as possible (avoiding the need for hundreds of millions of cycles if a point further to the right is chosen).

A peculiarity of the fatigue performance of steel provides further insight into why testing with "just" a few million cycles can provide great confidence in real-life performance with a billion load cycles or more. For ferrous alloys such as steel, as the applied cyclic stress on an S-N curve reduces from a high level, the number of cycles to failure increases – but when the applied stress reaches a certain limiting value, the number of cycles to failure suddenly appears to approach infinity. This value is known as the material's fatigue limit. In other words, at stresses below the fatigue limit, fatigue failure will never occur – and the S-N curve becomes horizontal at the fatigue limit, as can be seen in Figure 5. Therefore, it makes sense to conduct testing, where possible, with parameters that are taken from the flat part of the S-N curve. Such testing, in the so-called "infinite life regime", indicates that an infinite number of load cycles could be applied without failure as long as loading levels do not exceed the corresponding value that has been applied in testing.

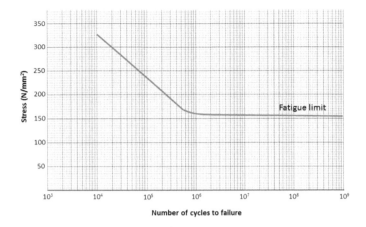

Figure 5.   S-N curve for a typical steel.

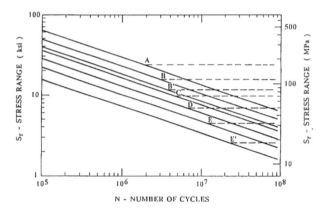

Figure 6.   AASHTO S-N curves for all detail categories (Figure 2.5 of NCHRP Report 402).

This understanding of fatigue performance and testing of materials is of great use when applied to structures or devices such as expansion joints. Modular expansion joints, for example, are generally manufactured predominantly from steel, with welded or bolted details such as the centerbeam to support bar bolted stirrup connection of the Tensa-Modular joint. The fatigue performance of the material is adequately covered by standard material and welding specifications, so the joint-specific assessment of fatigue performance focuses on the welded and bolted details. In most fatigue design specifications for structures, the fatigue resistance of details is reflected in so-called detail categories, which can be thought of as a ranking of the severity of the stress concentration associated with the detail geometry, with each detail category being a grouping of components and details having essentially the same fatigue resistance. AASHTO bridge design specifications define Categories A to E', Category A being the best, and represents the fatigue performance of each by means of an S-N curve (see Figure 6). As can be seen from the curves, the number of cycles (N) that can be withstood by a detail at any particular stress range increases rapidly as the detail category improves. The dashed lines on the graph indicate a limiting value of stress range, known as the Constant Amplitude Fatigue Threshold (CAFT), at which the number of cycles to failure suddenly approaches infinity – much like a material's fatigue limit as described above. The S-N curves of steel and of details manufactured from the steel are thus analogous in this respect.

The S-N curves for detail categories provide an insight into why the figure of two million load cycles has often been applied in the past in laboratory fatigue testing. As for testing of materials, it is sensible to select the parameters for testing from the point on the appropriate S-N curve where the curve becomes flat (horizontal), to minimise the number of cycles while benefitting from the infinite life regime aspect. For relatively uncomplicated expansion joint types, such as cantilever finger joints, Category A can be considered to apply, so the number of cycles, N, has often been set at two million (with the corresponding CAFT of approximately 165 MPa).

5   THE APPLICATION OF AASHTO FATIGUE TESTING REQUIREMENTS TO THE TENSA-MODULAR JOINT

It is desirable for an expansion joint's details to be recognised as belonging of a high category, as it provides confidence in the long-term performance of the joint and enables fatigue design requirements to be satisfied by less cumbersome, more easily installed and maintained expansion joints. Category A is typically only applicable for very simple details such as base metal with no welds or structural connections, so Category B is the best that can be realistically hoped for in relation to connections of any sort. In fact, the NCHRP Report 402 specifies that a centerbeam/support bar connection with a stirrup should be classified as Category D (unless a higher category is proven by testing).

The manufacturer of the Tensa-Modular joint, having committed to carrying out fatigue testing in accordance with AASHTO requirements, made arrangements to do so at America's leading institute in this field, the ATLSS Engineering Research Center of Lehigh University, Pennsylvania, USA. Testing of an entire expansion joint is not required, but of full-size parts which contain all relevant fatigue-sensitive details and elements. In the case of the Tensa-Modular joint, the requirements could be fulfilled by testing specimens consisting of just the critical stirrup connection between a section of centerbeam and a section of support bar beneath. Being convinced that Category B was achievable and appropriate for this connection, the company decided to conduct the testing with the objective of proving this. After extensive discussions with ATLSS, involving complex technical considerations such as real-life deviations from the idealised S-N curves mentioned above and considering the specifications of various American states, it was concluded that testing should consist of 6 million load cycles for each specimen. Although the S-N curve for Category B indicates a figure of three million at the point where the curve becomes horizontal, a factor of two is applied to this to reflect the effect of a statistical bell-curve distribution. In order for just 5% of the results represented by a normal distribution to fall below the figure of three million indicated by the S-N curve, that figure is increased by a factor of two times the standard deviation, which is evaluated by a factor of two. In effect, this introduces a much higher degree of statistical certainty to the testing; a bell-curve centred on the target value of three million load cycles (as it would be if that was the number of cycles chosen for testing) would allow 50% of the values to fall below the target figure, and thus to fall within the "finite life regime", but a bell-curve centred on a target value of six million cycles evaluates just 5% to fall below the target figure, with 95% falling within the "infinite life regime". In statistics, and thus statistics-based testing, an allowance for 5% (1 in 20) of values to fall outside proposed limits without compelling the statistician to consider these values particularly significant has been a key aspect of probability theory since Sir Ronald Fisher, the father of modern statistics, first championed the standard in 1925. In relation to the fatigue testing of modular expansion joints, the factor of two is specified, for example, by Washington State Department of Transportation, one of America's leading authorities in this field.

In accordance with AASHTO requirements, at least ten S-N data points are required to confirm that values consistently fall above the appropriate curve on the S-N graph. Four expansion joint specimens were tested, two at a time, each with three centerbeams, allowing for twelve data points in total. Testing was carried out between June 2012 and September 2013 (Figures 7 and 8), with almost continuous use of one of the industry's most elaborate testing facilities – facilities which were, in fact, used to conduct the original research relating to the development of NCHRP Report 402. The ATLSS laboratory is one of the largest of its kind in North America, with a 100 foot (30.5 m) by 40 foot (12.2 m) strong test floor, bordered on two adjacent sides by a monolithic rigid

Figure 7.   Fatigue testing of Tensa-Modular joint –
testing rig at ATLSS/ Lehigh University.

Figure 8.   Fatigue testing of Tensa-Modular joint –
one specimen.

reaction wall that is up to 50 foot (15.2 m) high. The laboratory is equipped to generate multi-
directional (multi-axis) static and time-varying loads, with hydraulic power systems that operate
at up to 3,500 psi (24.1 MPa). These systems serves numerous, computer-driven servo-controlled
hydraulic actuators simultaneously and independently using a system of six 40 gpm (150 liters/min)
independent hydraulic service manifolds.

The specimens, featuring noise-reducing surface plates ("sinus plates"), were tested under constant amplitude fatigue loading, with 70% of the total load range applied downward and 30%
applied upward, acting at the center of each centerbeam span. The centerbeam to support bar
bolted stirrup connection was tested for a nominal stress range of 110 MPa (16 ksi), corresponding
to the CAFT for AASHTO Category B. The testing was completed successfully, with the fatigue
resistance of all details being verified by testing of ten specimens as specified, each subjected to
6×10^6 load cycles without any fatigue cracking (run out, i.e. no failure). Special aspects such as
field splicing are subject to ongoing examination.

6   CONCLUSIONS

Expansion joints are arguably the parts of a bridge upon which the highest demands are placed,
being relatively light compared to the rest of the structure, yet highly stressed and subject to intense
fatigue loading. This is especially true of the most advanced modular joints, due to their exceptional
flexibility and complex movement capabilities. The described fatigue testing of such a modular
joint in accordance with AASHTO specifications – the most demanding by any major authority
worldwide – demonstrated adequate fatigue performance, in the "infinite life regime", at a level
of testing which is unprecedented in the industry for any type of expansion joint. It thus set a
new benchmark for what can be, and arguably should be, expected by bridge owners in terms
of independent verification that the modular expansion joints to be used on their structures will
provide good long-term performance.

REFERENCES

American Association of State Highway and Transportation Officials (AASHTO). 2004. LRFD Bridge
Construction Specifications, SI Units, 3rd Ed. Washington DC.
Dexter R.J., Mutziger M.J. and Osberg C.B. 2002. Performance Testing for Modular Bridge Joint Systems
(NCHRP Report 467), Transportation Research Board, National Research Council. Washington, DC.
Dexter R.J., Connor R.J. and Kaczinski M.R. 1997. Fatigue Design of Modular Bridge Expansion Joints
(NCHRP Report 402), Transportation Research Board, National Research Council. Washington, DC.
Spuler, T., Loehrer, R. & O'Suilleabhain, C. 2012. Life-cycle considerations in the selection and use of bridge
expansion joints. *Proc. 18th IABSE Congress*, Seoul, South Korea.

Chapter 13

Design of depth critical steel bridge superstructures

R. Schaefer & G. Ricks
HNTB Corporation, Parsippany, NJ, USA

ABSTRACT: The design of "depth-critical" steel superstructures for even simple span bridges is an emerging design concept for many engineers today and is inadequately explained in the AASHTO LRFD Bridge Design or Construction Specifications. Traditional empirical design assumptions as presented in the AASHTO Specifications do not necessarily apply to slender beams which require large cambers and specialized sequences of construction. Shallow girder design will only continue to rise in popularity for situations in which designers are forced to provide additional roadway underclearance where functionally obsolete bridges are replaced, or where construction of a new bridge requires spanning longer over widened roadways without the addition of a pier. The content of this paper addresses the unique concerns of designing steel bridges for minimum superstructure depth through the discussion of a shallow single span bridge which was successfully designed and constructed.

1 INTRODUCTION

Overpass bridges built during the first half of the 20th century were often designed with vertical clearance between underside of the bridge and the riding surface of the below roadway as an afterthought, at best, and sometimes with clearances of $12'-0''$ or less. As these bridges have reached the end of their service lives, bridge owners have typically demanded that additional clearance be provided to accommodate modern trucking standards. Current federal underclearance requirements mandate a minimum of $16'-0''$ and many state and independent bridge owners mandate $15'-0''$ or more.

In tight urban areas where it is not possible to either raise the above roadway or lower the below roadway, the only available solution is to provide a replacement superstructure with a shallower depth. In some cases, this can be easily accomplished by shortening the bridge spans by adding additional pier supports, or by the use of adjacent precast prestressed concrete slabs or box beams, if the span lengths are short enough. In other cases, it is not possible or practical to add piers to create shorter spans, or the span of the superstructure is too long to accommodate the precast prestressed concrete options. In these situations, steel girder type superstructures are the next most practical solution to achieve the maximum possible clearance under the bridge.

Recent versions of the AASHTO LRFD Bridge Design Specifications allow for the design of extremely shallow bridge superstructures by making deflection criteria (Article 2.5.2.6.2) and span-to-depth ratios (Article 2.5.2.6.3) optional for designers. The elimination of these previously mandated provisions creates design and constructability complexities that are not discussed in-depth in the AASHTO text. It is the intent of this paper to discuss these concerns so that designers and bridge owners may have a more complete understanding of the complexities of these structure types from a design and constructability point of view. With this understanding, it is noted that the otherwise conservative assumptions included within the current AASHTO design specifications may not be conservative for these bridge types.

The complexities are discussed below in the order in which a design engineer is expected to encounter them in the process of designing and preparing contract plans for this type of bridge. This guidance, where differing or expanding upon established design and construction practice, has

Figure 1.   Photo of finished bridge.

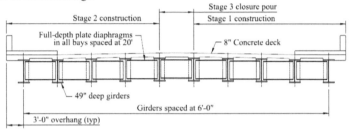

Figure 2.   Basic cross section of final bridge.

been validated through the construction of a long and shallow single span bridge, which is described below. The contents of this paper are not validated against continuous span negative moment regions, but the general design principles should be considered for positive moment regions of continuous girder arrangements.

2   SUBJECT STRUCTURE

The subject structure used to validate the concepts within this paper is a 22 degree skewed (as measured from a line normal to the girders), single span girder bridge with a span of 157'-8³/₃". The steel girders supporting the bridge consist of 5/8" × 45" web plates, 2¹/₂" × 24" bottom flanges, and 2" × 24" top flanges. The girders were spaced at 6'-0" and constructed of conventional ASTM A588 Grade 50 weathering steel. Full depth plate diaphragms were located along the length of the girders in staggered arrangement, spaced at a maximum limit of 20'-0" on-center. The deck was a conventional cast-in-place 8" thick slab using 4000psi High Performance Concrete (HPC). See Figures 1 and 2.

The original as-designed bridge shown on the original contract plans was configured with shallower girders consisting of 5/8" × 36" web plates, 2¹/₂" × 24" bottom flanges, and 1¹/₂" × 24" top flanges. The girder was to be of hybrid construction with ASTM A709 grade 50W web plates and grade 70W flange plates. However, the final girder size was revised and deepened to work with ASTM A588, as ASTM A709 grade 70W was not available in the correct plate sizes at the time of construction. The contractor (Anselmi and DiCicco) and their engineer (Garden State Engineering) value engineered the girder sizing to accommodate the lower grade A588 material in available plate sizes. Pot type bearings were selected for this bridge, as well.

3   DESIGNING FOR DEFLECTION

3.1   *Live load design parameters*

Traditionally, bridges have been designed for stress limits using allowable stress design (ASD) or for ultimate strength using load factor design (LFD) provisions published in the AASHO or

AASHTO codes from the first (circa1931) to seventeenth edition (2002), with the LFD provisions being added in the 1973 11th edition. These publications included provisions to place a lower bound on the overall stiffness of the bridge by the limiting the ratio of the span length (L) divided by a factor (800) for pure vehicular traffic or L/1000 for vehicular plus pedestrian traffic). The anecdotal history of these provisions was presumably to mitigate driver and pedestrian discomfort (Wright and Walker 1971), with additional concerns noted for durability of bridges and decks subjected to "excessive" flexure (ASCE, 1958). Many bridge owners still maintain this criterion in their own design guidelines, often without a complete understanding of the need of this provision or its effects on their bridges.

In evaluating the validity of the provisions, both of the above noted concerns were researched with no conclusive documentation found to support either supposition. If anything, it is worth pointing out that the inverse may in actuality be true for driver and pedestrian discomfort.

There is no human ability to directly measure the small amplitude service deflection of a bridge. At a computed deflection of L/1000, a 100' long bridge is permitted to deflect up to 1.2 inches, per current AASHTO optional provisions. It is implicitly unreasonable to expect a significant difference in perception if a pedestrian were to stand in the middle of the example bridge, look back 50' to the bridge abutment, and then note a deflection of, say 1.5" and unacceptable.

The human inner ear, however, has an excellent ability to detect changes in the ambient rate of acceleration. Changes in the rate of acceleration are more noticeable to pedestrians and occupants of stationary vehicles on bridges. Higher rates of acceleration generally cause more discomfort than lower accelerations.

In the field of psychophysics, the ability perceive a change in stimulus is referred to as the "just noticeable difference" as described by Weber's Law, which in mathematical terms, is defined as the percentage change in the stimulus which is noticeable to the observer. For example adding 1 lb. of weight to a 100 lb. weight (1% change) would likely not be noticed by the person holding that weight. However, adding 1 lb. to a 10 lb. weight (10% change) likely would be noticed. The "just noticeable difference" varies from person to person and among the 5 senses, as well. However, from a structural engineering perspective, the human ability to notice a deflection of L/1000 (0.1%) should carry far less influence than the human ability to notice a mild vertical acceleration of 0.1 g (10%).

Based on the above discussion, high rates of acceleration stemming from bridges with higher first resonant frequencies (natural vibration frequency), with small deflections, are more likely to cause more discomfort than lower rates of acceleration stemming from bridges with lower first resonant frequencies, with large deflections. The Ontario Bridge Design Code codified this relationship at least as far back as their 1991 3rd edition, where acceptable deformation limits of bridges were established via a graphic relationship between bridge deflection and first resonant frequency, Figure 3.

The Ontario Bridge Design Code (1991) was the most relevant published specification for the subject bridge insofar as consideration of natural frequency of the structure was included with the criteria for pedestrian discomfort on the bridge.

As can be seen from the above Figure 3, it is entirely possible to have two bridges of identical deflection limits, but with different overall first flexural frequencies (stiffness) resulting in either 'acceptable' or 'unacceptable' vibrations for pedestrians. A stiffer (deeper) bridge would cause pedestrians greater discomfort at a lower vertical deflection than a comparable but less stiff (shallower) bridge with a greater deflection, counterintuitive to current practice. This conclusion is significant, as it suggests that limiting deflection will not necessarily improve pedestrian comfort.

For reference, the subject bridge has a first flexural frequency (natural frequency) on the order of 1 Hz, which would have allowed up to 6" (152 mm) of acceptable deflection for "little pedestrian use' based on the above chart. Our L/800 deflection limit for the structure was 2.3" (58 mm), well within this criteria.

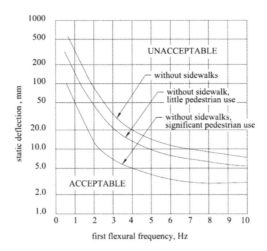

Figure 3.   1991 Ontario Bridge Design Code Deflection Limitation Chart; Clause 2-6.2.2.2.

3.2 *Constructability parameters*

Excessive deflections are still a concern during design, particularly during the steel erection. No research was available at the time of design to describe the anticipated behavior of a shallow depth steel girder bridge that was a highly flexible structure, expected to undergo large deflections at each stage of construction. These large deflections expected during construction, especially for the concrete pours, required equally large cambers. After conferring with peer review in the construction industry, it was determined that some design limitations should be observed for total camber of beams to facilitate conventional bridge construction practices. No guideline for maximum camber was determined to be 'unbuildable', but the following issues were identified as concerns when performing erection of shallow girders with large cambers:

- Correct alignment of field splices: the structural steel erector must 'follow the curve' of the girder and cannot rely on a straight line view of the side profile of the girder. Camber was listed on the girder at 9 increments along its length (10th points of span) including the field splice location in order to prevent misalignment.
- Stability: Shallow girders may have large cambers that approach or even exceed the depth of the girder web, creating concerns regarding lateral stability, particularly during plastic concrete deck placement. For this reason, full-depth diaphragms were used at tighter spacing along with wide girder flanges to stabilize the girders from out-of-plate distortion, similar to what would be done for a curved girder bridge. Overall tolerable camber was arbitrarily limited to the depth of the girder as an upper bound.
- Screed Rail Placement: The use of a hard dimension from the underside of the girder top flange to the top of the bridge deck would be required to allow the contractor to set his deck placement screed rails along the cambered girders.

The bridge girders were designed to consider the above constructability concerns, not live load deflection. The final stiffness of the girders when designed based on the above parameters provided for a design live load deflection of L/800. The live load deflection calculation relied on the base composite girder section property as well as the concrete sidewalk slab for added structural stiffness, which is permitted in the current AASHTO language. Parapets were not included in the stiffness calculation, as they were designed to include regular open joints explicitly to prevent structural interaction and possible cracking.

4   DESIGNING WITH HIGH STRENGTH STEEL AND CONCRETE

As with the traditional design methods, the dead load and live load deflection of the designed girders will be governed by the elastic moment of inertia property of the girder, which is based entirely on the geometric size of component members, and their respective Young's modulus. Since the value for Young's modulus of steel does not change for different bridge structural steel strength grades, the use of higher strength steel does not affect the elastic moment of inertia or the computed deflection of the composite or non-composite girder. More specifically, decreasing the size of the plates by using a higher strength steel will actually increase the deflections, despite maintaining the same strength capacity.

Increasing the specified strength of the concrete in the deck does however affect its Young's modulus through the well-known equation for normal weight concrete:

$$Ec = 1820(f'c)1/2 \quad \text{(f'c in thousands of pounds per square inch (ksi))} \tag{1}$$

As can be seen in Equation 1 above, the Young's modulus of the concrete varies with the square root of its specified strength and has a limited effect to the stiffness of the composite section where the transformed area of the concrete is proportional to that of the steel via the modular ratio (n) of their respective Young's moduli. This ratio of steel-to-concrete transformation can vary from a little as 1/9th for f'_c of 3 ksi, and as high as 1/6th for f'_c of 5 ksi and higher, up to 10 ksi. The modular ratio is capped at 1/6th, per the AASHTO LRFD Bridge Design Specifications. The effects of varying concrete and steel strength on a notional composite section will, at most, increase the stiffness of the composite section approximately 10% at the outer limit of f'c of 8 ksi–10 ksi. Given that use of concrete deck mix designs of greater than 5 ksi–6 ksi are still fairly unusual for cast-in-place heavy highway construction, it should be considered slightly conservative to design to the industry standard 4 ksi limit.

Where the superstructure depth is governed by deflection, the most efficient way to add stiffness to the bridge is by using larger steel flange plates for the girder, particularly at the bottom flange in positive moment regions, or by adding more girders at a tighter spacing. Since the deflection of the girder is driven by its geometric properties as discussed above, the larger plate sizes required for stiffness will generally result in composite girder sections that are 'overdesigned' for live load capacity. In fact, in all shallow composite girder designs performed to date by the authors, the resulting live load capacity of the girders is always greater than required to carry the minimum AASHTO LRFD 'HL-93' design live load, sometimes by as much as 50%. For most bridges, the increased cost due to the additional structural steel quantities was worth the benefit of the additional underclearance created. In any event, maximum plate available must be always be considered. Grade 50 steel is typically available in thicknesses of up to 4″. Grade 70 steel is typically not available in plates exceeding $2\frac{1}{2}″$, but increased interest in the higher strength material may justify larger plate sizes in the future.

In summary, high strength steel has little utility in deflection controlled shallow superstructures as they typically easily rate for required live loads, yet high-strength concrete bay be useful to reduce live load deflections, somewhat.

5   DESIGNING FOR CAMBER

Perhaps the greatest concern for designing and building shallow girder bridges is the design and inclusion of accurate camber into the fabricated girders. For single span bridges with span-to-depth rations approaching L/40, large total camber values should be expected. In some cases, 24″ or more of camber is possible. This creates a number of design related issues, as traditional bridge design relies on simple prismatic beam deflection calculations based on published values for Young's modulus of both concrete and steel members. In reality, these values are, at best, approximate, and do not account for natural variations in the material properties. For typical cambers on the order of 4″ to 6″, a 10% variation in actual stiffness, and thus variation of the camber (∼1/2″), is irrelevant, as it can easily be accounted for in the concrete haunch area between the steel girder top flange and the underside of the concrete deck slab. For a shallow girder with cambers on the order of 24″, a 10% variation could result in a 2½″ error in camber. This can be highly problematic for a traditional bridge where the haunch between the girder and the deck has been set at about 1″ minimum. The result is either an overly thick unreinforced haunch and potentially sagged girder or negative haunches where the girder top flange is embedded within the deck rather than supporting it from underneath.

On the subject bridge, cambers were set using 3D analysis in GTSTRUDL for dead load distribution rather than relying on traditional classical assumptions. Traditional design practice would have assumed that all girders carry their plastic uncured concrete dead load independently, while the incurred load from subsequently added sidewalks and parapets would have been distributed uniformly across all girders within that particular construction stage. The latter of these two assumptions were suspected to be inaccurate, and analysis suggested that the fascia girder may carry substantially more of the parapet and sidewalk weight than the interior girders due to the eccentric loading of the bridge parapet on the deck overhang. A 50% distribution of these loads was placed on the fascia girder, whereas traditional design practice would have uniformly distributed these loads across all girders in the staged section, or 100%/5 girders = 20%.

After completion of analysis, it was anticipated that the girders would each carry their traditional tributary plastic concrete dead load from the deck. However, for each construction stage, the analysis suggested that the superimposed dead loads would not be equally distributed to each girder. The results suggested that adjacent to the supporting bearings, the parapet and sidewalk loading placed on the deck overhang may actually exceed 100% on the fascia girder due diaphragm continuity and the counterbalancing uplift at the first interior girder. However, at mid-span where all girders were uniformly flexible, the parapet/sidewalk loading was anticipated to be equally distributed across all girders, as it was assumed that they would all be forced to deflect relatively equally due to the connectivity of the deck and diaphragm system.

During construction, the contractor recorded the cambers and reported them to the Engineer. It was found that the fascia girder deflected approximately 2″ more than anticipated, and the interior girders deflected approximately 1″-2″ less than anticipated. See Figure 4 below for graphical representation of the deformations. This variation was accommodated within the girder haunch, but was noteworthy in that the behavior of the bridge so strongly invalidated the traditional guidance of distributing superimposed dead loads to all the girders. The unanticipated extra deflection at the fascia girder would suggest that the flexibility of the system at mid span is not enough to share load carrying capacity across the width of the structure, and the fascia girder may carry a disproportionate amount of superimposed dead load weight for its full length, not just at the bearings. It is speculated that the observed dead load distribution is likely present to some extent in all girder bridges, regardless of section depth. Future designs should consider load distribution carefully when designing girders for camber. For highly flexible shallow structures, fascia girders may carry as much as 100% or more of the actual parapet and tributary sidewalk dead load on eccentric overhangs, regardless of inferred load sharing ability of the cured deck.

6   SPECIALIZED CONSTRUCTION METHODS

6.1   *Large displacements and end girder rotations*

In addition to the design concerns, it is also important to address the construction of the bridge within the contract plans so that the contractor fully understands how the anticipated behavior of the bridge differs from traditional expectations. For example, the ends of highly cambered girders can experience large rotations and longitudinal translations at the bearings during placement of the concrete deck. This condition is commonly encountered at bridges with long simply supported spans and/or large cambers which can rotate more than 0.02 radians as camber is removed. This

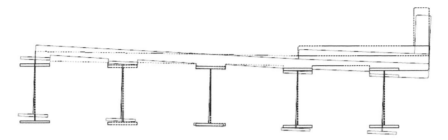

Figure 4.   Half section of bridge shown with anticipated post-construction orientation in dashed lines and actual post-construction deformations in solid lines (exaggerated for clarity).

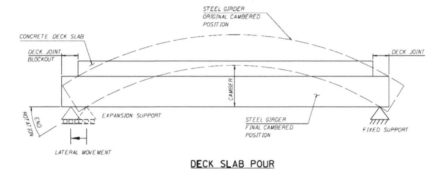

Figure 5.   (Slightly) exaggerated cambered simple span, prior to deck concrete placement.

rotation value is significant, as it is as much or more than the live load rotation that most girders see under daily live load conditions. In the case of the subject bridge, the expansion end of the girder also translated more than 1″ during deck slab placement. These large end rotations and translations would not typically be anticipated and would be difficult to accurately predict by contractors who would otherwise set their joint armor at the end of the girders, prior to deck slab placement.

For comparison, in traditional design practice, where short simple spans or continuous spans are common, the steel girders were sized such that camber is anticipated to be very small, and the ends of the girders were very nearly plumb in the unloaded condition. The contractor therefore incorrectly assumes that the ends of the girders will experience only minimal rotations as the deck concrete is poured. Due to the flexibility of shallow depth girders, this assumption cannot be relied upon for accurate results when the girders are expected to go through large deflections and have been designed with large cambers.

Designers are recommended to pay special attention to their bearing designs and either include the additional construction rotation into the design of reinforced elastomeric bearings, or publish the anticipated rotations at varying stages of construction where proprietary bearings are specified for use, such as high load multi rotational (HLMR) bearings.

Figure 5 above shows a sketch of a single-span steel girder as it would appear before and after the plastic concrete for the deck has been placed. The girder camber and end rotations are exaggerated for illustration purposes. In the arched cambered condition, the only loads acting on the girder are its own self-weight and the diaphragm dead loads. At this point, traditional construction practice in the north east region would include installing the joint armor at the ends of the girder and using it as a bulkhead for the plastic deck concrete placement. This method requires the contractor to accurately predict the displacement of the girder deflection as the deck concrete is placed. Should the contractor fail to properly anticipate the behavior of the end of the beam, the joint armor would be misaligned in the final condition.

While this was avoided on the subject bridge, the misaligned joint armor condition has been witnessed in other bridges. To prevent error in subsequent projects of similar type, contractors were required to leave a 19″ block-out at the end of the girder which did not receive concrete during placement of the plastic deck concrete. Smaller block-outs may be possible, depending on joint hardware and end diaphragm geometry.

The alignment and installation of the joint armor after the deck pour prevented the possibility of error and gross misalignment in the joint assembly. While there is some additional cost in performing multiple deck concrete placements, the combination of providing nigh-guaranteed aligned joints and a smooth riding surface is highly desirable as best possible fit of joint hardware implicitly reduces direct joint impact from vehicular traffic. This was considered more valuable to the durability of the bridge than the minor additional costs incurred.

In addition to the above, it is important to note that during the first deck concrete pour, the expansion bearings of the subject bridge were left unattached to the girder bottom flanges and wood blocking was placed between the backside of the bearing sole plate and the backwall of the abutment, to prevent the sole plate from over-translating, which otherwise may have damaged the bearings as the girder displaced under the concrete dead load. The girder bottom flange was then forced to slide on top of the wood-blocked bearing sole plate. After the concrete deck slab had been placed, the sole plates were welded to the bottom flanges of the girders.

6.2   Staged construction

For where an existing bridge is to be replaced in multiple stages, additional construction provisions must be made to accommodate new bridge framing where highly flexible superstructures with large cambers maybe used. Consider a typical staged construction where half the width of an existing is demolished while active traffic is maintained on the remaining half. After the first half of the new bridge is constructed and traffic is shifted onto it, the remaining portions of the old bridge will be demolished, and the remainder of the new bridge superstructure steel will be erected adjacent to the completed portion of the new bridge.

For the subject bridge which had similar staging to the above, the new superstructure steelwork with only its self-weight acting on the camber was approximately 24″ higher than the adjacent completed half of the bridge at mid-span. It was not be possible to connect the diaphragms between the completed Stage 1 half of the bridge and newly erected bare steel for Stage 2 half, nor was it be possible to install stay-in-place deck pan forms between the adjacent superstructure stages. These elements could not be installed until the new deck slab had been placed on the bare steel girders erected in Stage 2 and the girders allowed to deflect into their final position. After the deck had been poured in Stage 2, the diaphragms and deck formwork could then be installed and the final stage completed, where the last portion of the deck between the two new bridge stages was placed (closure pour). The addition of this third and final stage allowed the camber to be relieved to permit the installation of the connecting diaphragms and deck formwork. For the purposes of this bridge, the closure pour was placed under live traffic. The effect of live load on the quality or cracking of the closure pour was considered negligible since the first resonant frequency of the span did not appreciably change from the overall structural stiffness increase attributable to the addition of the closure pour. In other words, the closure pour concrete constituted only a slight change in the stiffness of the superstructure and attracted negligible force effect as it cured. In addition to this, robust full-depth diaphragms at tight spacing forced the two adjacent stages to deflect equally, reducing the possibility of damage from differential deflection. It's important to note that while a fairly flexible superstructure may deflect more than a stiff superstructure, it is conversely fairly straightforward to force two adjacent flexible superstructures to deflect in tandem with rigid connections.

As discussed in Section 5 above, for highly flexible shallow superstructures, superimposed dead loads added to the composite section of the bridge do not distribute equally to all girders. The analysis for the closure pour placement should also consider this occurrence. The girders at the stage limits on both sides of the closure pour must be cambered specifically to carry the actual

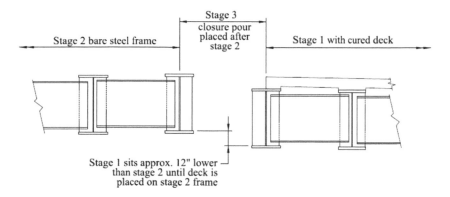

Figure 6. Staged construction condition. Difference in elevation shown to scale.

weight of the deck to be placed upon them in the first stage of construction, which is one half of the plastic deck weight within the deck span to the adjacent girder. When the concrete deck is cured, its deflection from carrying the tributary weight of the closure pour must be based on the composite section of that girder assuming only the portion of the cured deck it supports as contributing to the stiffness of the member. Since the closure pour concrete will harden in less than 24 hours, use of the short term 1n section is recommended for this computation, as opposed to the long term (3n) section typically used for superimposed dead load. Weight of the closure pour concrete should be assumed to be borne solely by the staged girder, similar to what was noted above for parapet weight supported by the fascia girder. See Figure 6 below.

7 CONCLUSIONS AND RECOMMENDATIONS

It is both possible and practical to design and construct bridges that are significantly shallower than traditional methods would suggest, even under staged construction. Increasing the span length to superstructure depth ratios by 30% or more is feasible, where adjusting roadway profiles and/or shortening of span lengths is not possible. However, shallow bridge designs are inherently far more flexible than traditional depth bridge designs, and special considerations must be included in both the design and the construction documents for a successful project. The following considerations should be considered mandatory for a successful design:

- Current AASHTO optional criteria for deflection limitations are overly conservative and do not actually serve their intended purpose of ensuring bridge user comfort. Amplitude of allowable deflection has limited, if any, appreciable effect on the comfort of pedestrians, and in some cases, overly stiff bridges may actually exacerbate pedestrian discomfort. Exceeding the L/1000 suggested limit should be considered to be acceptable. Exceeding the L/800 suggested limit may be practical, but designers should focus on constructability rather than pedestrian or driver comfort as a primary concern when doing so. Following the guidance of the Ontario Bridge Code (1991) provided more reliable estimates of pedestrian discomfort and it is recommended that similar procedures be followed when pedestrian comfort on a highway bridges is a concern.
- Shallow girder design should be performed with an understanding of the applicable engineering materials and properties. Thicker girder plates of lesser strength steel will provide better deflection control than thinner or like thickness plates of higher grade steel. Higher strength concrete only offers a nominal increase in deflection control.
- Lateral bracing and substantial top flanges are essential. With expected cambers/deflections sometimes matching or possibly exceeding the depth of the girder, the potential for lateral bucking is heightened during deck pour. Additional consideration should be given to more substantial bracing, similar to that required for curved girders with full-depth diaphragms to adequately support the full height of the web and the flanges.

- Computed distribution of eccentric superimposed dead loads such as parapets and sidewalks should be performed via rational analysis such as 3D model or at least a 2D grid model, not via conventional uniform distribution. Construction experience has been that the fascia girders may take the majority of these loads directly, without equal contribution from interior girders. When in doubt, provide additional camber at the fascia girders, as they will be the only ones noticeable from on-coming traffic. Residual camber will only be noticed as architectural element, if at all.
- Classic elastic equations for prismatic beam deflection do not have the accuracy required to adequately reflect the range of actual stiffness that may occur due to real variances of material properties for construction grade concrete and steel. For a shallow girder with a large camber, a small discrepancy between computed and actual deflection may result in an error of several inches. Where possible, additional room for adjustment should be factored into the haunch to avoid potential negative haunch depths. The subject bridge displayed camber discrepancies on the order of 10% of the camber values. It is recommended that at least 10% of the calculated camber should be provided as a minimum haunch depth.
- Deck joint areas should be detailed to be placed after the main portion of the deck has been placed. This will allow the contractor to more accurately align the deck joint despite the large anticipated rotations and displacements of the girder ends as the camber is removed.
- Girder bearings should be blocked to prevent unwanted translation as the girder camber is removed or, if the bearing design can tolerate the anticipated construction displacement, the bearings should be reset to relieve them from construction deformations. Where large rotations are anticipated, ensure that bearings are adequately designed to accommodate the significant construction rotations and the temporary condition where partial dead load is borne eccentrically on the bearing sole plate.

Lastly, it is recommended that further research be considered to revisit the optional deflection criteria noted in the current edition of the AASHTO LRFD Bridge Design Specifications, and consider elimination of it as a consideration solely for user comfort. It is recommended that a frequency or acceleration based criteria be developed for user comfort, instead. Total deflection, if considered from a design aspect, should be focused on controlling the constructability concerns noted in this paper. Additional research is also recommended to determine a more appropriate supplemental dead load distribution for girder design, particularly fascia girders. Development of an empirical formulation for supplemental dead load distribution of common overpass bridges is highly recommended, as constructing 2D or 3D models for these routine structures should be unnecessary. Until such time as an empirical formulation can be developed, it is suggested that designers consider applying up to 100% of the bridge parapet load upon the supporting fascia girders for new bridge design, with consideration of the specific features of the bridge.

REFERENCES

AASHO. 1931. *Standard Specifications for Highway Bridges*, First Edition, American Association of State Highway and Transportation Officials, Washington, DC.

AASHTO. 2002. *Standard Specifications for Highway Bridges*, Seventeenth Edition, American Association of State Highway and Transportation Officials, Washington, DC.

AASHTO. 2014. *2014 LRFD Bridge Design Specifications*, Seventh Edition, with interims through 2015, American Association of State Highway and Transportation Officials, Washington, DC.

Wright, R.N., and W.H. Walker. 1971. Criteria for the Deflection of Steel Bridges. *AISI Bulletin*, No. 19, November 1971, Washington, DC.

ASCE. 1958. Deflection Limitations of Bridges: Progress Report of the Committee on Deflection Limitations of Bridges of the Structural Division. *Journal of the Structural Division*, Vol. 84, No ST 3, May 1958, American Society of Civil Engineers, New York, NY

Ontario Ministry of Transportation. 1991. *Ontario Highway Bridge Design Code*, Third Edition, Ministry of Transportation, Quality and Standards Division, Structural Office, Toronto Ontario.

Chapter 14

Proportioning and design considerations for extradosed prestressed bridges

S.L. Stroh
AECOM, Tampa, FL, USA

ABSTRACT: An extradosed prestressed bridge is a girder bridge that is externally prestressed, using stay cables over a portion of the span. Extradosed prestressed bridges can provide an economical bridge solution for spans in the transition range from conventional girder bridges and cable stayed bridges. This paper discusses initial proportioning guidelines for this bridge type, based on work by the author in developing the design for the first extradosed prestressed bridge in the US, the Pearl Harbor Memorial Bridge in New Haven Connecticut, and from reviewing over 60 extradosed prestressed bridge designs world-wide. Propositioning guidelines are discussed for items such as efficient span layouts, depth ratios for the deck, tower height, efficient span ranges, and stay layouts. The paper also addresses strength and fatigue design guidelines for the stay cables that are specific to the extradosed bridge type. The results of these guidelines are demonstrated with respect to the design for the Pearl Harbor Memorial Bridge.

1   INTRODUCTION

The introduction of "extradosed prestressed bridges" is a new and exciting development in bridge engineering, extending the application of prestressed concrete bridge principles into new areas. The extradosed prestressed bridge has the appearance of a cable stayed bridge with "short" towers, but behaves structurally closer to a prestressed girder bridge with external prestressing.

The earliest documented discussion of an extradosed prestressed bridge concept in the literature is by J. Mathivat in a 1988 FIP Journal article, "Recent Developments in Prestressed Concrete Bridges" (Mathivat 1988). Mathivat describes a common cable layout scheme consisting of two types of prestress for box girder type bridges erected in a balanced cantilever technique:

- Semi-horizontal prestress internal to the concrete and arranged within the area of the upper flange of the deck and countering the cantilever moments, and
- Prestress external to the concrete but within the concrete box girder void, placed after mid-span closure, running from pier diaphragm to pier diaphragm and deviated by means of special arrangements and countering the positive moments.

This type of system represents a mixed system, with a combination of internal and external prestress. Mathivat proposed to substitute for the first type of prestress, cables placed above the running surface of the deck and deviated by stub columns or towers above the deck. He calls this type of construction "extradosed prestress", and suggests that this type of construction would offer an economical transition between traditional concrete box girder structures built by cantilevering, and cable-stayed bridges.

The definition of an extradosed prestressed bridge must make a fundamental distinction from a cable stayed bridge and from a girder bridge in the structural behavior. Mathivat suggested that the tower height as a differentiating feature between the two bridge types. Cable stayed bridges were defined by tower height (H) to span (L) ratios of H/L of approximately 1/5. He suggested that extradosed prestressed bridges are defined by H/L ratio of approximately 1/15.

Since the introductory work by Mathivat, more than 60 extradosed bridges have been constructed in more than 25 countries. Based on work by the author in developing the design for the first extradosed prestressed bridge in the US, the Pearl Harbor Memorial Bridge in New Haven Connecticut, and from reviewing existing extradosed prestressed bridge designs world-wide, this paper discusses initial proportioning guidelines for this bridge type. Propositioning guidelines are discussed for items such as efficient span layouts, depth ratios for the deck, tower height, efficient span ranges, and stay layouts. The paper also addresses strength and fatigue design guidelines for the stay cables that are specific to the extradosed bridge type. The results of these guidelines are demonstrated with respect to the design for the Pearl Harbor Memorial Bridge.

2   GENERAL PROPRTIONING FOR EXTRADOSED PRESTRESSED BRIDGES

2.1   *Applicable span ranges*

Extradosed Prestressed bridges can be considered in the transition region of span lengths between traditional girder bridges and the longer span bridge types such as truss, arch and cable-stayed. Sources in Japan, where most of the extradosed bridges have been constructed, have set the applicable span range for Extradosed Prestressed bridges to be generally between 100 and 200 meters (328 and 656 feet) (Kasuga 1994 and Komiya 1999). Although even in Japan, a number of these bridges have been constructed outside of this span range.

Stroh (2012) summarizes 63 extradosed prestressed bridges worldwide where span length information is available. The spans for extradosed prestressed bridges range from 172 feet to 902 feet, however several of the longer spans are a hybrid design, with a steel middle section of the main span. The longest all-concrete extradosed bridge has a span of 886 feet. The mean span length for extradosed bridges is 435 feet. The standard deviation of the range of span lengths is 171 feet. Assuming a normal distribution of a random variable, this means that within one standard deviation each side of the mean (giving a span range of 265 to 606 feet) we capture 68% of the data. Based on this data a span range from 300-600 feet would seem a common span range for typical bridges of this type.

Figure 1 expands recommendations from Poldony (1994) to include extradosed bridges. These bridges fill an important niche between girder bridges and the longer span bridge types of arch, truss and cable stayed, giving designers another option for bridge type

2.2   *Side span ratios*

The ratio of span length between the main span (L) and side spans (L_1) has influence on the vertical reactions or anchoring forces at the anchor pier, the moment demands on the deck (positive moments in main span vs. side span, and negative moments at the tower), and stress changes in the stay cables. A good choice of the ratio between main and side spans is important for a good design. This ratio is commonly expressed as the ratio of side span to main span (L_1/L).

For cable stayed bridges, Leonhardt (1980) provides recommendations on economical span ratios in graphical form based on a function of dead load to live load ratio of the bridge, main span length and live load change in stay cable stress (fatigue stress). For the common case of a steel cable stayed bridge the L_1/L ratio works out to about 0.35. For a heavier concrete cable-stayed bridge this ratio works out to about 0.42.

For three span concrete girder bridges the side/main span ratio should range from about 0.8 for conventional cast-in-place-on-falsework construction to about 0.65 for balanced cantilever construction (Poldony 1982).

Stroh (2012) summarizes the span lengths data for 50 extradosed bridges worldwide.

The L_1/L ratios for extradosed bridges varied from 0.33 to 0.83 with a mean of 0.57. The standard deviation is 0.12, so one standard deviation each side of the mean gives a range for L_1/L of 0.45

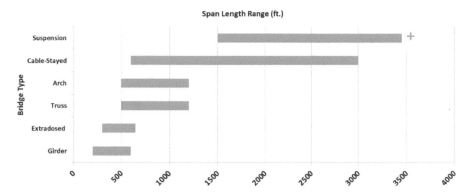

Figure 1.   Recommended span ranges for various bridge types.

to 0.69. This places extradosed bridges essentially between the envelope of concrete cable stayed bridges, at 0.42, and balanced cantilever constructed concrete girder bridges at 0.65.

It should be noted that when evaluating these span ratios, for some bridges geometric and site constraints set the span ratios rather than structural efficiency. The data shows that these shorter or longer side span ratio can be accommodated in the design of extradosed bridges without a major impacts. The data of existing bridges indicates a wide range of side span ratios.

Based on the data examined a reasonable recommendation for side to main span ratios for an extradosed bridge is about 0.6, unless geometric or site constraints would require otherwise.

2.3   Multi-span bridges

Crossings of wide rivers many times have poor foundation conditions, deep water, large vessel impact considerations, large navigation clearances, or other factors driving the decision to use a long span bridge. For very wide rivers or waterways, several long spans may be required in order to span the waterway.

Cable stayed bridges are typically either a two-span or three-span arrangements. These span arrangements are ideal for cable stayed bridges because back-stay cables can be provided from the anchor pier to the top of tower to provide stiffening of the tower. The Structurae (2011) website documents more than 1200 examples of cable stayed bridges world-wide, and of these, only seven cable stayed bridges are multi span bridges (more than 3 spans). Design of a multi-span cable stayed bridge presents a special challenge, in that for the central spans there is no opportunity for backstay cables, and special design considerations must be made to address the resulting flexibility of the structural system (Leonhardt 1980). Solutions include the provision of very stiff towers, as was done for the Rion-Antirion Bridge in Greece (Figure 2), or providing crossing backstay cables that are anchored to adjacent towers multiple main spans, as was done for the Ting Kau Bridge in Hong Kong (Figure 3).

Extradosed prestressed bridges do not rely on backstay cables. So, unlike cable stayed bridges, multi-span extradosed bridge arrangements do not require special measures Stroh (2012) examined 63 bridges extradosed bridges built around the world to date and 19 of these had 4 or more spans (representing some 30% of the bridges built). The extradosed bridge type is well suited to long multi-span bridge arrangements, and provides a viable bridge alternative for this design condition. An example multi span extradosed bridge is shown in Figure 4.

2.4   Curved alignments

Modern highway construction frequently required bridges that conform to curved roadway alignments. For longer span bridges, this becomes a challenge for designers from both the viewpoints of structural demand and accommodation of the curved geometry.

Figure 2.   Rion-Antirion Bridge in Greece – multi-span cable stayed bridge with stiff towers (photo courtesy structruae.de, photographer Inge Kanakaris-Wirti).

Figure 3.   Ting Kau Bridge in Hong Kong – multi-span cable stayed bridge with crossing cable from central tower (photo courtesy structurae.de, photographer Baycrest).

Figure 4.   Kiso River Bridge in Japan, a multi-span extradosed prestressed bridge (photo by author).

From a structural demand viewpoint, a curved bridge sees torsional demands resulting from the vertical loads. For significant torsional demands, as would result from tight curvatures or long span bridges, a closed cross section is significantly more efficient in carrying these torsional demands, and is the preferred structural system (Menn 1986).

The geometrics of design must also accommodate the curvature. This can be a challenge for some bridge types. For example, a curved cable stayed bridge deck must be detailed so that the stay cable avoids conflicts with the roadway traffic, considering the stay cable is essentially a straight line from the top of tower to the connection at the deck level. This can result in geometric conflicts on the outside radius of the curve and can require the bridge to be widened along the outside curve to accommodate stay clearances.

The extradosed bridge type can accommodate a modest curvature without special consideration of the structural system. The girder for extradosed bridges is typically a concrete box girder section, which can efficiently resist the torsional demands by shear flow around the closed cross section. The stay cables for extradosed bridges are typically only provided over a limited region of the span, and do not extent all the way to midspan of the main span or to the anchor piers in the side spans. Therefore the geometry conflicts between the stay cables and traffic are minimized. Extradosed prestressed bridges offer an added opportunity for the longer-span bridges in that they can accommodate at least a modest curvature of the roadway alignment in an efficient manner. Five of the 63 existing extradosed bridges examined by Stroh (2012) were constructed on curved alignments with radius of curvature as small as 1,312 feet.

2.5 *Tower height*

An important parameter for extradosed bridges, and one that differentiates them from cable stayed bridges, is the tower height. The tower height directly influences several other design parameters, such as the stay stress variation under live load (fatigue range), the cable inclination, and the proportion of loads shared between the deck and the cables. A fundamental distinction between a cable-stayed bridge and an extradosed bridge is the role of the stay cables. The basic role of the cables in a cable stayed bridge is to develop elastic vertical reactions. In an extradosed bridge they are used to prestress the girder.

The taller the tower, the smaller the size of cable is required to carry a given load. As discussed by Leonhardt there is a limit to the economical tower height because even though the cable cost reduces with higher towers, the tower cost increases. Leonhardt (1980) places the optimal ratio of the tower height (H) to main span (L) for cable-stayed bridge between 1/4 to 1/5.

For extradosed bridges, the role of the cables is to act as external post-tensioning tendons and provide prestress to the deck. For an extradosed bridge, the post tensioning is elevated above the cross-section of the girder using a short tower, and therefore provides a much larger eccentricity, and therefore more efficient use of the prestressing steel. However, if we continue raising the tower, at some point the vertical component of the cable reaches a force level that starts to significantly carry the vertical live load of the structure. This also means that the fatigue stress in the cable becomes more significant, and the bridge starts to behave more like a cable stayed bridge, rather than an externally prestressed girder. According to Mathivat, the optimal ratio to tower height to span length should be on the order of 1/15 (Mathivat 1988). Although Mathivat did not provide a basis for this recommendation, one may derive an approximation for the tower height limit based on a simple relation of the stay geometry and target fatigue limits.

If we assume a geometric distribution of stays as shown in Figure 5, we can determine, based on a tower height to span length ratio of 1/15, that the vertical component of stay force, equal to about 17% of the total stay force (the sin of the steepest stay angle).

$$Tan\ \alpha = H/0.375, \quad Where:\ H/S = 1/15$$

$$Therefore:\ \alpha = 10.08°$$

$$Sin\ \alpha = 0.17 = 17\%$$

We can also establish a limit on the vertical component of the stay force based on a target fatigue limit. In simple terms, (AASHTO 2010) provides a nominal 18 ksi fatigue stress limit for conventional prestress (i.e., strand stressed to the 0.6 f's limit – similar to target stress for extradosed cables). We can express this as a fraction of the total stay force by dividing by the maximum permissible stay force of 60% f's, which gives a live load limit to compare with the vertical load limit based on geometry in the preceding paragraph. However, we need to recognize that the fatigue truck is lighter than a conventional live load truck in AASHTO, and it has a lower impact factor (AASHTO 2010). So we need to increase the 18 ksi fatigue stress target by the difference in load factors for service vs fatigue loading (1.0/0.75) and by the ratio of the service vs

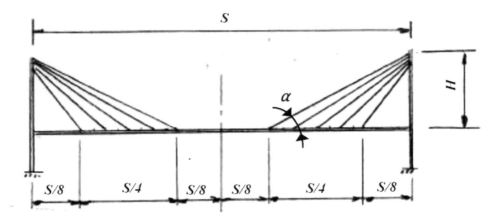

Figure 5.   Assumed distribution of stay cables along span.

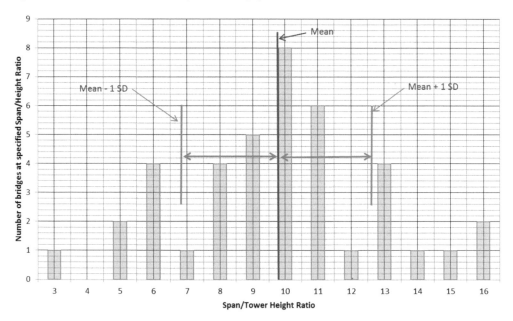

Figure 6.   Tower height/span length comparison for extradosed prestressed bridges.

fatigue impact factors (1.3/1.15). Therefore we can calculate the target vertical limit on live load as a fraction of total stay force as:

$$(18 \ ksi \ (1/0.75)(1.3/1.15))/(0.6 \ (270) \ ) \ =0.17$$

This is the same limit on the vertical component of live load that results from Mathivat's limit of span to tower height of 1/15.

Stroh (2012) examined 40 existing extradosed bridge where there was sufficient data to study the tower height parameters (Figure 6). The numerical calculation for mean and standard deviation is based on a discrete random variable calculation using the referenced data.

The mean tower height ratio for extradosed bridges is 1/9.75 and going one standard deviation each side of the mean gives a range of tower height ratios from 1/6.9 to 1/12.6. This shows that the population of existing extradosed bridges has not followed Mathivat's original suggestion that a 1/15 height/span ratio would be the optimal value. Based on existing bridges tower heights have been used are slightly taller that recommended by Mathivat and a suggests typical H/L ratio of 1/10 would appear appropriate, or a H/L range between 1/7 to 1/13, based on the data from the population of existing extradosed prestressed bridges.

2.6   *Girder depth/girder haunch arrangement*

For cable stayed bridges, the stay cables carry most of the dead and live loads and the deck structure (or girder) is proportioned with adequate strength and stiffness to span between and carry any local load effects between the stays, to carry the overall global flexural loading from the entire stay-girder structural system deflections and to carry the horizontal compression from the stays. For cable-stayed bridges the depth to span ratio can range from values to 1:50 to more than 1:250, representing very flexible decks (Poldony 1982).

For a girder bridge the loads are carried by flexure and shear in the girder, and the girder depth is proportioned for strength and stiffness to carry these loads. The proportioning of girder bridges is directly dependent on the construction method, and whether the girder is constant depth or variable depth (haunched). For balanced cantilever erection methods, recommended girder depth/span ratios are as follows (from Menn 1986):

For constant depth sections   depth/span = 1:22
For variable depth sections   depth/span = 1:17 at supports
                               depth/span = 1:50 at midspan

Extradosed prestressed bridges are typically constructed in balanced cantilever and their behavior is similar to a girder bridge constructed in balanced cantilever, but with more efficient external prestressing. Therefore we would expect a reduction in the structural depth at the support. Mathivat (1982) recommended a depth/span ratio between 1:30 and 1:35 for extradosed bridges.

Stroh (2012) examined 29 existing extradosed structures that have sufficient data available to examine the depth to span ratios. For those bridges, there is a wide range of deck depth/span ratios ranging from 1:13 to 1:40 with a mean ratio 1:28.2. Based on an assumption of a normal distribution of the data, the standard deviation of the data is 8, giving a range to depth/span ratios from 1:20 to 1:36 for one standard deviation each side of the mean.

In general, we would expect that the depth/span relation should be nearly a constant, based on the efficient design of the structural system. An example is for concrete box girder bridges, where there was shown to be strong correlations between girder depth and span length (FHWA 1982). However for extradosed bridges, when the data is plotted for the depth/span ratio as a function of span length, there is a clear trend for increasing depth span ratios for longer spans (Figure 7).

This indicates that for extradosed bridges, the structural proportioning is under control of the designer, meaning that the designer can control the load distribution between the girder and cable system. For longer-span extradosed bridge the stay system is controlled more by fatigue demands, placing more demand on the girder and trending towards a slightly deeper girder for longer span bridges.

If we limit the span range for the data to the 300 to 600 foot span range that is considered most common for extradosed prestressed bridges, then the existing depth span data is plotted as a function of span length as shown in Figure 8. This is only slightly shifted from Mathivat's recommendations of a depth/span range of 1:30–1:35 for extradosed bridges. Based on the existing bridges a span/depth ratio of 1:30 (or a range of 1:25–1:35) is recommended.

The girder depth at mid-span for an extradosed bridge should be similar to a girder bridge constructed in balanced cantilever. The moment and shear demands of both systems are similar at the mid-span location. As previously noted Menn (1986) recommends a depth/span ratio of about 1:50 for the mid span region of continuous girder bridges constructed in balanced cantilever. The mean mid-span depth ratio for variable depth extradosed bridges is 1:46 from Stroh (2012), agreeing closely with Menn's recommendations. Therefore, a recommended mid-span depth/span ratio for variable depth extradosed bridges is 1:50.

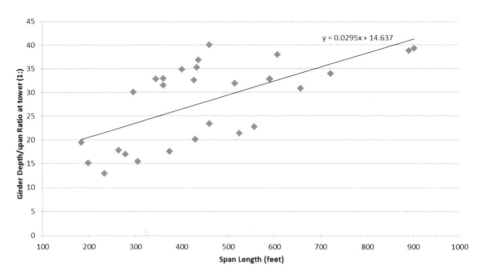

Figure 7.   Girder depth/span length comparison for full range of extradosed bridges.

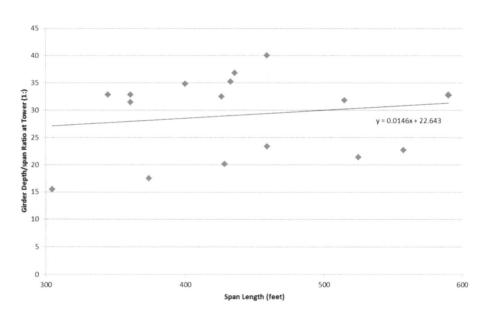

Figure 8.   Girder depth/span length comparison for 300–600 ft. span range of extradosed bridges.

About 10% of the existing extradosed bridges use a constant depth girder, rather than a haunched, variable depth arrangement. As previously noted, for a constant depth girder bridge a depth/span ratio of about 1:22 would be expected. The depth/span ratio used for extradosed bridges ranges from 1:25 to 1:40, with a mean ratio of 1:32. It is noted that a constant depth girder was typically used for shorter span length extradosed bridges, with a mean span length of only 285 ft. for the constant depth bridges.

In general a variable depth girder section would be expected for extradosed bridges, in recognition of the higher negative moment demand at the towers. However for short-span extradosed bridges (less than 300 ft. span) a constant depth section may be appropriate.

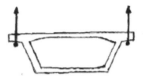

Figure 9.   Narrow extradosed bridge.

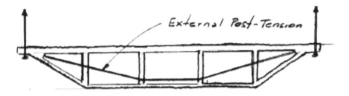

Figure 10.   Wide extradosed bridge.

2.7   *Bridge width*

Extradosed prestressed bridges have been used for a wide range of bridge deck widths, ranging from 30 to 112 feet wide. The mean deck width from the data examined by Stroh (2012) is 65 feet and there is a relatively uniform distribution of deck width variations between the two extremes.

It is expected that the extradosed bridge concept would function better with relatively narrow deck widths, due to more direct load path from the girder webs to the cables on a narrow bridge (Figure 9). For a wide bridge, either a strong diaphragm or an external arrangement of post-tensioning must be provided to transfer the intermediate web loads to the stay cables (Figure 10). Either of these adds cost and complexity to the design and construction.

However, even with the added complexities, there is a relatively even distribution of constructed bridge deck widths for extradosed bridges. It is the author's opinion that this is probably a reflection of the construction of bridges to the required roadway widths in response to traffic demands, as opposed to structural efficiency, and that given the option, a narrow extradosed bridge that allows a direct force transfer between the box girder webs and the stay cables, is preferable.

3   STAY CABLE CONSIDERATIONS

3.1   *General*

The previous section reviewed a variety of general proportioning and detailing factors that "define" extradosed bridges, among which was the tower height. The tower height is a particularly important parameter because it influences how the loads are shared between the cables and the girder. Specifically how the live loads are carried, and how much change in live load, or fatigue, the cable is subjected to. The cable fatigue demand is central to the definition of extradosed prestressed bridges because the fatigue capacity of a cable is directly related to the maximum stress limit in the cable. For cable stayed bridges, the allowable maximum stress on the cable is set in order to provide an appropriate fatigue range. That is, the maximum stress is the cable is set low enough that there is an appropriate fatigue range available for live load variation. For an extradosed we can set the maximum stress of the cable higher, because there is less fatigue demand. This provides a more efficient use of the cable material for extradosed bridges, and consequently, cost savings.

The tower height alone does not sufficiently control the fatigue stress range in the cables to a level of accuracy to safely and consistently establish appropriate maximum stress limits. In order

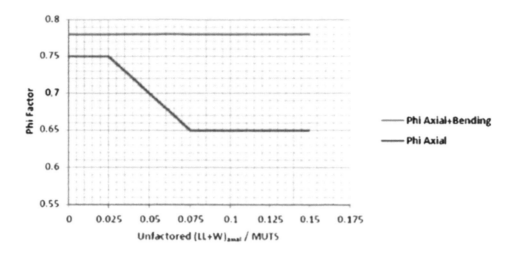

Figure 11.   Strength Resistance Factor (ϕ) (PTI 2006).

to provide a more comprehensive assessment of extrasosed bridges, there needs to be extradosed-bridge-specific criteria on the design of the stay cables, specifically, the establishment of the maximum allowable stress and the establishment of appropriate fatigue ranges based on that maximum stress. The Current PTI Specifications, (PTI 2006) address an adjustment to the Phi factor for stay cable design to recognize the lower fatigue demand for extradosed prestrressed bridges.

3.2   *PTI approach to stay cable fatigue*

The PTI Recommendations for Stay Cable Design, Testing and Installation 6th Ed. (PTI 2006) includes provisions that address bridges with relatively low live load demands on the stay cables in a general manner. The term "extradosed bridges" is not used in the specifications, however the intent of the specifications is to accommodate this bridge type. This is accomplished by allowing a transition in the stay capacity for low demand cables, expressing this as a function of the total live load plus wind stress to maximum ultimate tensile strength ratio (Total LL + W/MUTS) for the cable. The procedure modifies the material phi factor (ϕ) to recognize the lower live load fatigue demand on bridges that have relatively small live load demand on the cables. For the total LL + W/MUTS ratio over 7.5% the material factor is 0.65 as for a normal cable stayed bridge. For a Total LL + W/MUTS ratio of 1.0% a ϕ factor of 0.78 is allowed, essentially allowing a maximum stay stress of 0.6 f's. A linear transition is permitted between these two limits (Figure 11).

The result of this variable ϕ can be illustrated by a simple example. For a low live load fatigue demand extradosed bridge we would anticipate designing under a service load condition to an allowable cable stress of 0.6 f's. That is:

$$DL + LL \leq 0.6 f's \dots\dots\dots\dots\dots\dots\dots\dots\dots\dots\text{Eq. 1}$$

Based on the PTI 6th edition provisions, the cables would be sized for strength Group I loading (the load group typically governing the strength design of the cables) (AASHTO 2010). Inserting the Group I load factors into Equation 4–2 and using the ϕ-factor of 0.78, determined from Figure 11 for a bridge with a low LL + W/MUTS ratio (<1%) gives:

$$1.25(DL) + 1.75(LL) \leq 0.78 f's \dots\dots\dots\dots\dots\dots\dots\text{Eq. 2}$$

In order to compare equations 1 and 2, we need to consider a typical ratio of dead to live load forces in the cables. Experience from the authors design for the Pearl Harbor Memorial Bridge (Stroh

2012) would place this ratio at about 92% dead load and 8% live load. Inserting these ratios in the left side of equation 2, we can compute a blended load factor (x) on (DL + LL) as follows:

$$1.25(DL) + 1.75(LL) = x(DL+LL)$$

Inserting the DL and LL ratios:

$$1.25(.92) + 1.75(.08) = x(1.0)$$

$$x = 1.29$$

Inserting this blended load factor into equation 2 gives:

$$1.29(DL+LL) \leq 0.78 f's$$

And dividing through by 1.29 gives:

$$DL+LL \leq 0.605 f's$$

This gives essentially the same result as designing as a service level stress of 0.6 f'.

The variable ϕ-factor approach serves to provide a simple means of transitioning the cable design criteria between the typical cable-stayed stress limit of 0.45 f's to a stress limit of 0.6 f's that would represent a low fatigue demand condition.

4   APPLICATION OF CRITERIA TO A PROTOTYPE DESIGN

4.1   *Project description*

The Pearl Harbor Memorial Bridge in New Haven Connecticut is the first extradosed prestressed bridge designed in the United States, and as such provided a test bed for the adaption of the extradosed bridge concept to the United States (Figure 12). This includes the adaptation of design codes and standards, the application of new design criteria where none previously existed for this bridge type. The Pearl Harbor Memorial Bridge is a three span continuous cast-in-place segmental concrete box girder structure with 515 foot main span and 248.8 foot side spans. A vertical clearance of 60 feet is provided. The superstructure is supported on pot/disk bearings at the towers and end piers. The structure is fixed against longitudinal translation at Tower 3 and free to move at the other locations.

The northbound and southbound roadways are carried on separate parallel structures, accommodating 5 lanes a tapering auxiliary lane and 10 foot shoulders on a deck that varies in width from 95.4 feet to 107.6 feet. Each deck is a 5-cell concrete box girder section. The depth varies through a parabolic haunch from 9.84 feet at midspan to 16.4 feet at the towers. The superstructure box section is post-tensioned both longitudinally and transversely. Longitudinal tendons are internal to the concrete. Transverse slab tendons are internal to the concrete. Draped external transverse post tensioning is provided at each stay anchorage. This external tendon is anchored near the stay and deviates near the bottom of the inner two webs.

The stays are anchored at the edges of the cross section in reinforced edge beams. The tower legs are spaced slightly outside the superstructure, to allow the deck to pass through, and the stay cables are therefore slightly inclined outward from the vertical plane.

The twin decks are supported by a common tower, each comprised of with three pylons above deck and two additional intermediate columns below deck. The tower legs are constant cross section, elliptical in shape and hollow in cross-section. The stay cables are anchored in steel frames erected prior to pouring the tower sections. Foundations are 8 foot diameter drilled shafts founded on rock.

Figure 12.   Computer image of pearl harbor memorial bridge (adjacent Tomlinson Bridge removed from view for clarity) (Image courtesy of AECOM).

4.2   *Side span proportioning*

As previously noted desirable main span/side span length ratio for extradosed bridges is 0.6, with typical extradosed bridges in the range of 0.45 to 0.69. This ratio for the Pearl Harbor Memorial Bridge is 0.48, which is at the low end of the range. The span length was selected for the Pearl Harbor Memorial Bridge based on geometric constraints. There is a horizontal curvature on the bridge approaches, and the side span was set to avoid the horizontal curvature on the cable supported bridge. Also the roadway continues to taper (widen) on the approaches, and the side span length was set in order to provide a reasonable design width for the bridge.

In final design, the result of this relatively short side span was that there was an uplift condition at the anchor piers under certain live load conditions. This uplift was about 5% of the maximum reaction at the anchor pier, or about 570 kips. A concrete counterweight was cast inside the box girder to balance this uplift condition and result in a net positive reaction under all load conditions on the bearings. The use of a counterweight was considered preferable over a mechanical hold-down device (such as tie-down cables of a pinned bearing) because the hold down device would require ongoing future maintenance. There is also an issue of redundancy of hold-down devices, since their failure could lead to collapse of the bridge. There was however a negative consequence of the counterweight in that it adds mass to the superstructure which increases the seismic demands on the structure. However in this case the counterweight was not too large and the added mass was judged acceptable.

It is noted that if the side spans were shortened even more, this uplift condition would become a significant design issue. Therefore the lower range limit to the side span ratio is an important design parameter.

4.3   *Tower height*

The recommended tower/span (H/L) ratio is 1/10 for extradosed bridges, with a range of 1/7 to 1/13. The tower height selected for the Pearl Harbor Memorial Bridge (measured from deck level to the uppermost cable) is 60 feet. This gives an H/L ratio of 1/8.6. The tower height for the Pearl Harbor Memorial Bridge was selected with a slightly taller tower height than the suggested value, but well within the suggested range. This decision was made based on the very wide deck of the bridge, and the desire to reduce demand on the girder system. The consequence of this is that the stay system will be somewhat stiffer, due to the slightly steeper cable inclination, which will place more demand on the cables (especially for fatigue). This was closely coordinated with the stay design.

4.4 *Deck depth*

As previously noted, most extradosed bridges use a variable depth cross section. The recommended girder depth/span ratio at the tower is in the range of 1:25–1:35 and the recommended depth span ratio at midspan is 1:50.

For the Pearl Harbor Memorial Bridge a variable depth girder was chosen. The haunched section maximizes the girder capacity for the cantilever construction prior to installation of the first stay cable, while reducing the section size to save weight and cost for the lower moment demand sections near mid-span. A depth 16.4 feet is selected at the towers, which was selected in order to provide adequate negative moment capacity for the cantilever construction of the girder prior to installation of the first stay. This gives a depth:span ratio of 1:31.4, in the middle of the recommended range. At mid-span a depth of 11.5 is selected. This depth was in part chosen to provide a 6.5 foot internal clear height within the box girder for inspection access purposes. This depth gives a depth:span ratio of 1:45, close to the recommended range.

4.5 *Stay cable design*

The Pearl Harbor Memorial Bridge was designed prior to publication of the PTI code that addressees a variable phi factor approach for low fatigue demand cables. Rather, it was designed on an allowable stress basis based on an evaluation of the single strand fatigue performance of strand loaded at 0.55 of ultimate, with a corresponding fatigue range.

The Pearl Harbor Memorial Bridge was re-evaluated using the 6th Edition PTI (PTI 2006) the variable ϕ-factor. The results of this analysis show that for most of the cables all stay stresses are within the factored resistance limits using the variable ϕ-factor. There are 7 cables that are overstressed up to 6.2% for Group I loading (dead plus live load).

It is noted that the variable ϕ-factor calculation is based on the ratio of live load + wind stress divided by the maximum ultimate tensile strength of the cable (MUTS). Reviewing the load summaries, the reason these seven cables are overstressed is related to their wind loading, and its effect on the ϕ-factor. For these cables, the wind stress is relatively high, and this results in a reduction of the ϕ-factor to nearly that of a conventional cable stayed bridge, and hence the overstress.

AASHTO has historically assessed wind loads for a fewer number of fatigue cycles than for live loads. Wind fatigue is typically assessed at few hundred thousand cycles, whereas live loads cycles are based on actual traffic loading and the service life of the structure, often reaching 50 million or more cycles of fatigue. By including the wind stress in the determination of the ϕ-factor, the wind effects on cable fatigue are treated the same as live load effects, which is not the case. If we re-calculate the ϕ-factor as a function of live load/MUTS (leaving the wind stress out of the equation), then the overstress would not occur and the design would be acceptable.

It is the authors opinion that the approach to the variable ϕ-factor presented in the draft PTI specifications (PTI 2006) are unnecessarily conservative in the inclusion of wind stress in the determination of the variable ϕ-factor, and shown for the Pearl Harbor Memorial Bridge, and will in some cases result in designs being needlessly controlled (albeit to a small degree) by the wind provisions.

5 CONCLUSIONS

Extradosed prestressed bridges have been shown to provide a viable and economical option for bridges in the 300 to 600 foot span range by successful completion of more than 60 of this bridge type worldwide. At least one example of this bridge type has been constructed in at least 25 countries. Based on review of existing bridges and evaluations made in conjunction with the design

Table 1.   Recommended proportioning guidelines for extradosed prestressed bridges.

Common Span Range	300–600 feet
Side Span/Main Span Ratio	0.6
Girder Depth at Tower (Girder Depth/Main Span Length)	1:25–1:35
Girder Depth at Midspan (Girder Depth/Main Span Length)	1:50
Tower Height (Tower Height/Main Span Length)	1/10

of the Pearl Harbor Memorial Bridge, recommendations on proportioning parameters for extradosed bridges are shown in Table 1. The extradosed bridge type is also shown to be especially applicable to multi-span bridges (four or more spans), and can accommodate modest curved alignments. Extradosed prestressed bridges have been constructed for a wide range of bridge widths. The paper also provides discussion on stay cable design requirements following the PTI specifications related to strength and fatigue considerations, (PTI 2006).

ACKNOWLEDGEMENTS

The author gratefully acknowledges the Connecticut Department of Transportation, Federal Highway Administration and AECOM for their assistance and permission to use the information from the Pearl Harbor Memorial Bridge in this paper.

REFERENCES

AASHTO (2010). American Association of State Highway and Transportation Officials, AASHTO LRFD Bridge Design Specifications, 5th Edition, 2010.

FHWA, 1982. Feasibility of Standards for Segmental P/S Box Girder Bridges, Offices of Research and Development Report No. FHWA/RD-82/024 dated July 1982.

Kasuga, Akio, Shirono, Yoshiaki, Nishibe, Gou, and Okamoto, Hiroaki, 1994. Design and Construction of the Ordawa Port Bridge – The First Extradosed Prestressed Bridge, proceedings of the XIIth FIP International Congress, Washington, D.C., May 29-June 2, 1994, pp. F56–F62.

Komiya, Dr. Masahisa, 1999. A Characteristics and Design of PC Bridges with Large Eccentric Cables (PC Extradosed Bridges), Japan & Structure Institute, Inc., report dated 16 March 1999.

Leonhardt, F. and Zelner, W., 1980. Cable-Stayed Bridges, *International Association for Bridge and Structural Engineers, IABSE Surveys S-13/80, IABSE Periodica 2/1980*, ISSN 0377-7251, pp. 21–48.

Mathivat, J., 1988. Recent Developments in Prestressed Concrete Bridges, *FHP Notes, Quarterly Journal of the Fédération Internationale de la Précontrainte*, 1988/2, pp. 15–20.

Menn, Christian, 1986. *Prestressed Concrete Bridges*, Birkhäuser Verlag.

Podolny, Walter, Jr., 1994. "Cable-Stayed Bridges State-Of-The–Art in the United States", Proceedings of A Seminar Series on Cable-Stayed Bridges held in Miami, Florida, October 17–18, 1994. Edited by Ahmad H. Namini, University of Miami, Coral Gables, FL.

PTI, 2006. Recommendations for Stay Cable Design, Testing and Installation, 6th Ed., Post Tensioning Institute.

Stroh, Steven L., 2012. On the Development of the Extradosed Bridge Concept, A dissertation submitted in partial fulfillment of the requirements for the degree of Doctor of Philosophy, Department of Civil and Environmental Engineering, College of Engineering, University of South Florida, February 8, 2012.

Structurae, 2011. (http://en.structurae.de/structures/stype/?id=1002), website data retrieved October 2011.

Chapter 15

Bridge-weigh-in-motion for axle-load estimation

E. Yamaguchi & M. Kibe
Kyushu Institute of Technology, Kitakyushu, Japan

ABSTRACT: Good maintenance of a bridge requires the information on traffic loads. A method of estimating the traffic loads is bridge-weigh-in-motion (BWIM). The conventional BWIM is based on the strains of main girders. To obtain supplemental information of truck velocity, the strains of transverse stiffeners are measured additionally. Our previous research has shown that truck loads could be estimated solely by the strains of transverse stiffeners, provided that the transverse stiffeners were chosen carefully. The method is superior, since the strain measurements of main girders are no longer necessary. The method involves the integration of time-history response of strain and is called BWIM by Integration Method with Transverse Stiffeners (BWIM-IT). In the present study, BWIM-IT is extended to the estimation of axle loads. Good results are obtained if the influence of the truck in the adjacent lane is removed from the strain data.

1   INTRODUCTION

The information on actual traffic loads travelling on highways is essential to determine maintenance requirements of the highways. Because of this, the techniques that weigh trucks in motion without disturbing traffic flow have attracted many highway engineers and researchers. Those techniques are called collectively weigh-in-motion.

One of the weigh-in-motion techniques is based on the deformation of a bridge and is called bridge-weigh-in-motion (BWIM). The technique was first explored by Moses (1979). Because of its lower cost than conventional pavement scales, much attention has been drawn to BWIM in Japanese bridge engineering community (Matsui & El-Hkim 1989, Ojio et al. 2001, Miki et al. 1987, 2001, Ishio et al. 2002), as maintenance work is getting very important in recent years in Japan.

The authors have studied BWIM since 2002, too (Yamaguchi et al. 2004). It was related to a project of investigating actual truck weights on National Highway 201, Japan. A problem encountered in this project was that a two-span continuous bridge with skew was the only bridge available to the project while a short, simply-supported steel bridge with no skew is ideal for BWIM. Therefore, a careful preliminary field test with trucks of known weights was conducted, having shown that even this type of bridge could be used for BWIM. The method for this BWIM is based on that employed by Miki et al. (1987, 2001). The estimation of truck weight is carried out by the strains of main girders. But those strains are insufficient to conduct BWIM. The supplemental information such as truck velocity is required and needs be acquired by the strains of transverse stiffeners.

After the project, the possibility of estimating truck loads solely by the strains of transverse stiffeners was explored (Yamaguchi et al. 2010). It has been shown that the approach can give a good estimation of truck loads, provided that the transverse stiffeners are chosen carefully. In short, the transverse stiffeners whose deformations are influenced considerably by trucks running on the adjacent lane should not be employed for this approach. For example, the transverse stiffener close to the cross frame or the cross girder is not appropriate. The method is named BWIM by Integration Method with Transverse Stiffeners (BWIM-IT). BWIM-IT can lower the cost since the required measurements are much less and the data manipulation is much simpler. The objective of present study is to extend BWIM-IT further to the estimation of axle loads. It is noted that the axle loads are important information for bridge maintenance, as they are closely related to the durability of the floor slab of a bridge.

2   BWIM-IT

2.1   *Fundamentals*

The present approach is based on the BWIM technique presented by Ojio et al. (2001). It involves the integration of time-history response of strain. It was originally associated with the strains of stringers, but herein it is with the strains of transverse stiffeners. The supplemental information such as truck velocity is needed as well, which can be obtained in the same way as that of Yamaguchi et al. (2004).

The strain at the point of interest is a function g(x) of the position of a load x. The area A surrounded by the x-axis and g(x) in the x-g(x) graph, which is called the influence area in the present study, is proportional to the magnitude of the load. Hence, if the influence area A_c due to a truck of known weight W_c is obtained, the weight of a truck W can be estimated from the corresponding influence area A as follows:

$$W = \frac{A}{A_c} \cdot W_c \qquad (1)$$

Strains are usually obtained as the time-history response. When a truck runs with constant velocity V, the influence area A can be computed as

$$A = V \int_{-\infty}^{\infty} r(t)dt \qquad (2)$$

where $r(t)$ is the time-history response of the strain.

The truck velocity is obtained by measuring the strains of two transverse stiffeners in each traffic lane. To that end, the identification of the strain responses due to the same truck is required. This is done by using the auto-correlation function.

2.2   *Extension to estimation of axle loads*

In case of the axle-load estimation, the influence area due to each axle load needs be identified. Which requires the information on the strain induced by a single-axle load. That strain is called the strain influence line herein.

With the strain influence line, the number of axles in a truck, axle loads and distances between the axles, the strain caused by the running truck can be computed. The estimation of axle loads is its inverse problem: the axle loads and the distances between the axles are to be computed, while the strain influence line, the number of axles and the measured strain are given.

In the present study, the strain influence line is obtained by the three-dimensional finite element analysis of a bridge. The number of axles can be determined from measured strain. Then by assuming axle loads and distances between the axles, strain can be computed right away. This computation is to be conducted with various axle loads and distances. The axle loads and the distances between the axles that yield the influence area A equal to the influence area A of the measured strain would be identified, which are the estimated axle loads and distances, the answer to the inverse problem mentioned above, in the present study.

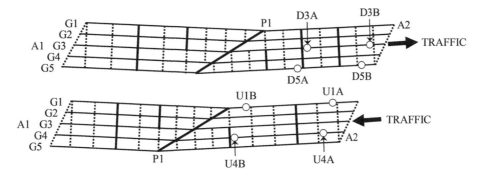

Figure 1.   Sasaguri Bridge: up-lane bridge (bottom) and down-lane bridge (top).

3   ESTIMATION OF AXLE LOADS

3.1   *Bridge for BWIM-IT and field test*

The bridge used in the project of National Highway 201 is Sasaguri Bridge, which actually consists of two parallel bridges, the up-lane bridge and the down-lane bridge. Each bridge has two lanes, each of which is for traffic in one specific direction, up or down. The up-lane bridge is about 70 m long while the down-lane bridge is about 80m long. Otherwise, they are very similar to each other. Both are a two-span continuous plate-girder steel bridge with five main girders. The plan is presented in Figure 1. The solid lines and the dotted lines in the direction perpendicular to the bridge axis indicate cross girders and cross frames, respectively. Each bridge has a side walk: it is located above the G5 girder in the up-lane bridge while it is above the G1 girder in the down-lane bridge. The bridges cross a river and the pier is inclined severely to the bridge axis, making the bridges quite skew. While the axes of the girders are composed of line segments, the floor slabs and thus the traffic lanes are curved, the distance between a running truck and the girders vary along the bridge axis.

At the initial stage of the project, the accuracy of BWIM was the major concern, since Sasaguri Bridge is by no means ideal for BWIM. Therefore, preliminary field test was carefully conducted, in which the strains of transverse stiffeners were measured for obtaining supplemental information. Three trucks of known weights were used in this test. The strains of the transverse stiffeners recorded in the test are utilized in the present study to carry out BWIM-IT.

The circles in Figure 1 show the locations of the transverse stiffeners whose strains were measured in the preliminary field test. There are two measured transverse stiffeners in each lane so as to evaluate the truck velocity. The strains measured at U1A, U4B, D3A and D5A are made use of for the axle-load estimation. Therefore, the strain influence line at each of these four transverse stiffeners is obtained by the three-dimensional finite element analysis of Sasaguri Bridge with the single axle load of unit magnitude, 1kN, travelling on the lane above the transverse stiffener. Nastran (MSC 2003) is used for the analysis.

The preliminary field test was conducted in May, 2002. Using three calibration trucks, the following four patterns of traffic situation were considered in the preliminary field test:

Pattern 1: Only one truck runs
Pattern 2: Two trucks run in the same lane
Pattern 3: Two trucks run side by side
Pattern 4: Two trucks run in the same lane while the other runs in the other lane

Figure 2 illustrates the four patterns. Within each pattern, various combinations of trucks were considered and the test of the same traffic situation was conducted three times. The number of the truck-running tests amounted to 60. It is noted that in an effort to enhance the reliability of the

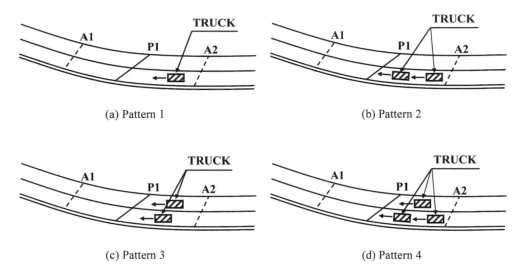

(a) Pattern 1   (b) Pattern 2

(c) Pattern 3   (d) Pattern 4

Figure 2.   Traffic situations in preliminary field test.

Table 1.   Average percentage error.

	Pattern 1	Pattern 2	Pattern 3	Pattern 4
Up-Lane Bridge	3.4 (8.6)	4.1 (9.4)	3.8 (7.8)	3.5 (7.2)
Down-Lane Bridge	4.4 (8.8)	4.5 (10.8)	50.7 (88.9)	33.9 (79.6)

( ): Maximum error.

field test data, traffic was controlled so that no public vehicles would run on the bridge during each truck-running test.

For the estimation of truck weight, A_C in Equation 1 needs to be evaluated first. The measurements in Pattern 1 are used, to this end. Since multiple truck-running tests were conducted within Pattern 1 and each test yields its value of A_C, the medium is taken for the value of A_C to be applied for the present BWIM-IT. It is however noted that the variation of A_C is quite small.

3.2   *Axle loads*

The results of the axle loads estimated by BWIM-IT are shown in Table 1 in terms of percentage error. BWIM-IT yields satisfactory accuracy in the case of all the patterns of the up-lane bridge and Patterns 1 and 2 of the down-lane bridge: even the largest error is only 10.8%. However, the error is much greater in the case of Patterns 3 and 4 of the down-lane bridge, the maximum error being up to 88.9%. To be interesting, the number of axles and the distances between the axles have been identified rather accurately even in the cases when the estimation of axle loads is not good.

By closely observing the strain measurements, the reason for the low accuracy in the down-lane bridge has been identified as the influence of the adjacent truck on the strains of transverse stiffeners. For instance, when a truck runs in the cruising lane of the up-lane bridge, the strain at U4B is much larger than that at U1A. On the other hand, when a truck runs in the cruising lane of the down-lane bridge, transverse stiffeners in both of the cruising and passing lanes deform: the strain at D5A is not negligible in comparison with that at D3A. The influence of a truck running on the adjacent lane is not expected in BWIM-IT and is considered a source of the error of BWIM-IT in the down-lane bridge.

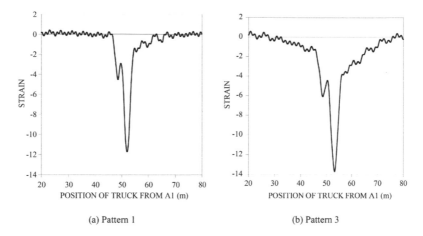

(a) Pattern 1   (b) Pattern 3

Figure 3.   Measured strains.

4   IMPROVEMENT

The discussion in the previous section indicates the necessity of removing the influence of the truck load on the adjacent lane so as to improve the accuracy of axle-load estimation for Patterns 3 and 4 of the down-lane bridge.

The strain influence lines are required for the improvement. They are obtained by the three-dimensional finite element analysis of the bridge in the same way as those in 3.1, except that the single axle load of unit magnitude moves on the adjacent lane instead of the lane above the strain measurement.

Since the number of axles of the adjacent truck and the distances between the axles can be identified before the removal of the influence of the adjacent truck, the issue at this stage is to estimate the axle loads of the truck on the adjacent lane and the distance between the two trucks in the different lanes. Once those pieces of information are obtained, the strains induced by the truck on the adjacent lane can be evaluated, using the strain influence line.

Strain occurs when the axle load is located within a certain distance from the transverse stiffener of the strain measurement. Such influence regions are observed in Figure 3. The strains shown there are the measured values at D3A. It is pointed out that the influence regions are different in Patterns 1 and 3. This is also due to the presence of the truck on the adjacent lane. When the influence of the adjacent truck is removed from the strains of Pattern 3, the influence region should be the same as that of Pattern 1.

Making use of the above point regarding the influence region, the axle loads of the truck on the adjacent lane and the distance between the two trucks are estimated. To be specific, by assuming various values of the axle load and the distance, evaluating the influences of the adjacent trucks, and modifying the measured strain each time, the values that correct the influence region are looked for. Figure 4 shows the strain thus modified from the measured strain in Figure 3b.

Based on the modified strains, the axle loads of Patterns 3 and 4 of the down-lane bridge are estimated. The errors involved are summarized in Table 2. The average error is significantly reduced down to around 5%, and the largest is to 18.5%. The estimation can be considered good enough in practice.

5   CONCLUSIONS

BWIM-IT was applied to the estimation of axle loads. The results were good except in Patterns 3 and 4 of the down-lane bridge. The reason for the low accuracy in the down-lane bridge was

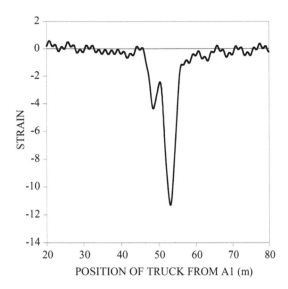

Figure 4.   Modified strains of Pattern 3.

Table 2.   Average percentage error with modified strain.

	Pattern 3	Pattern 4
Down-Lane Bridge	5.2 (12.1)	5.4 (18.5)

( ): Maximum error.

identified as the influence of the truck in the adjacent lane upon the strains of transverse stiffeners. A remedy was then proposed. Its validity has been demonstrated, confirming that BWIM-IT can achieve an acceptable accuracy for the estimation of axle loads.

REFERENCES

Ishio, M., Nakamura, S., Tamakoshi, T. & Nakasu, K. 2002. About the WIM system using the influence line. *Proc. of 57th Annual Conference of The Japan Society of Civil Engineers*: 1447–1448.

Ojio, T., Yamada, K. & Kobayashi, N. 2001. Development of bridge weigh-in-motion system using stringers of steel plate girder bridge. *Journal of Structural Engineering, JSCE* 47(A): 1083–1092.

Matsui, S. & El-Hkim, A. 1989. Estimation of axle loads of vehicles by crack opening of RC slab. *Journal of Structural Engineering, JSCE* 35(A): 407–418.

Miki, C., Murakoshi, J., Yoneda, T. & Yosimura, H. 1987. Weighing trucks in motion. *Bridge and Foundation* 87(4): 41–45.

Miki, C., Mizunoue, T. & Kobayashi, Y. 2001. Monitoring system of bridge performance with fiber-optic communications. *Journal of Construction Management and Engineering* 686/VI-52: 31–40.

Moses, F. 1979. Weigh-in-motion system using instrumented bridges. *Transportation Engineering Journal of ASCE* 105(TE3): 233–249.

MSC.visual Nastran, professional 2003.

Yamaguchi, E., Matsuo, K., Kawamura, S., Kobayashi, Y., Mori, M., Momota, K. & Nishinohara, T. 2004. Accuracy of BWIM using two-span continuous steel girder bridge. *Journal of Applied Mechanics, JSCE* 7: 1135–1140.

Yamaguchi, E., Naitou, Y., Matsuo, K., Matsui, Y., Takaki, Y. & Kawamura, S. 2010. BWIM by transverse stiffeners of steel I-girder bridge: *Journal of Civil Engineering, JSCE* F66(2); 251–260.

Seismic analysis of bridges

Chapter 16

Post-earthquake stability of Gerald Desmond Bridge

P. Banibayat, M. Carter, M. Nelson & T. Chandler
ARUP, New York, NY, USA

ABSTRACT: The Gerald Desmond Bridge is a vital link in the nation's trade infrastructure and a major commuter corridor which connects Long Beach with Terminal Island. The Port of Long Beach intends to replace the existing deteriorating bridge with a 2000 ft long cable stayed bridge with 1000 ft main span and two 500 ft side spans. The 515 ft tall towers, which provide primary means of vertical support to the cable-stayed bridge, are relatively slender tall hollow reinforced concrete sections thus requiring assessment of possible buckling. The buckling resistance of the towers comes from the global structural system with the stay cables providing restraint to the top of the towers and the viscous dampers providing restraint at deck level. The octagonal tower shape at the connection to the pile cap tapers to a diamond in the upper part of the tower. This non-standard tower geometry is a signature architectural statement and can be constructed with fairly simple formwork; however, the constantly variable cross-section is an analytical challenge. The slenderness evaluation captured the unique boundary conditions of the bridge as well as the irregular tower geometry. This paper provides an in-depth discussion of this analysis. The bridge is located in an area of extreme seismic hazard. The non-linear time history analysis of the bridge includes simultaneous tri-axial earthquake accelerations as well as gravitational acceleration. This paper discuss the explicit nonlinear time history analysis performed for tower stability during an earthquake and show the tower remain stable and elastic after Functional Evaluation Earthquake (FEE) and 1000 year return period Safety Evaluation Earthquake (SEE) event.

1 INTRODUCTION

The Gerald Desmond Bridge is a vital link in the nation's trade infrastructure and a major commuter corridor, which connects Long Beach with Terminal Island in Southern California. The new Gerald Desmond is a cable stayed bridge with Main Span of 1,000 ft and two side spans of 500 ft. The bridge is supported by two approximately 115 ft mono-pole towers, cables are connected to the outside of edge girders. Superstructure is 156 ft wide carrying three traffic lanes on both direction and a shared bike path-pedestrian on south side. Figure 1 shows an overview of main span and approaches of the new Gerald Desmond Bridge.

1.1 *Tower geometry and key dimensions*

The towers comprise of reinforced concrete tapering hollow towers. The tower is octagonal at the connection to the pile cap. The taper is achieved by reducing the width of four of the eight sides whilst keeping the other four sides constant such that the cross section is diamond shaped in the upper part of the tower. The transformation to a diamond shape rather than a more traditional square facilitates the anchorage of the stay cables with each fan of cables projecting from one face of the tower and no interference between the stay cable anchor tubes and the corners of the tower. The variation of tower wall thickness is shown in the Figure 2. Tower thickness is 3 ft for the lower 70 ft of the tower and 2 ft for the upper part. The tower is generally doubly symmetric and is prismatic in

Figure 1.   Gerald Desmond cable stayed bridge.

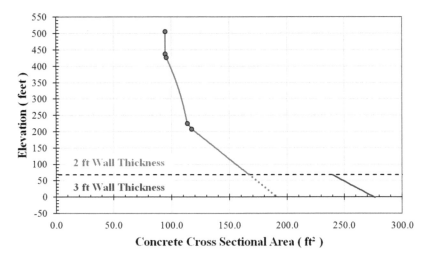

Figure 2.   Variation in tower cross sectional area.

the upper part where the stay cables are anchored. The upper eight sets of stay cables are anchored in a fabricated steel box. The lower two sets of stay cables are sufficiently vertical so as to be anchored on reinforced concrete corbels.

The towers are provided with a bi-linear taper below the stay cables with setting out dimensions at the top of the taper, superstructure level and top of pile cap. The inclination of the taper below superstructure level is the same on both the west and the east tower such that the slightly shorter east tower is also slightly narrower at the connection to the pile cap. A substantial diaphragm is provided at superstructure level to transfer loads into the tower from the viscous dampers. Structural steel skin plates are also provided at superstructure level.

1.2   *Concrete tower shaft*

The tower will be provided with two ductile zones which are analogous to the plastic hinges provided in traditional Caltrans ductile design. The extent of the ductile zones is indicated in Figure 3 and Table 1. The skin plates at deck level provide additional reinforcement to the tower which defines the position of the upper ductile zone.

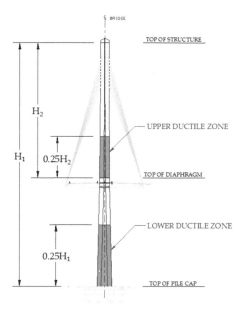

Figure 3.   Tower ductile zones.

Table 1.   Tower ductile zone.

Location		H (ft)	0.25xH (ft)
West Tower	Lower	507.00	126.75
	Upper	281.46	70.37
East Tower	Lower	486.00	121.50
	Upper	281.46	70.37

Since the maximum wall slenderness ratio of the tower is 6.8 the towers can be designed in accordance with AASHTO LRFD (AASHTO 2009) without compensation for local buckling effects in the panels. Similarly, local buckling effects do not need to be considered in the time history analysis model.

2   TOWER SLENDERNESS EVALUATION

The towers provide the primary means of vertical support to the Main Span Bridge. The towers are relatively slender tall hollow reinforced concrete sections. The buckling resistance of the towers comes from the global structural system with the stay cables providing restraint to the top of the towers and the viscous dampers providing restraint at deck level. Slender evaluation is focused on the West Tower since it is slightly taller and therefore slightly more susceptible to slenderness effects than the East Tower.

2.1   *Axial loading in towers*

Below the stay cable anchor zone the axial force is reasonably uniform, increasing slightly towards the base due to the self-weight of the tower as indicated in Figure 4. Also, Figure 5 shows the

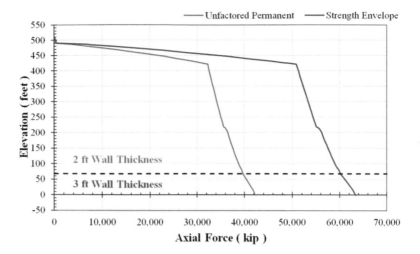

Figure 4.   Axial force in tower.

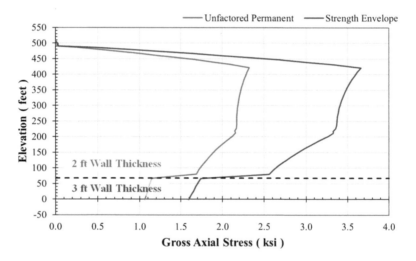

Figure 5.   Gross axial stress in tower.

variation in axial stress. The maximum gross permanent axial stress of 2.3 ksi occurs immediately below the stay cable anchorage zone and the stress is between 2.2 and 2.3 ksi in the region of the tower between deck level and the stay anchorage zone. Below deck the gross permanent stress decreases to a value of 1.5 ksi at the pile cap due to the increasing cross section area.

The maximum permitted axial force in the tower is determined from AASHTO LRFD (AASHTO 2009) Equations 5.7.4.4-1 and 5.7.4.4-3. Rearranging these equations, the maximum allowable strength limit state axial stress in the tower can be expressed in terms of the longitudinal reinforcement percentage as shown in Equations 1 and 2:

$$\frac{P_\gamma}{A_g} = \frac{\varphi P_n}{A_g} = \frac{\varphi \times 0.80 \times \left[0.85 f'_c (A_g - A_{st}) + f_y A_{st}\right]}{A_g} = 0.75 \times 0.80 \times 0.85 \times [7 \times (1 - \rho_{st}) + 60 \times \rho_{st}] \qquad (1)$$

$$\rho_{st} = \frac{A_{st}}{A_g} \qquad (2)$$

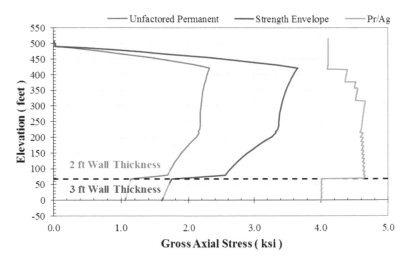

Figure 6. Tower axial stress vs limit state.

Figure 6 compares the maximum strength limit state axial stress with the factored allowable stress. The demand is less than the capacity over the full height of the tower. The critical section is just below the stay cable anchorages.

3 METHODOLOGY FOR EVALUATING BUCKLING/SLENDERNESS

3.1 *Procedure*

The slenderness of the towers is evaluated using a fully non-linear analysis approach as explained in the following

– Apply ultimate loads to the global analysis model
– Run an elastic critical design buckling analysis for the towers to determine significant buckling modes
– Deform the geometry of the model into the same mode-shape as the result of the buckling analysis scaled to achieve a target "initial imperfection".
– Run a cracked stiffness-P-δ analysis for each significant ultimate load combination. This requires an iterative approach to determine a compatible set of displacements, member forces and EI_{eff}.
– Carry out sectional verification using the forces from the final iteration
– Run a second elastic critical Eigen buckling analysis to determine the elastic critical buckling load factors with cracked stiffness.

Depending upon the results of the verification the reinforcement pattern could be adjusted and the above procedure would be repeated.

3.2 *Critical ultimate loads combinations*

In this analysis ultimate limit states STRENGTH I, III, IV and V are considered. By inspection STRENGTH II will not be critical since the load factor on live load is reduced and the permit vehicle is a localized load effect which will not govern the tower design.

Preliminary analysis has shown that three configurations of live load are critical:

• All (A) Loading applied to the entire bridge
• Main (M) Loading applied only to the main span
• South (S) Loading applied only to the southern half of the bridge

In addition SERVICE I is used for verifying the effectiveness of frictional bearing restraint at the end bents after the Functional Evaluation Earthquake (FEE). Wind in STRENGTH III and V is applied from the north since this is in the same direction as the out of balance forces, while this is conservative assumption since the wind climate analysis showed that north winds are generally lower. The live load applied for STRENGTH V is Main (M) since this has been found from the STRENGTH I analyses to be the most critical loading for the tower. These combinations have been selected since they provide the highest lateral displacements and bending moments (worst case destabilizing loads) in the towers.

3.3   Global analysis model

Two global analysis models are used to perform the slenderness evaluation. The In-service buckling analysis model is established in Oasys GSA. The superstructure is modeled with a pair of edge beams in order to achieve analytical convergence during the buckling analysis while at the same time accurately modeling the torsional restraint at the end bents. This model is used to carry out Strength Limit State Combination I P-δ and Strength Limit State Combination IV P-δ analysis as well as elastic critical Eigen buckling analysis for all combinations.

The In-service and temporary stages wind buffeting analysis model is also used to carry out Strength Limit State Combination III P-δ analysis and Strength Limit State Combination V P-δ analysis. In each analysis the tower member is split into 47 elements ranging in length between 2.1 ft and 6.5 ft. Figure 7 shows an overview of main span cable stayed bridge model.

3.4   Deformation of global model

To achieve a target "initial imperfection" the global analysis model was deformed to match the modal shapes determined from the initial elastic critical Eigen buckling analysis. The minimum value of initial imperfection is considered to be 6". Elastic deformations as a result of the erection sequence are taken into account directly without the need to introduce deformed nodal geometry. However, since the buckling analysis does not directly include creep and shrinkage the initial imperfection is increased to allow for this effect. The critical direction for creep and shrinkage is in the transverse direction and the additional deformation is less than 2".The total initial imperfection should therefore be a minimum of approximately 8" in the transverse direction and 6" in the longitudinal direction. However, an initial imperfection of 12" has conservatively been used for the in-service analysis cases in both the longitudinal and transverse directions. Since this exceeds the minimum requirement it is acceptable to be adopted for design.

New models were created by taking the displacements from the given mode shape and applying a factor so that the largest displacement in the tower matched the design value of imperfection. These factored displacements were then added to the original geometry and fed back to the model to create a deformed geometry model as shown in Figure 8. Transversely the towers are always displaced to the south, which is more critical since the elastic deformations and creep and shrinkage are always towards the south side which is heaver due to the bike path.

The models were derived from the first transverse mode and the first longitudinal mode. Preliminary analyses have shown these two mode shapes to be critical in terms of moment demand and buckling load factor. For the determination of the critical buckling load factors the geometry of the model is reset such that the deformed geometry of the P-delta analysis is "locked in" as shown in Figure 9. If the structural stop is engaged in the P-δ analysis then a rigid restraint (very stiff spring) is added at the location of the structural stop.

3.4.1   Moment curvature analysis

A moment-curvature analysis is carried out for each element in the tower based on the reinforcement arrangement shown on the plans. The moment curvature analysis is carried out following the assumptions prescribed for strength limit state in AASHTO LRFD (AASHTO 2009). The objective of the moment-curvature analysis is to determine the secant stiffness of the tower for every element

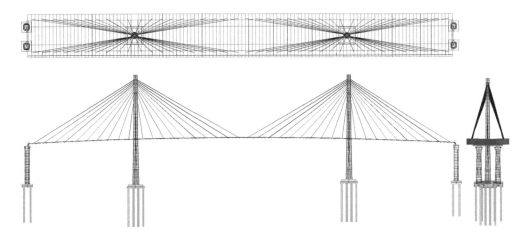

Figure 7.   Finite element model for slenderness evaluation.

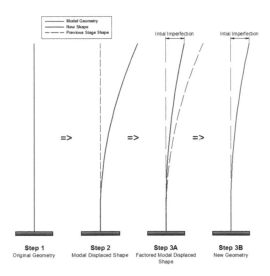

Figure 8.   Buckling Procedure – Stage 1.

to allow a full non-linear cracked section properties analysis. This variation of stiffness with demand is a more accurate model of the behavior than the use of a single "fully cracked" effective stiffness throughout the tower.

3.4.2   *P-δ analysis*

The P-δ analysis is carried out by applying factored loads to the global analysis model in order to determine second order moments. The process is iterative due to the non-linear behavior of the cracked concrete sections. A process diagram illustrating the iteration is provided in Figure 10.

The iteration process is deemed to have converged once the new bending moments are within 0.01% of the previous iteration bending moments. The final iteration case is then used to create a new model geometry. This new model is used for the Eigen buckling analyses, which determines the cracked stiffness elastic critical buckling load factors.

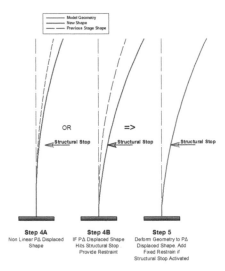

Figure 9.   Buckling Procedure – Stage 2.

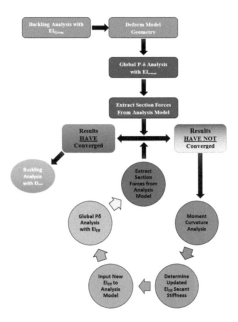

Figure 10.   P-δ process diagram.

3.5   *Calculation of elastic critical buckling load factors*

The key features of the Eigen buckling analysis are:

- Factored loads are applied to the deformed geometry model
- The tower element stiffness are based on the cracked stiffness P-δ analysis including deformed initial geometry
- Engaged structural stops are modeled as ridged elements (very stiff springs)

Table 2.   In-service critical buckling factors.

Case	First Longitudinal Mode (λ_{crit})	First Transverse Mode (λ_{crit})
STRENGTH I (A)	8.2	6.3
STRENGTH I (M)	8.1	7.4
STRENGTH I (S)	9.0	8.0
STRENGTH III	9.9	8.2
STRENGTH IV	8.3	7.3
STRENGTH V (A)	8.4	6.1
STRENGTH V (M)	8.4	6.2
STRENGTH V (S)	9.0	8.1

The Eigen buckling analysis determines the elastic critical load factor (λi). This is the factor by which the Strength Limit State loads in the model would have to be multiplied before Euler buckling would occur.

4   RESULTS AND DISCUSSION

4.1   *Preliminary evaluation*

Preliminary investigations where conducted to determine which load cases are critical for tower buckling. The first investigation looked at the axial load demands in the tower. This investigation shows that the highest axial demand in the tower is from STREGNTH I and the lowest is from STRENGTH III. Although STRENGTH III produces the lowest axial demand in the tower it does produce a significant transverse moment in the tower. For this reason STRENGTH III is deemed to be significant for buckling in the transverse direction.

The second investigation looked at the critical buckling factor (λ_{crit}) for the in-service base model, (i.e. no non-linear iterations). Generally all first modes are under 10 and there is no significant variation between the different load cases, hence all load cases will be considered.

4.2   *In-service slenderness evaluation*

The minimum elastic critical buckling load factors and mode shapes are as tabulated in Table 2.

The worst case transversely is STRENGTH V (A), which produces high axial load on the tower combined with a significant transverse load. STRENGTH I (M) is the worst case longitudinally. The critical mode shapes are displayed on Figure 11 and Figure 12.

4.3   *Post-earthquake slenderness evaluation*

As has been mentioned previously, the tower has significant reserves of capacity under strength limit state loads. However, an evaluation must be carried out to investigate whether the effects from the Functional Evaluation Earthquake (FEE) or the Safety Evaluation Earthquake (SEE) ground motions could result in the tower being unable to maintain its vertical load carrying capacity. This study addresses two main questions; first, to what extent will the tower be damaged in an earthquake and does this affect the load carrying capacity? and second, what are the implications of the fuses being activated?

4.3.1   *Influence of damage to tower*
The tower is designed to remain essentially elastic under the SEE ground motions. Based on that performance requirement it is expected that damage experienced during the SEE will be negligible and will have no influence on the ability of the tower to maintain its load carrying capacity during

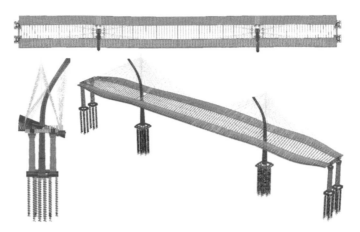

Figure 11.   In-Service Buckled Shape – Transverse Mode 1.

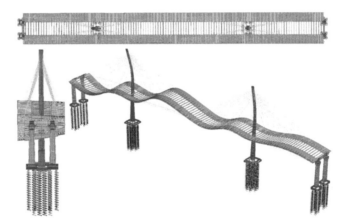

Figure 12.   In-Service Buckled Shape – Longitudinal Mode 1.

service conditions. Calculations show the concrete strains under the SEE event to be significantly less than 0.003 meaning there is no damage in concrete. (Carter et al. 2013) Also the current analyses have demonstrated that the tower remains essentially elastic. Therefore, tower may be considered to be undamaged as a result of the SEE ground motions.

4.3.2   *Implications of transverse fuses being activated*
The fuses at both end bents and towers are designed to fail (activate) in both the FEE and SEE earthquake events thus allowing the dampers to function. The elastic critical buckling load factor of the tower is sensitive to the lateral restraint at the end bents. The activation of the fuses changes the lateral restraint conditions at the end bent. However, structural stops are provided at both the towers and end bents. Therefore, any instability in the tower due to lack of lateral restraint would not cause collapse but would cause deformations of the tower so as to engage the structural stops and restore lateral restraint to the system.

Initially, under small movements, there is no lateral restraint in either the transverse or longitudinal direction between the superstructure and substructure except for the stay cable system. However once the movements reach the limit of the structural stops (i.e. 18 inches transversely at end bents or 30 inches longitudinally at towers) they will introduce lateral restraint to the system.

Table 3. Post-earthquake slenderness parameters summary.

			Post FEE Service	Post FEE Strength	Post SEE Strength
			Un-factored Nominal Properties		
Effective Stiffness					
Initial Imperfection			12"	12"	18"
Load Cases			Service I (A, M, S)	Strength I, III, IV, V	Strength I
Restraint	End Bent	Transverse	Fixed/5" offset	Fixed/18" offset	Fixed/18" offset
Conditions		Longitudinal	Fixed/5" offset	($\Delta < 40$")	($\Delta < 40$")
	Tower	Transverse	($\Delta < 30$")	($\Delta < 30$")	($\Delta < 30$")
		Longitudinal	($\Delta < 30$")	Fixed/30" offset	Fixed/30" offset

Table 4. Post-earthquake critical buckling factor.

	First Transverse Mode (λ_{crit})				First Longitudinal Mode (λ_{crit})			
Case	POST FEE		POST SEE		POST FEE		POST SEE	
STRENGTH I (A)	**1.7**		4.1	ET*	**2.0**		**2.0**	
STRENGTH I (M)	1.8		4.2	ET*	7.4	TL*	7.4	TL*
STRENGTH I (S)	2.0		**2.0**		2.4		2.4	
STRENGTH III	5.1	ET*	5.1	ET*	3.0	ET*	3.0	ET*
STRENGTH IV	1.9		4.3	ET*	2.5		2.5	
STRENGTH V (A)	4.3	ET*	4.3	ET*	2.2	ET*	2.2	ET*
STRENGTH V (M)	4.4	ET*	4.4	ET*	2.3	ET*	2.3	ET*
STRENGTH V (S)	4.6	ET*	4.5	ET*	2.5	ET*	2.5	ET*

Note: *first letter "E-" indicates that structural stop in the end bent has been engaged
first letter "T-" indicates that the tower structural stop is engaged
last letter "-T" indicates the transverse stop is engaged
last letter "-L' indicates the longitudinal stop is engaged

A two-step approach is used to determine the buckling load factor. The first step is to assume there is no lateral restraint in either the transverse or longitudinal direction between the superstructure and substructure except for the stay cable system ignoring the additional geometric restraints which exist. The relative displacements between the superstructure and substructure are evaluated. If these displacements are greater than the displacement allowed by the structural stop and the structural stop is active then a spring is added to model the structural stop. A second iteration process is then conducted to alter the spring stiffness to achieve the desired displacement.

4.3.3 *Slenderness evaluation*

The Post Earthquake slenderness evaluation has been split into three parts; Post FEE – Service, Post FEE – Strength, and Post SEE – Strength. Table 3 outlines the basis of the slenderness evaluation for each part.

The minimum elastic critical buckling load factors are displayed in Table 4. If the end bent transvers structural stop is engaged it will provide lateral restraint and hence stabilize the system.

It was observed that the structural stops have minimal effect on the mode shapes for both Post FEE and Post SEE. The effective stiffness of the tower for Post FEE and Post SEE is very similar, in that the base of the tower sections remain uncracked and the upper part of the tower has some minor cracking. There is only one exception that is Post SEE – Strength 1 (M). This is because the tower longitudinal structural stop is engaged transferring force from the deck to the tower and hence increasing the tower demands.

5 CONCLUSIONS

The 515 ft hallow tall towers are relatively slender which requiring assessment of possible buckling. The constantly variable cross-section of the tower is an analytical challenge. The slenderness of the towers was evaluated capturing the unique boundary conditions and the irregular tower geometry.

To understand the tower behavior at earthquake events it was investigated that to what extent will the tower be damaged in an earthquake and does this affect the load carrying capacity. Calculations show the concrete strain under SEE event to be significantly less than 0.003 meaning there is no damage in concrete. Also the current analysis demonstrated that tower remain essentially elastic.

Also, implications of fuses at towers and end bents being was studied, it was observed that the structural stops have minimal effect on the mode shapes for both Post FEE and Post SEE. The effective stiffness of the tower for Post FEE and Post SEE is very similar, in that the base of the tower sections remain uncracked and the upper part of the tower has some minor cracking.

ACKNOWLEDGEMENTS

The authors would like to acknowledge and thank other colleagues within Arup as well as Caltrans, the Port of Long Beach and the Shimmick/FCC/Impregilo JV for their invaluable contributions in developing the design of the project.

REFERENCES

AASHTO 2009, Guide Specifications for LRFD Seismic Bridge Design. Washington: American Association of State Highway and Transport Officials.
Carter, M., Matusewitch, P. & Carstairs, N. 2013. Gerald Desmond Bridge Replacement – Seismically Driven Design of the Main Span Bridge, *Proceedings of the 7th New York City Bridge Conference*, New York, NY.

Chapter 17

Impact of secondary fault findings on the design of Izmit Bay Bridge in Turkey

O.T. Cetinkaya
IHI Infrastructure Systems, Osaka, Japan

M. Yanagihara & T. Kawakami
IHI Infrastructure Systems, Tokyo, Japan

J. Chacko
Fugro, Istanbul, Turkey

ABSTRACT: Izmit Bay Bridge with a 1550 m main span will be the world's 4th longest span suspension bridge when completed. The bridge crosses the Izmit Bay in western Turkey from North to South and situated at very close premises of North Anatolian Fault. Therefore, from the early stages of design, detailed geophysical and geotechnical surveys have been conducted to find out the faulting, and the consequent earthquake risk in the region to lay the required basis for detailed design. In this paper, the conceptual bridge design, the detailed geophysical studies and the implication of the faults revealed by those geophysical studies on the conceptual design are introduced.

1 INTRODUCTION

Izmit Bay Bridge will be the most important component of Gebze-Orhangazi-Izmir Motorway which is currently under construction through the route shown in Figure 1. The Gebze-Orhangazi-Izmir Motorway with a total length of 420 km road construction including the Izmit Bay Bridge was tendered by the General Directorate of Highways (KGM) of Republic of Turkey as Build Operate Transfer basis project. The project was awarded to a consortium company named as OTOYOL AS for an operation period of 22 years and 4 month in 2010 September.

Having been appointed for the highway construction the OTOYOL AS opened an international tender for the construction of Izmit Bay Bridge in 2010. According to the conditions of tender every bid owner was required to propose a conceptual design that will satisfy the tender specifications. For the purpose of enabling the tenderers to propose a conceptual bridge design across the Izmit Bay, OTOYOL AS has conducted a detailed geophysical and geotechnical on-shore, near shore and offshore studies in an area not limited to the bridge axis but covering a wide region in the Izmit Bay. Later the outcomes of those studies have become a part of tender documents according which the tenderers developed the conceptual design proposals. In 2011, a consortium formed by IHI and ITOCHU, two Japanese companies, was awarded the contract based on their conceptual design, hereafter the "Tender Design" which is a suspension bridge with 1550 main span to be constructed in 38 months after a 6 months of preparatory work period which will be the period for the developing the detailed design (Kawakami et al, 2014).

After receiving the "Notice to Proceed" IHI-ITOCHU has started the Preparatory Works Period of which the first task was to evaluate the validity of the tender design by carrying additional soil investigations at the proposed locations of anchorages and towers. Detailed near-shore geophysical

Figure 1.   The Gebze - Orhangazi-Izmir Motorway route and the location of Izmit Bay Bridge.

investigation was initiated (Fugro, 2011) at the south anchorage location which had been indicated as "Secondary Fault" location by the Tender Specifications. As the result of those additional geophysical investigations several secondary faults are identified with their characteristics. The newly identified secondary faults caused the tender design to be drastically changed prior to the development of detailed design i.e. shortening of total bridge length in suspension and the change of south anchorage location.

2   TENDER DESIGN

2.1   *Location characteristics in terms of fault hazards*

The tender design of IHI-ITOCHU was prepared based on the results of the soil investigations conducted in the pre-tender stage. In 2010 IHI-ITOCHU conducted detailed geophysical and geotechnical soil investigations and made assessment for fault hazard at the greater bridge location. According to these studies the greater bridge location was classified into three fault regions as illustrated in Figure 2.

The primary fault zone is identified as the main strike zone of North Anatolian fault, where typical horizontal displacement at the surface level is expected from 300 cm to 600 cm and vertical

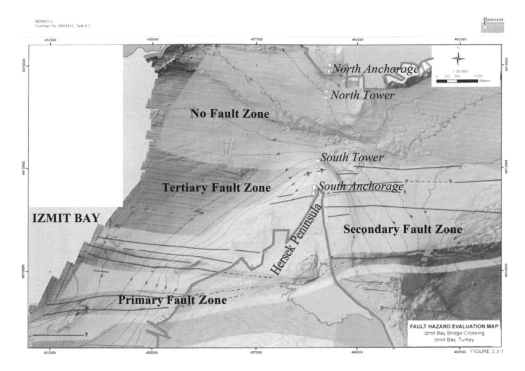

Figure 2.   The fault hazard map prepared at the 2010 geophysical investigations.

displacements may occur between 15 cm to 30 cm. This region is located about 2 km south of proposed south anchorage location. Apart from this main fault, secondary and tertiary fault zones are identified to the northern direction, which are interpreted as numerous traces of the North Anatolian Fault Zone. At the south, on both sides of the Hersek Peninsula, a secondary fault zone is indicated, where a secondary fault which is capable of causing $50 \sim 100$ cm horizontal displacement and $15 \sim 30$ cm vertical displacement can exist at any location within the region. Additionally a tertiary fault zone is identified as well, where faults capable of causing $15 \sim 30$ cm horizontal and vertical displacement can be found.

The tender design envisaged a suspension bridge with 1550 meter main span. The towers are located at deep offshore locations which led the side spans to be suspended as well. As a result, a bridge profile as shown in Figure 3 was generated where the total bridge length in suspension was formed as 2800 meters.

The North and South Towers are located at "No Fault Zone" as shown in Figure 2, whereas the south anchorage is located at the secondary fault zone which is susceptible to have an actual secondary fault nearby.

2.2   *Other characteristics*

One of the distinguishing characteristic of tender design is the reinforced concrete side span piers located at the ends of the suspended deck. The main cable is deviated on these side span piers and anchored to the two gravity type anchorages. By this way the splay saddle is taken down from the deck level which led to a smaller size of anchorages. Both at the north and south side the anchorage and the side span piers where the deviation saddle and splay saddle are located are planned as structures with discrete foundations. At the North the anchorage and the side span pier are located at a no fault zone. The South Anchorage structures of which the profile is given in the Figure 4 is located at a location where there is risk of existence of secondary faults. Therefore prior to the

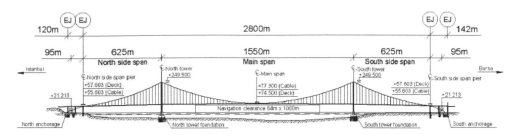

Figure 3.   The profile of tender proposal.

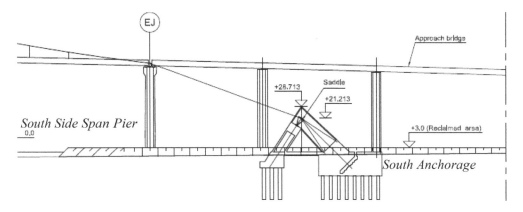

Figure 4.   The South Anchorage Structures during the tender design.

detailed design works, for the purpose of revealing the secondary fault hazard at the proposed south anchorage structures, detailed geophysical investigations are conducted as condition of the tender design for its validity to be used for the detailed design development.

During the tender design the tower height is designated as 251 m in a way to allow a main cable sag of 1:9. The towers (Yamane et al, 2014) are designed as steel single box composed of 4 stiffened panels located on deep off shore foundations. The foundations (Yamamoto et al, 2014) are designed as reinforced concrete caissons and located on gravel bed over improved ground (Refer to Figure 5) according to the seismic isolation concept. According to the seismic isolation concept the connection of foundation to gravel bed is designed as a fuse which will fail during strong ground motions to give way to foundation to slide on the gravel bed causing energy dissipation. Therefore, during a strong earthquake event the tower foundations will act relatively to the ground so that the superstructure will oscillate in longer periods causing only partial force transfer from foundation to the superstructure. The ground under the gravel bed is reinforced by inclusion piles deep into 35 meters to endure the shear forces and prevent subsequent liquefaction during an earthquake event. Both the north and south tower foundations are located at no fault zone indicated in Figure-2.

3   DETAILED DESIGN

The works of detailed design was initiated on September 2011, after IHI-ITOCHU has received "Notice to Proceed" subsequent to the signing of contract. First step was to carry out additional geophysical and geotechnical investigations at the locations of south anchorage structures, which were located at the secondary fault zone. Through these investigations it is aimed to find out the exact locations of secondary faults and make necessary changes on the tender design, if actually exist.

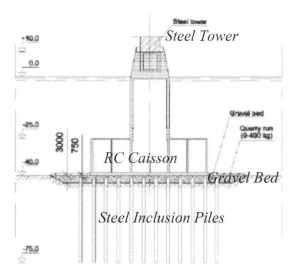

Figure 5. Tower Foundation.

3.1 *Additional geophysical investigations*

The tender design had located the south anchorage structures at a near shore location about 50 meters away from the tip of Hersek Peninsula (See Figure 2) at a very shallow sea section. Only a limited geophysical study could have been carried out at this location by the Employer at the pre-tender stage in 2010 due to insufficient depth of water (Average depth is about 1.5 ~ 2 meters) which limits the boat operations. With the start of detailed design additional geophysical investigation at the near shore locations was re-executed in 2011 to indicate the soil profile at the lines shown in Figure 6. In order to overcome and/or mitigate the problems associated with boat operation and obtaining good quality near-surface seismic reflection data in shallow water the additional geophysical survey was designed to use bottom laid cables and very shallow draft vessels. Use of a bottom laid cable eliminated the operational problems associated with towing a surface cable in shallow water. Energy source is adopted as either air gun or boomer. Approximately 70 lines of either boomer or air gun data were collected using 15 bottom cable lays. An additional 5 lines were obtained using a towed cable from a small skiff for shallow water areas. Most of the survey lines for the 2011 survey are shown in red in Figure 6.

As a result of these surveys thrust and strike-slip fault models are interpreted all along the survey lines. The thrust fault interpretation lead to the following results.

a) In addition to the Fault 15, which was also known by the 2010 Surveys, four typical faults are found to the north of the Fault 15 as shown in Table.1 and Figure.7, as the result of additional geophysical surveys.
b) A zone between the Faults A and B can be evaluated as a less risk area than a zone between the Faults A and 15, where the south anchorage was located in the Tender design. Because, the zone between the Faults A and B shows the apparent continuity of horizon with a thickness of 40m and is elder than the zone between the Faults A and Fault 15.
c) Even no apparent faulting was seen between Fault A and Fault B this region is also considered as a region where relative movement can occur due to its closeness to other faults.

Strike-slip fault interpretation lead to same results for Faults B, AB and 15. Fault-A and Fault A South for their locations. However, rather than interpreted as listric they are interpreted as vertical and near vertical, which was the only difference in two interpretations. As a result owing to the

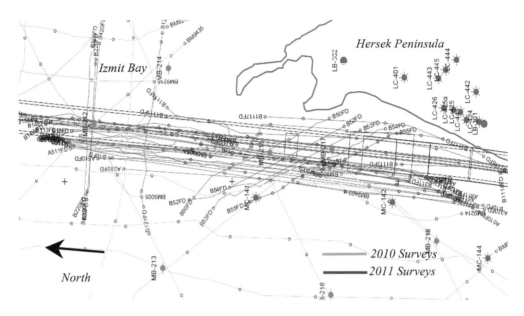

Figure 6.   South Anchorage Geophysical Investigation Site.

Table 1.   The faults found as result of additional geophysical investigations (Thrust Fault Interpretation).

Fault A-south	Some 70 m to the north of the Fault 15, located under the anchorage in the Tender design. This fault does not appear on the seabed but up to 40 m below the seabed. It is shown in red in Figure-7. This fault appears to join the listric (decreases in slope with depth) trace in the north.
Fault A	Some 160 m to the north of the Fault 15, located under or close to the side span pier in the Tender design. This trace is interpreted to be listric at depths of 75 to 100 meters and associated with the south dipping reflector that offsets north dipping reflectors in the nominal depth range of 115 to 150 meters. It is this apparent listric appearance that resulted in the initial interpretation of a thrust fault.
Fault AB	Some 220 m to the north of the Fault 15. This fault does not appear to reach the sea bed located at least 80 meters below the bottom. This fault does not appear to offset the T3 horizon which seems to be older than Holocene, yet can be considered as inactive.
Fault B	Some 320 meters north of Fault 15. It is shown with gold in Figure-7. This fault appears to be the northern limit of coherent seismic reflections in the survey area.

matching of results of the two interpretations it was considered that the revealed faults are valid to be used for engineering purposes.

3.2   *Design implications of secondary faults*

As a result of the findings of additional geophysical surveys it was considered that the tender design anchorage structure locations are coinciding with the high risk zone. Especially, the south side span pier and south anchorage residing on separate foundation imposed a great risk due to the 'Fault-A South' passing through these structures. The main cable between the deviation saddle of the side span pier and the splay saddle of the anchorage is shortest compared to other spans of main cable making this location its most rigid part. Therefore, any relative displacement between these saddles will impose very high forces which is difficult to be accommodated by detailed design. Therefore the foundations of side span pier and anchorage was unified on a monolithic foundation

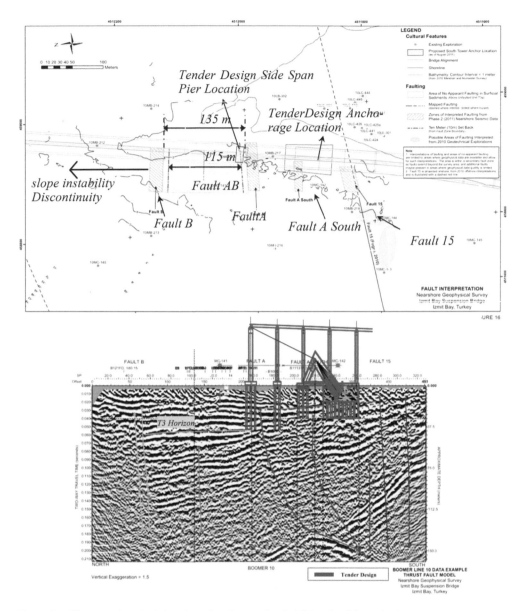

Figure 7.   The secondary fault locations found as result of additional soil investigation.

(See Figure 8). The other modification made on the tender design due to secondary fault findings are indicated below:

a) The south anchorage and the south side span pier are moved by 138.25 m and 118m toward the north so that the both are located in "No Fault zone" between the Faults A and B.
b) The concept in the Tender design is basically maintained as it is;
    – Main span length is 1550m.
    – Deviation saddle is placed between the tower saddle and the splay saddle.
    – The suspension bridge is symmetric within expansion joints of the suspended girder

Figure 8.   Modified South Anchorage Structure.

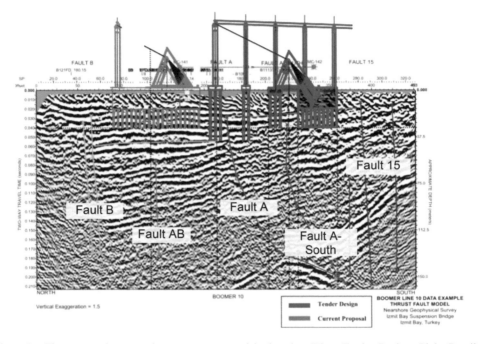

Figure 9.   The proposed new anchorage structure and its location (Blue: Tender Design; Pink: Detailed Design).

c) The bored piles under the anchorage and side span piers are cancelled since the required lateral resistance can be achieved by the friction between the foundation's bottom surface and soil, due to enlarged volume of foundation.

The shifting of south anchorage to North direction will make the anchorage structures closer to the steep slope under the sea (Refer to Figure 2) which is considered to be instable during seismic events due to the apparent traces of historic land slide on it. Therefore, the northern bound of the south anchorage structures are designated as Fault-B. The suitable location of anchorage structures is then selected as the area of no apparent fault zone between Fault-A and Fault-B with a total length of 115 meters by keeping sufficient margins from edge of the faults.

The location of the modified anchorage location and its impact to overall bridge design is shown in Figure 9 and 10. Here, the North and South Towers are shifted with an equal distance of 59 meters to the North Side to keep the center span unchanged as 1550 meters and keeping the bridge profile symmetric. The side spans are reduced to 566 meters. This modification also helped the south tower foundation to be located further away from the instable steep slope found at the south of it.

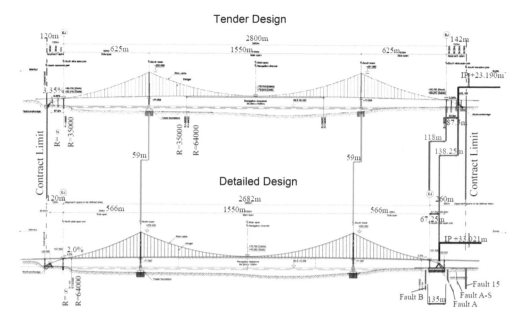

Figure 10.   The shifting of bridge layout.

4   CONCLUSION

The south anchorage structures are located at a no fault zone between Fault A and Fault B which have been identified after the Tender design by carrying out additional geophysical investigations. However, even the area between Fault A and Fault B is considered to be prone to occurrence of relative ground displacement during a seismic event due to the closeness to the nearby secondary faults.

In the detailed design, the design verification of south anchorage was performed (Avar et al) by the soil-structure analyses in which the relevant design parameters such as the magnitude of postulated offsets and the depth at which the offsets are applied are taken into account. It was verified by this analysis that the postulated fault displacements can be accommodated by local soil distortion and/or soil failures that find surface expression by circumventing the structure.

REFERENCES

Avar, B.B., Augustesen, A.H., Steenfelt, J.S., Kasper, T. & Foged, B. 2015. Behaviour of an Anchorage Block of Izmit Bay Suspension Bridge Subjected to Fault Movement. *SECED 2015 Conference: Earthquake Risk and Engineering towards a Resillient World, 9–10 July 2015, Cambridge UK.*
Kawakami, T., Yanagihara, M., Yamasaki, Y., Ozturk, A.N. & Zeybek, F. 2014. Izmit Bay Suspension Bridge – Overview of the Project. *IABSE Madrid Symposium Report,* Vol.102: Page 2081–2088.
Nearshore Geopyhsical Survey and South Anchorage Area Fault Study, Izmit Bay Suspension Bridge. 2011. Fugro.
Yamamoto,Y. & Koga, A. & Dizman, U. & Ikoma, N. 2014. Izmit Bay Suspension Bridge: Construction of Tower Foundation. *IABSE Madrid Symposium Report,* Vol.102: Page 2065–2072.
Yamane, M. Kudo, M., Asai, N., Mutaguchi, T., Kan'o, M. & Seki, S. 2014. Izmit Bay Suspension Bridge, Erection of Steel Tower. *IABSE Madrid Symposium Report,* Vol.102: Page 2057–2064.

Chapter 18

Poplar Street Interchange replacement – seismic design

L.E. Rolwes

HNTB Corporation, St. Louis, MO, USA

ABSTRACT: The existing Poplar Street Interchange at I-55/I-64 in downtown St. Louis, MO adjoins the approach viaduct to the Poplar Street Mississippi River Bridge. A major seismic retrofit initiative was completed on the approach viaduct within the last 15 years. The ramps were not part of the retrofit project and are currently being replaced. The eastbound approach viaduct will also be widened to accommodate an additional lane. A design response spectrum was developed through the use of site specific probabilistic rock accelerations in conjunction with at-depth analysis of ground motion. Using the refined spectrum it was demonstrated, without time-consuming modeling, that the widening and new ramps would not appreciably affect the existing viaduct structure and associated retrofit details. Design of the new ramps for the seismic hazard was also greatly simplified. The refined seismic spectrum ultimately translated into time savings during a compressed design schedule and reduced structure costs.

1   PROJECT OVERVIEW

The existing Poplar Street Interchange Ramps at I-55/I-64 in downtown St. Louis, MO adjoin the approach viaduct to the Poplar Street Mississippi River Bridge, which carries two interstates into Illinois: I-64, and I-55. The existing configuration of the interchange is shown in Figure 1. It consists of two ramps exiting the westbound direction of I-64 heading in the northbound and southbound directions. Similarly, there are two ramps adjoining the eastbound direction from the northbound and southbound directions. The existing ramps are comprised of cast-in-place multi-cell concrete box spans constructed in the 1960's. These structures have been deteriorating quickly in the last 10 years, requiring constant maintenance and emergency repairs to keep them open to traffic.

The regional transportation system was enhanced with the opening of a new river crossing approximately two miles north of the project location. The bridge was opened to traffic in 2014 and carries Interstate 70, which formerly was serviced by the Poplar Street Bridge. Removal of this traffic from the interchange allowed for modifications to its configuration.

1.1   *Interchange improvements*

Three of the four existing ramps will be demolished and replaced, while the fourth movement will be permanently removed. The new ramps will provide two 16-foot lanes and improved horizontal and vertical geometry in comparison with the existing structures, which each carry only one lane of traffic.

The proposed configuration of the new interchange is shown in Figure 2. The westbound ramps are to be replaced with a single ramp that bifurcates to service the northbound (Ramp 1) and southbound (Ramp 3) directions. The northbound to eastbound ramp (Ramp 2) will likewise be replaced approximately on the existing alignment. The southbound to eastbound structure will not be replaced. This movement was deemed unnecessary with the opening of the new I-70 crossing.

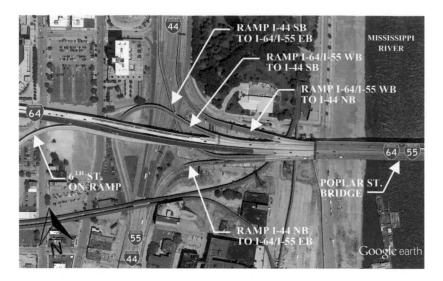

Figure 1.   Aerial view of existing Poplar Street Interchange.

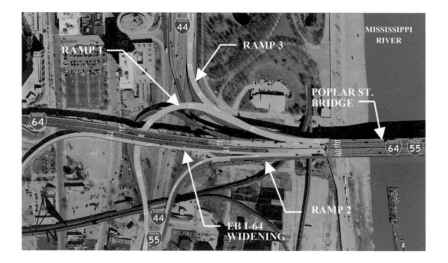

Figure 2.   Proposed configuration of interchange ramps and widening of eastbound viaduct.

1.2   *Existing viaduct*

The existing viaduct approaching the Poplar Street Bridge consists of parallel structures comprised of steel girder spans on concrete bents. These structures were also built in the 1960's. A major retrofit initiative was completed on this segment of the viaduct within the last 15 years to address numerous seismically-vulnerable details. Some of the common retrofit details are depicted in Figure 3 and include seat extenders, column wraps, post-tensioned cap beams, and lockup devices. The ramps were not part of the retrofit project as illustrated in Figure 4, which shows the tie-in condition between the viaduct and adjoining ramp. The last span of the ramp is supported on coped bearing seats at the ends of steel girders that are cantilevered over the adjacent bent and which form part of the viaduct framing. Seat length requirements are not met nor has any form of restraint been provided to prevent the ramp from becoming unseated.

Figure 3.   Typical retrofit details.

Figure 4.   End condition between existing ramp and viaduct.

As depicted in Figure 2, the eastbound approach viaduct will also be widened to accommodate an additional lane across the Poplar Street Bridge. The widening will start at the end of the 6th Street on-ramp and extend to the first pier of the Poplar Street Bridge. The Poplar Street Bridge is to be widened under a separate project.

1.3   *Project approach*

A primary objective that became evident early in the design development phase of the interchange was to limit impacts to the existing viaduct. Significant changes in load and/or stiffness could render the retrofit details obsolete. Modeling the entire complex would quantify the impacts of the proposed changes on the existing viaduct. If required, however, such a modeling effort would have been extremely time-consuming, considering the size and complexity of the viaduct, and the various retrofit details and modifications. This would have also added to the challenge of meeting a very aggressive design schedule covering about six months.

These challenges were partially tackled through the project seismic design strategy. Specifically, a design response spectrum was developed through the use of site specific probabilistic rock accelerations, in conjunction with at depth analysis of ground motions. The refined spectrum aided in addressing the concerns centering on the viaduct structure and simplified the design of the interchange in general.

2   INPUT GROUND MOTION

2.1   *General procedure*

The interchange improvements were designed in accordance with the AASHTO Guide Specifications for LRFD Seismic Bridge Design (2011) (Guide Spec). The general procedure for establishing the design response spectrum involves selecting three base spectral acceleration values for the site and applying response modification factors to account for the effects of the soil. These three acceleration values then provide the basis for constructing the design response spectrum.

Using the charts provided in the Guide Spec, base response spectrum parameters for the site (38.62 latitude; −90.19 longitude) were determined: PGA = 0.174 g, Ss = 0.352 g, and $S_1 = 0.099$ g.

The soil strata are fairly consistent across the site and generally consist of layers of loose silty-clay and silty-sand overlaying limestone bedrock. The overburden thickness varies considerably, however, from about 15 feet at the eastern limits of the project site to more than 60 feet at the western limits. The average N-values within the overburden are generally less than 20 blows/ft. Using AASHTO Guide Spec Table 3.4.2.1-1, the site may conservatively be classified as Site Class D. Applying the corresponding site factors to the basic spectral data results in the acceleration spectrum shown in Figure 5. The corresponding displacement spectrum is shown in Figure 6.

The retrofit details for the mainline structures were developed based on a design hazard in accordance with AASHTO Standard Specifications, Division IA (1996) (Standard Spec). A peak ground acceleration of 0.10 g was used in conjunction with Site Class II subsurface conditions

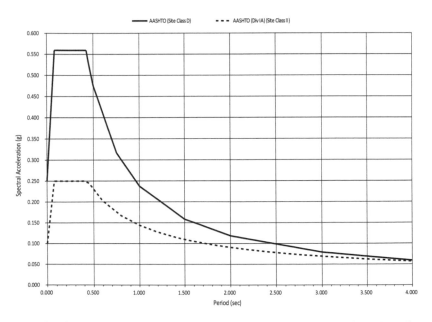

Figure 5.   Acceleration response spectrum based on AASHTO general procedure. Retrofit spectrum (AASHTO Div. IA) is shown for comparison.

(Banan et al 2001). The corresponding acceleration and displacement spectra are shown in Figures 5 and 6.

The Poplar Ramps site is located just two miles south of the I-70 crossing and within the same rock formation. Given the close proximity of the two sites, the changes in the attenuation of the source accelerations would be negligible. It was therefore deemed reasonable to apply the I-70 site specific hazard UHS to the Poplar Ramps project. Figure 7 shows the acceleration spectrum constructed using the site specific rock accelerations with the AASHTO site modification factors for Site Class D. The general acceleration spectra based on the Guide Spec and the Standard Spec are shown for comparison. The Guide Spec places a lower bound on spectral accelerations developed through a site specific approach, stipulating they shall not be less than two-thirds of those based on the general procedure. The limiting spectrum is also shown in Figure 7. As shown, the site specific accelerations are less than those per the general procedure but still greater than those per the Standard Spec. Further reductions in spectral acceleration may be possible through refinements in the site effects.

It is important to note that the design hazards targeted by the Guide Spec and Standard Spec are not identical in the level of safety they provide. The spectral accelerations per the Standard Spec have a probability of exceedance of 10% in 50 years, whereas those based on current AASHTO specifications have a probability of exceedance of 5% in 50 years. It is therefore not surprising that the spectral accelerations and displacements based on current design standards exceed those developed from the Standard Spec at all periods. The Guide Spec spectral accelerations are nearly double those based on the Standard Spec around a period of 0.5 second. The difference steadily diminishes to nearly zero at a period of 4 seconds. The displacement spectrum paints a similar picture. In general, the spectral displacements based on the Guide Spec are about 1 inch larger than those given by the Standard Spec over the expected range of periods within which the bridge would be anticipated to respond (0.7 to 2.5 seconds). Finally, the Seismic Design Category (SDC) for the new structures, which is established based on the spectral acceleration at 1 second, would be SDC B ($0.15\,g \leq S_{D1} = 0.238\,g < 0.30\,g$).

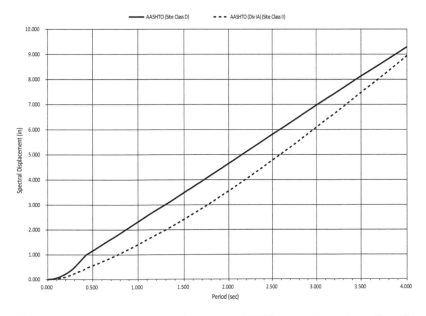

Figure 6.   Displacement response spectrum based on AASHTO general procedure. Retrofit spectrum (AASHTO Div. IA) is shown for comparison.

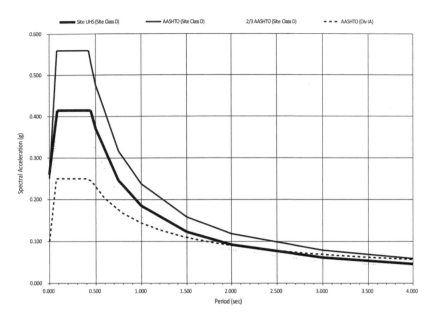

Figure 7.   Site specific acceleration spectrum for Site Class D compared with Guide Spec and Standard Spec Spectra.

2.2   *I-70 MRB site specific hazard acceleration spectrum*

In lieu of the general procedure described above, the Guide Spec allows for a refined site specific approach to be employed that encompasses a site specific hazard analysis (rock accelerations) and/or a site specific ground motion analysis (soil response). This refined approach was adopted for the recently completed I-70 Mississippi River Bridge (I-70 MRB) in St. Louis, MO. The probabilistic hazard analysis was prepared by the University of California at Berkeley under the direction of Norm Abrahamson. The hazard was found to be primarily dominated by two sources: a nearby source located in the Wabash Valley (about 12 miles east) which was responsible for most of the shorter period responses; a far field source at the New Madrid seismic zone (about 120 miles south) which was responsible for most of the longer period responses (Abrahamson 2009). A uniform acceleration response spectrum (UHS), having a 5% probability of exceedance in 50 years, was developed based on the input source motions.

2.3   *Evaluation of site effects*

Site soil classification was based on the weighted average of N over the top 100 feet of the soil profile. Taking the soil profile from existing ground surface, the site class would be assessed as a D as previously discussed.

The site classification and the corresponding site effects can be improved upon considering the depth of motion for deep foundations as discussed in Article 3.4.2.2 of the AASHTO Guide Spec. Figure 8 illustrates the depth of motion concept. For spread footings or pile footings in which the piles are flexible (e.g. H-piles), the depth of motion will tend to be centered at the level of the pile cap as shown in Figure 8(a). In this case, the upper elevation of the soil stratum should be taken at the ground surface when evaluating site effects. If the piles are relatively stiff (e.g. drilled shafts), or for columns supported directly on single drilled shafts as shown in Figure 8(b), the depth of motion will tend to be centered at the depth of fixity of the foundation, with the soil above this level having negligible contribution to the input accelerations to the structure.

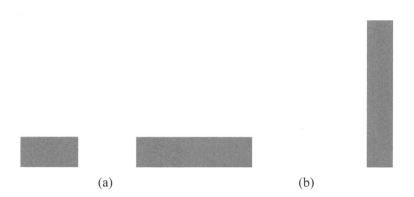

(a)   (b)

Figure 8.   Depth of motion concept: (a) flexible pile with stiff footing; (b) column supported on single shaft.

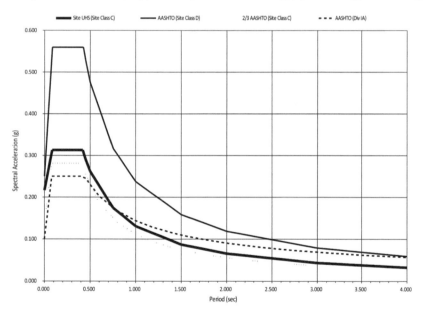

Figure 9.   Site specific acceleration spectrum for Site Class C compared with Guide Spec (Site Class D) and Standard Spec Spectra.

All new substructure units throughout the project will be supported on drilled shafts socketed into rock. The existing pile foundations of the viaduct were enlarged as part of the seismic retrofit. The enlarged footings are supported on drilled shafts. Taking this into consideration, and discounting the soil above the anticipated depth to fixity for new and existing foundations (15 to 20 feet) when evaluating N, the site soil class would be assessed as a C. The site factors would correspondingly decrease, further reducing the design spectral accelerations.

2.4   *Project design seismic hazard spectrum*

The site specific UHS for Site Class C is shown in Figure 9. The general procedure spectrum for Site Class D, and the retrofit spectrum are provided for comparison. The lower bound spectrum

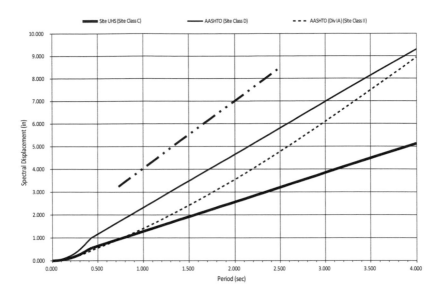

Figure 10.   Site specific displacement spectrum for Site Class C compared with Guide Spec (Site Class D) and Standard Spec spectra. The heavy dashed at the top shows the lower bound displacement capacity of the existing bents for the expected range of natural periods within which the existing structure would respond.

in accordance with the Guide Spec is also shown. Figure 10 provides displacement spectra for the same.

The site specific UHS for Site Class C demonstrates clear improvements over the general procedure as the magnitude of displacement is decreased at all periods. The site specific spectral displacements are less than the retrofit displacements at periods greater than approximately 0.70 seconds. Considering the makeup of the substructure units of the mainline structures, the predominant dynamic response will be between 0.7 to 2.5 seconds. The displacement demands on the retrofitted structure based on the site specific UHS would therefore be less than those for which the structure was designed. Furthermore, the spectral acceleration at a period of 1 second is 0.13 g, which falls below the upper bound for Seismic Design Category A of 0.15 g (see Figure 9).

3   DESIGN OF I-64 EASTBOUND WIDENING

3.1   *Widening scheme*

A typical section through the eastbound viaduct depicting the widening scheme is shown in Figure 11. A new steel girder will be added outboard of the existing exterior girder and supported on new single column bents. The bents will be in line with, but independent of, the existing viaduct bents, and will be supported on single drilled shafts socketed into rock.

3.2   *Expected dynamic behavior of viaduct*

The fundamental period of the existing structure would be expected to fall between about 0.7 to 1.5 seconds in the transverse direction and about 1.5 to 2.5 seconds in the longitudinal direction. The retrofit details at a minimum correspond to SPC C detailing requirements of the Guide Specification. A lower-bound estimate of displacement capacity may therefore be calculated using Equation 4.8.1-2 from the Guide Specification. For shorter period responses, the displacement capacity is at least 3.50 inches, while for longer period responses, the displacement capacity is at

Figure 11.   Widening scheme at eastbound viaduct.

least 8.25 inches. In either case, the displacement capacity is larger than the demand defined by the site specific UHS as shown in Figure 10.

The modifications required to accommodate the widening will impact the dynamic response of the structure. The additional superstructure components will contribute an approximate increase in total structure mass of about 33%. This in turn will constitute a period shift of about 15% such that the upper bound period of the structure will be slightly less than 3.0 seconds in the longitudinal direction. The latter corresponds to a spectral displacement demand of about 4 inches, which is less than that anticipated for the longer period response as discussed above. This excludes the additional stiffness that would be introduced by the new bents and which would tend cause a reduction in the fundamental period of the structure. Widening the viaduct will therefore not significantly increase the displacement demands on the existing retrofitted bents beyond their current capacity. This alleviates the need to model the existing structure to determine new displacement demands for the existing and proposed bents.

3.3   *Design approach*

The new elements of the widening were designed for compatibility with the existing structure with respect to ductility and displacement capacity. Target displacement demands were selected considering the displacement capacities of the existing retrofitted bents and the maximum anticipated displacements based on the site specific UHS. The new bents were then designed to have a displacement capacity greater than the target values. Detailing followed the requirements for Seismic Design Category B. Columns were detailed to hinge near the top of the drilled shafts, which in turn were designed to resist the corresponding over-strength loads elastically. The drilled shafts are continuously cased. The steel casing was only considered for confinement. The shear capacity of the columns and drilled shafts was checked against the over-strength shear demand. The cross frames over bents and bearing connections were similarly designed to transfer the over-strength shear force into the new bents.

4   DESIGN OF NEW INTERCHANGE RAMPS

Design of the new ramp structures was simplified through use of the site specific UHS. Based on the site specific UHS, the structures could have been classified as SDC A ($S_{D1} = 0.13\,\text{g} < 0.15\,\text{g}$).

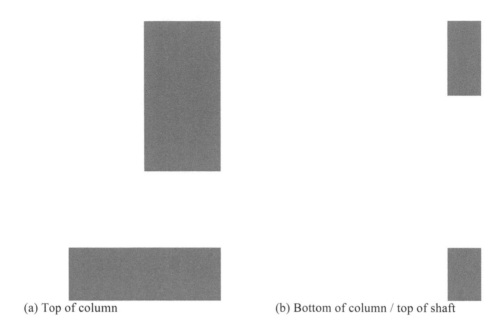

(a) Top of column   (b) Bottom of column / top of shaft

Figure 12.   Typical details at top and bottom of columns (hinge regions).

In order to provide seismic performance comparable with the existing viaduct, however, the ramps were detailed to meet the requirements of SDC B. A multi-modal analysis, however, was deemed unnecessary as the anticipated spectral displacement based on the site UHS would be less than the displacement capacity implied by the SDC details.

Specific details that were addressed included: hinge regions, seat lengths, and superstructure to substructure connections. Hinge regions at the top and bottom of columns were provided with confinement reinforcement as shown in Figure 12. Splices of longitudinal steel were only made outside these regions. The confinement reinforcement was extended into the cap to column connections. Steel casing was provided over the full length of the drilled shafts and was only considered as confinement to the concrete. Expansion bearings were designed to accommodate movements up to the maximum expected spectral displacement of 5 inches, while seat lengths were detailed to meet the requirements of SDC B. Cross frames over bents with steel girders and bearing connections were designed to transmit 25% of the tributary dead load for each restrained direction. At the steel spans, lateral seismic forces are transmitted from the superstructure to the substructure through combined resistance of bearing device anchor bolts and shear blocks. The shear blocks are concrete elements cast integral with the cap beams. A similar approach was taken at the expansion ends of the prestressed concrete units. Interior supports within the concrete units, however, were detailed with full depth continuity diaphragms. A series of steel reinforcing dowels that extend into the diaphragm are to be cast along the centerline of the cap beam between girders. Resistance to lateral seismic force is therefore provided through shear friction and dowel action.

The connection between the existing viaduct and new ramps was also addressed. The existing detail was provided in Figure 4, which demonstrated the limited seat length at the coped beam ends. Rather than attempting to modify the coped ends or provide longitudinal restraints, the first span of the new ramps will be comprised of simple span steel girders. The bearings at the existing coped seats will be fixed and provisions for expansion will be provided at the first new bent where adequate seat length will be provided. The fixed bearings were designed to transfer 25% of the tributary dead load in the restrained directions. The expansion bearings were detailed to accommodate the maximum spectral displacement of 5 inches. The expansion joint opening had to be set considering movement of the existing viaduct and new ramp.

5 CONCLUSIONS

Refinement of the design spectrum through the use of site specific rock accelerations and consideration of depth of motion when evaluating the site classification provided several benefits on the Poplar Ramps project.

1. It was demonstrated, without time-consuming modeling, that the widening and new ramps would not appreciably affect the existing viaduct structure and associated retrofit details.
2. Design of the new ramps for the seismic hazard was also greatly simplified.
3. The refined seismic spectrum ultimately translated into time savings during a compressed design schedule and reduced structure costs.

REFERENCES

Abrahamson, N. A. 2009. *Probabilistic Seismic Hazard Analysis for I-70 Mississippi River Bridge*; prepared for HNTB.

American Association of Highway and Transportation Officials (AASHTO) 1996. *Standard Specifications for Highway Bridges, Division IA Seismic Design*. AASHTO: Washington D.C.

American Association of Highway and Transportation Officials (AASHTO) 2011. *AASHTO Guide Specifications for LRFD Seismic Bridge Design*. AASHTO: Washington D.C.

Banan, M. R., and Capron, M. R. 2001. Seismic retrofit of the US-40/I-64 double-deck approach to Poplar Street Bridge in St. Louis, *Structures 2001: A Structural Engineering Odyssey;* Bridge Design and Rehabilitation, pp. 1–7.

Chapter 19

Seismic behavior of a long viaduct in Mexico DF: A combined FEM and SHM approach

J.M. Simon-Talero & A. Hernandez
Torroja Ingenieria S.L., Madrid, Spain

M. Ahijado
OHL Concesiones, Madrid, Spain

M. Santillan
Solvver Solutions S.L., Madrid, Spain

ABSTRACT: Some long viaducts have been constructed in Ciudad de Mexico in the recent past. As it is very well known Mexico City is quite vulnerable to earthquakes. Concerning the consequences of seismic events on this kind of structures is largely unknown. So, a fast and precise assessment would be extremely useful for continuing the operation of the infrastructure without compromising safety. This paper is proposing a methodology for tackling these problems, based on the combined use of Finite Element Modeling and Structural Health Monitoring practices. This methodology involves the implementation of precise FEM models combined with real-time, web-based remote monitoring systems providing dynamic data obtained from a sensor networks what can provide a deep and fast knowledge of the structures' and of the dynamic behavior and post-earthquake state. The results obtained from the monitoring implemented on 2014 and the comparison with the results of the FEM dynamic analysis performed are presented. A method for evaluating the real behavior of a long viaduct under the effects of a seismic load is presented.

1 GENERAL

Earthquakes are not unsual in Mexico City. The number and the intensity of seismic events are to be considered when designing urban infrastructures in the City. On the other hand, the city is under a large infrastructure program which includes the construction of very long viaducts that can cope with the increasing urban traffic. This is the case of the Bicentenario Viaduct and the Urbana Norte Viaduct (Figure 1), being the overall length of 23 km and 10 km, respectively, that are operated by OHL Concesiones under a DBFOM Contract.

The real behavior of these long urban structures in case of seismic events is largely unknown. This is because no monitoring of any of these type of structures have been extensively monitored. On the other hand, the consequences of the seismic effects when the operational issues are considered could be really important. Thus, having a fast and precise assessment of the structure condition after an earthquake would be a very useful tool for the operation, in order to make a quick decision about the safeness of the structure if continuing the operation after the earthquake is decided.

Additionally, this operational assessment can be complemented by the acquisition of more data and information about its dynamic behavior. This extra information would increase the knowledge about these structures and also be used to improve the design and maintenance techniques of similar viaducts (Hipley and Huang, 1989).

Figure 1.   Viaduct of the Autopista Urbana Norte (Ciudad de Mexico).

Following a methodology to address these problems is proposed. This methodology is based on the use of a SHM system, combined with the development and analysis of some Finite Elements Models (FEM) of the structure.

The main goals to be achieved by the implementation of the system are:

– Making a quick evaluation of the structures once the earthquake occurs, in order to decide whether to continue operating the infrastructure or not.
– Quickly reporting the damages caused by the earthquake.
– Generating a system of warnings, alarms and protocols in case a threshold (level of acceptance) has been achieved, and so some decisions and repairing actions could be quickly adopted and the operation could be reinitiated as soon as possible.
– Having a good knowledge of the real dynamic behavior of the structure in order to properly evaluate the condition of the structure in case a bigger earthquake occurs.
– Investigating the behavior of these kind of structures in order to get some more efficient designs in the future.

2   THE PROOF OF CONCEPT. OBJECTIVES

The total length of the Viaducto de Urbana Norte is nearly 10 km. Additionally, there are 4–5 km of elevated ramps. Because of its great length, the sensors system installed on the structure must be restricted, both in cost and quantity. Otherwise, it would be economically unfeasible. In doing so, the proposed monitoring will not cover the entire length of the viaduct and, so, some assumptions should be made in order to extent the conclusions of the monitored segments to the remaining segments of the structure that have not been monitored.

In addition it was also necessary to adequately choose the type of monitoring, in order to be sure that the selected method of monitoring and the features of the sensors were adequate. That is the reason a Proof of Concept, the so called "Pilot Project", was proposed to be made before the final implementation of the system was made.

The main objectives of the Pilot Project were:

– Adjusting the initial proposed monitoring to the particular situation of the structure.
– Selecting the technology and type of the sensors and of the data transmission system to be proposed for the final implementation.

– Discovering the potential problems and unefficiencies related to the installation of the sensors.
– Getting a precise and reliable estimation of the cost of the final implementation of the system.
– Adjusting the proposed algorithm for obtaining the movements of some parts of the structure from the measured accelerations.
– Analyzing the dynamic properties of a segment of the structure that was considered as representative one of the long portion of the structure.
– Calibrating a theoretical structural model using the results of the monitoring.

3   SYNOPSYS OF THE PILOT PROJECT

The Pilot Project involved the following tasks:

– The selection and testing of some sensors to be installed on the selected segment of the viaduct.
– The installation of the selected sensors on the structure.
– The creation of a software for adquiring, transmitting, and presenting the data of the sensors.
– The creation of a structural model of the said segment. The non linear response and the dynamic behavior of the structure should be analyzed using the proposed modelization.
– The calibration of the structural model using the result of the monitoring.

3.1   *Selection, testing and installation of the sensors*

The study of seismic events – eminently as a dynamic phenomena – makes necessary the installation of accelerometers in the structure, which provide such dynamic information (Arici and Mosalam, 2000). For these reasons, and for the particularities of the typology of the elevated viaduct – with high hyperstatic degree and with many singularities in its layout- the damage detection technologies based on the change of their dynamic charactcristics (Vibration-Based Damage Identification Methods, VBDIM) were rejected (Doebling, Farrar and Prime), (Wang and Chan).
   The following sensors were installed, Figure 2:

– 14 high precision accelerometers
– 2 clynometers
– 1 seismographe
– 7 displacement switches
– 2 LVDT sensors
– 1 temperature sensor
– 1 wind sensor

Once some sensors were selected they were tested. The Laboratory of the CENIM at the Polytechnical Industrial Engineering University at Madrid was used for making two kind of tests:

– Studying the sensitivity of the accelerometers at low frequencies (below 10 Hz).
– Studying the noise floor level at the frequency bands of interest.

The second objective is important because the sensor data obtained during the pilot project will be used to estimate the modal parameters of the bridge. For this purpose, high sensitivity sensors are desirable whereas for the final application of measuring the vibrations during earthquakes, the sensitivity must be limited in order to prevent saturation of the DAQ system from occurring. To find an economically priced sensor that satisfies both objectives is not an easy task.
   Manufacturers of accelerometers usually provide calibration charts, which cover frequency ranges starting at 10 or 20 Hz (Figure 3). For the present application however, the frequencies of interest are thought to be much lower, around 1 Hz. Taking into account that the acceleration data will be used to estimate displacements, a stable frequency response at low frequencies is of primary importance.

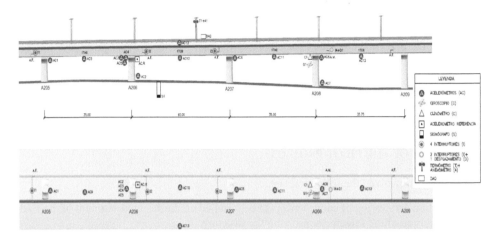

Figure 2.   Sensors installed on the pilot project segment.

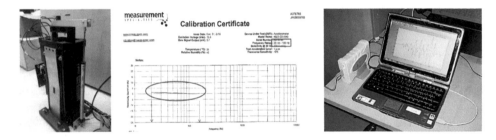

Figure 3.   Lab tests at the Polytechnical Industrial Engineering University at Madrid.

Figure 4.   Installation of sensors.

In addition to the constant frequency excitation the shaker used on the lab tests was programmed to reproduce a preset group of real earthquakes reaching peak displacements of around 6 cm. In order to calibrate the parameters of a double integration technique the measured acceleration data has been integrated twice and compared to the displacements registered by the potentiometric sensor. Figure 4 shows installed sensors.

3.2   System architecture of the SHM solutions

The stringent requirements of the project posed significant challenges to the design and implementation of the solution that come from two main requirements:

– Engineering specs imply data volumes in the order of tens of thousands of data-points per second, which need to be processed and stored for long periods of time.

- Operational requirements prescribe an end-to-end, web-based, real-time SHM systems architecture: data must flow continuously from the sensor readings, through the signal filters and engineering algorithms, up to the UI (User Interface) in order to provide the timely response needed after an earthquake.

In order to reduce the time-to-market, and improve the systems reliability, the SHM solution has been built on top of solvview, a registered web-based, soft-real time, cloud monitoring platform for engineering environments, which has been used as the core component of the architecture.
The data flow in the system is as follows:

- Data adquisition and processing stage.
- The system integrates a wide variety of sensing technology, which includes accelerometers, gyros, clinometers, displacement sensors, seismometers, as well as contextual information (temperature, wind, …). Due to the data quality requirements of the engineering algorithms implemented, sensor data streams undergo a series of signal processing routines, including analog and digital filtering, as well as real-time noise shaping techniques specifically designed to improve the Signal-to-Noise Ratio of the low-frecuency components most relevant for seismic analysis.In this stage, seismic detection algorithms are run. These algorithms include classical level triggering detection as well as energy window detection based on Short Term Average/Long Term Average approaches.
- Processed data is streamed to the SHM central system in soft real-time. Since the central system is deployed as a cloud service, standard Internet protocols are used for the live streaming of data. Moreover, backup mechanisms are in place to cope with possible communications interruptions during strong-motion seismic events.
- On the central system, data streams are stored for further analysis, alarms and events are detected and displacement calculations algorithms are run. This system will also be responsible for the orchestration of the operational workflows associated to earthquakes.
- Data flows are streamed to the end users' web interface live, ensuring the timely availability of data. Both technical operators and structural engineers receive these live streams of sensor data as well as alarms and indicators. The web nature of the UI makes the system easily accesible from any location, and only requires a standard Internet connection and valid credentials.
- External information can be integrated in the system in the form of seismic reports of the earthquakes, human comments and any other document that is relevant for the system.
- The information provided to the engineering teams is used to further calibrate the structural models, which in turn serve as a basis for the fine-tuning of the alarm thresholds and the detailed definition of the operational post-earthquake workflows.

Since the system manages huge volumes of information, providing an easy-to-use, simple-but-powerful UI has been a key design concern. Based on UX (User Experience) best practices, the interface (Figure 5) has been designed following a drill-down approach: a global dashboard provides an holistic, real-time view of the system, that can serve as a starting point for analysis, while specific modules provide detailed views on the different data streams, history, alarms and events. Direct download of data files enables researches to further analyze and process data with external, specialized tools.

3.3 *Structural models*

A linear elastic model (Figure 6) was used for determining the theoretical dynamic properties of the structure. This model incorporate all the geometric and mechanical properties of the structure, but the reinforcing and the prestressing of the piers was not considered.

The foundation of the piers is made of reinforced concrete cast in place piles that are 900 mm in diameter and some 30–40 m long. The interaction of the piles with the soil of the foundation was made using non linear springs, simulating the deformational properties of the soil.

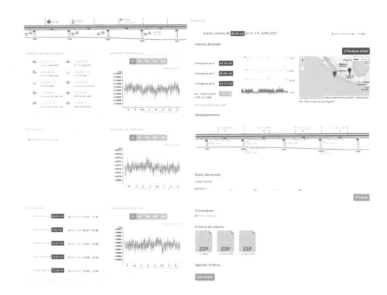

Figure 5.   SHM user interface. Global dashboard and seismic event summary.

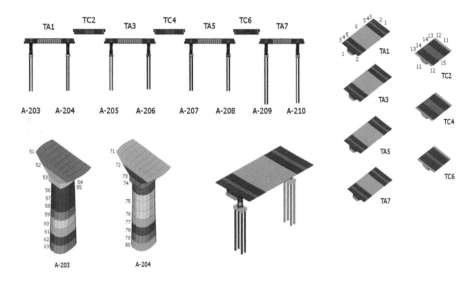

Figure 6.   FEM linear elastic model.

Another more sophisticated model was prepared for making dynamic calculations (Clough and Penzien, 2003). The geometry of the structure was defined in a more detailed way and, again, springs were used for simulating the connection of the piles to the terrain. Ansys software and Sophistik software was used to introduce the time dependent loads due to an earthquake to the structure. Some results of the dynamic calculations are presented later on this paper.

Some other more refined models were used for knowing the response of the structure to heavy loads.

Firstly, the resistance of the transverse section when loaded by a bending moment and by an axial force was calculated for several cross sections of the piers and of the piles. The exact position of every reinforcing bar and the effect of the prestressing forces were also considered. The cracking of the concrete and the non linearities of the concrete, of the reinforcing steel and of the prestressing

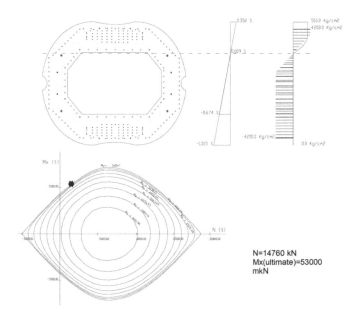

Figure 7.   N-M interaction diagram of a cross section of the pier.

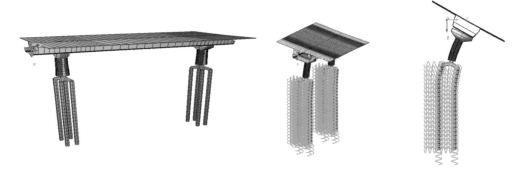

Figure 8.   Pushover modelization.

steel were also taken into account. In doing so, the interaction diagram N-M was calculated for every relevant cross section (Figure 7).

Then, a second order analysis of some piers was made (Figure 8). Again, the cracking of the concrete and the non linearities of the concrete, reinforcing steel and prestressing steel were taken into account. The displacement of the top of the pier depending on the horizontal forces was calculated. The load was increased and the displacement calculated till the ultimate load was reached (Figure 9). Three ultimate scenarios were studied: failure of the pier due to combined N-M effects, failure of the piles due to combined N-M effects and failure of the foundations due to a heavy tension force on the pile.

4   RESULTS OF PILOT PROJECT

The first result of the Pilot Project is related to the efficiency of the proposed and installed monitoring system. The accuracy and adequacy of the monitoring has been demonstrated by means of

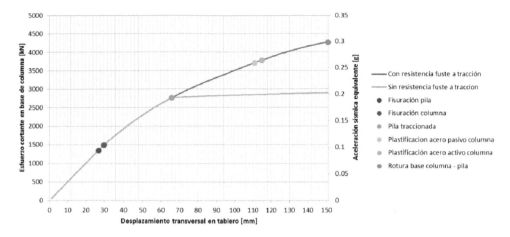

Figure 9.   Results of the pushover calculation of a relevant pier.

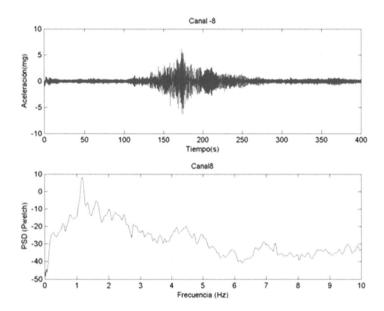

Figure 10.   July/29/2014 earthquake. Horizontal acceleration and frequency dependant response.

obtaining the dynamic response properties of the structure. Using the dynamic excitation created by the normal wind forces or by the effect of the traffic on the deck, the natural frequencies and the natural shape modes of the structure have been obtained. The transverse canonical movement has been found to be dominant being the measured natural frequency equal to 1.18 Hz.

The calculated natural frequency was 1.28 Hz: this means that the theoretical model is OK and it is accurately representing the existing structure. This is the second important result of the Pilot Project.

A strong earthquake measuring M6.3 on the Richter scale was registered in Veracruz, Mexico, on July 29th, 2014. The epicenter was located 418 km (260 miles) ES of Ciudad deMéxico. The installed sensors of the Viaducto de Urbana Norte at Ciudad de México were able to detect the dynamic load on the bridge. As the dynamic load was much greater than the load

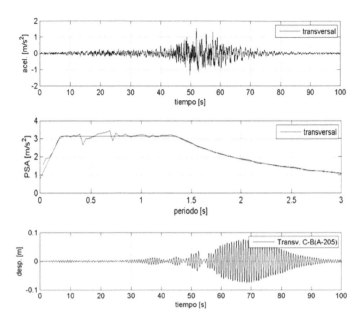

Figure 11.   Design seismic load. Horizontal acceleration, acceleration spectra and calculated displacement of the top of the pier.

produced by the said normal wind and traffic, this seismic load was used to calculate again the dynamic properties of the structure. The conclusions were exactly the same as the ones obtained before. That is, the relevant mode shape is the first canonical transverse modal form and the first natural frequency is the said 1.18 Hz. The frequency dependant diagrams that are presented following on Figure 10 are showing the said results.

Once the theoretical model was calibrated, it then can be used for obtaining the dynamic response of the structure to the effect of any seismic load (Lus et al, 1999), (Loh and Lee, 1997). For example, the horizontal acceleration produced on the said seismic was measured using the seismograph located close to one pier of the viaduct. The measured acceleration-time relationship was introduce to the model as an external load and the deflection and acceleration on the top of the piers was computed. Then, the results of the calculations were compared to the accelerations and deflections measured by the sensors installed on the top of the pier. The correlationship between the calculated and the measured deflections and accelerations was really good: the amplification factor on the transverse direction was some 5.7; that is, the acceleration of the top of the pier was 5.7 times the acceleration of the terrain on the bottom of the pier.

Some synthetic accelerograms were also produced for simulating the expected design seismic load. Figure 11 presents the a-t relationship of one of this dynamic load, the acceleration spectra of the said dynamic load and the resulting calculated deflection on the structural model. It is to be noted that the calculated maximum transverse displacement results some 9 cm. If we compare this movement with the movements resulting on the pushover calculation (Figure 9), we'll see that the concrete of the piers and of the piles will crack and that the piles will be tensioned but the concrete and the reinforcing steel is still on the elastic part of the sigma-epsilon diagram, with no yielding expected for the said loads. That is, the bridge is safe enough for the design seismic load.

5   CONCLUSIONS

Even though Ciudad de Mexico is not on or near any fault, it is quite vulnerable to earthquakes, because of the surface geology of the area, specially the downtown area. However, due to its huge population, as well as its economic relevance for the country, the city is developing a large infrastructure program that includes the construction of very long viaducts than can cope with the increasing urban traffic.

Concerning the consequences of seismic events on these kind of structures, the resulting state of these structures is largely unknown. In this context, a fast and precise assessment would be extremely useful for continuing the operation of the infrastructure without compromising safety. This paper is proposing a methodology for tackling these problems, based on the combined use of Finite Element Modeling and Structural Health Monitoring practices. This methodology involves the implementation of precise FEM models combined with real-time, web-based remote monitoring systems providing dynamic and static data generated by sensor networks composed of seismometers, accelerometers and other measuring instruments to jointly provide a deep and fast knowledge of the structures' dynamic behavior and post-earthquake state.

Also the data obtained from the monitoring implemented on 2014 and the comparison with the results of the FEM dynamic analysis performed are presented.

REFERENCES

Arici, Y. and Mosalam, K.M. System identification and modelling of bridge systems for assessing current design procedure. *SMIP2000 Seminar Proceedings*, pp. 77–95, 2000.

Clough, R. and Penzien, J. Dynamics of structures, *Computers & Structures*, Inc., Berkeley, 2003.

Doebling, S., Farrar, C. and Prime M. A Summary Review of Vibration-Based Damage Identification Methods. Los Alamos National Laboratory.

Hipley, P. and Huang, M. Bridge instrumentation and post-earthquake evaluation of bridges, *SMIP89 Seminar Proceedings*, pp. 55–71, 1989.

Loh, C.H. and Lee, Z.K. Seismic monitoring of a bridge: Assessing dynamic characteristics from both weak and strong ground motions. *Earthquake Engineering and Structural Dynamics*. pp. 269–288, 1997.

Lus, H., Betti, R. and Longman, R.W. Identification of linear structural systems using earthquake induced vibration data. *Earthquake Engineering and Structural Dynamics*. pp. 1449–1467, 1999.

Wang, L. and Chan, T.H.T. Review of Vibration-Based Damage Detection and Condition Assessment of Bridge Structures using Structural Health Monitoring.

Bridge rehabilitation & strengthening

Chapter 20

Titanium alloy bars for strengthening a reinforced concrete bridge

C. Higgins
Oregon State University, Corvallis, OR, USA

D. Amneus
Group Makenzie, Portland, OR, USA

L. Barker
David Evans and Associates, Salem, OR, USA

ABSTRACT: A visually distressed vintage conventionally reinforced concrete deck girder (RCDG) bridge was identified by routine inspection. Subsequent investigation showed the distress was due to a poorly detailed splice location for the flexural steel and the ratings determined the girders to be significantly understrength. The bridge was shored to allow it to remain in service until it could be strengthened. The bridge was strengthened using near-surface mounted (NSM) titanium alloy bars. Round titanium alloy bars with a unique deformation pattern were specially developed for this application. Experimental research was conducted to evaluate the behavior of the as-built poorly detailed girder and then to evaluate the performance of the strengthening approach. Realistic full-scale girder specimens were constructed, instrumented, and tested to failure. The specimens mimicked the in situ materials, loading interactions, and geometry. The as-built strength was verified to be very low and the distress observed in the tests priot to failure and were similar to those observed in the field. Two specimens were strengthened with NSM titanium alloy bars and exhibited much higher strength and deformation capacity. The observed strengths of the specimens with NSM titanium enable the bridge to carry legal and permit loads without restriction. The member strength was shown to be well predicted using the analysis program Response 2000 and was conservatively predicted using AASHTO-LRFD design methods. The approach and materials were applied to the actual bridge, and the bridge was restored to service without the need for shoring or posting. The first ever application of titanium alloy bars to a reinforced concrete bridge was completed at a 30% cost savings compared to alternatives.

1 INTRODUCTION AND BACKGROUND

Conventional reinforced concrete deck girder (RCDG) bridges were widely constructed in the middle of the last century. At the advent of standardized deformed reinforcing steel bars in the 1950's, designers used straight-bar terminations of the flexural reinforcing bars where they were no longer required by calculation and commonly terminated flexural steel in tension zones without special detailing provisions. Truck loads have increased from those in the past and these can result in cracking and distress of these poorly detailed bridge elements. One such RCDG bridge, called the Mosier Bridge, was recently identified.

Mosier Bridge is an overcrossing of the I-84 in the state of Oregon and is shown in Figure 1. The four span bridge serves a nearby quarry and is subjected to heavily loaded trucks with short axle lengths. The original three spans were built in 1953 and the bridge was lengthened in 1959 which added a fourth span (Oregon Department of Transportation (ODOT), 2013). Typical of practice in the 1950's, the bridge girders taper and are haunched which produces a smaller cross section at midspan compared to that at the interior support locations. During a biennial bridge inspection

Figure 1.   Mosier Bridge with highlighted Span 1 containing cracks at splice locations (Google Maps, 2014).

Figure 2.   View of crack with vertical offset on interior girder of Span 1 (ODOT 2013).

in 2013, wide vertical cracks (0.03 in. (0.762 mm)) were observed around cutoff locations of the longitudinal reinforcing steel in an interior girder as shown in Figure 2. The observed cracks had a vertical offset and identified the need for further investigation.

Analysis of the Mosier Bridge was conducted using a line analysis of the continuous span bridge and sweeping the ODOT prescribed rating truck models over the span. The moment distribution factor was 0.872 and the impact factor was 1.2. The maximum factored moment was computed as 284.9 kip-ft (386.3 kN-m) and was produced by the continuous trip permit truck (CTP3) having a live load factor of 1.3. The live load places the deck into flexural compression over the critical section. The service-level dead load moment (using a load factor of 1.0) at the section was computed as −68.9 kip-ft (93.4 kN-m) which puts the deck into flexural tension. Combining the factored live load moment and service-level dead load moment, the demand at the critical section was 219 kip-ft (297 kN-m) which is 46 kip-ft (63 kN-m) above the AASHTO design moment capacity (AASHTO-LRFD 2013). The corresponding shear resulting from factored live load and service-level dead load was 60.5 kips (269 kN), which results in a M/V ratio of 3.6 ft (1.1 m) at the critical section.

The very low bridge rating and observed visual distress required immediate shoring to maintain traffic until a strengthening approach could be implemented. Near-surface mounted (NSM) titanium, stainless steel, and CFRP were considered. However, due to geometric constraints, titanium alloy bars were selected as the most cost effective system.

2   EXPERIMENTAL PROGRAM

An experimental program was developed to evaluate the effectiveness of the strengthening approach. Full-scale specimens were constructed to simulate the as-built conditions of the Mosier Bridge girders that were under-strength. The specimens were constructed, strengthened, and tested to failure. The experimental program reports the specimen design, details, material properties, and test protocol of the Mosier specimens.

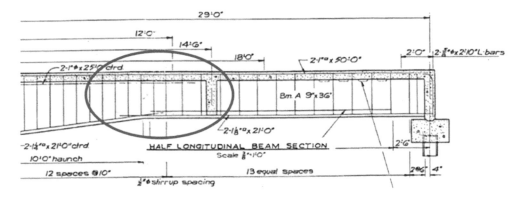

Figure 3. Original design drawing details for haunch transition of Span 1 (ODOT 2013).

Table 1. Concrete properties for Mosier specimens.

Specimen	f'_c (psi) [MPa]	Standard Deviation (psi) [MPa]	f_{ct} (psi) [MPa]	Standard Deviation (psi) [MPa]	Concrete Age (days)
Mosier 1	3038 [21.0]	76.9 [0.53]	348 [2.40]	25.2 [0.17]	33
Mosier 2	3629 [25.0]	244 [1.68]	275 [1.90]	216 [1.49]	63
Mosier 3	3344 [23.1]	426 [2.94]	353 [2.43]	16.1 [0.11]	58

2.1 Specimen design

The specimen design was based the details taken from the Mosier bridge design drawings. The girder used in this program was the interior girder of Span 1 shown in Figure 3. All specimens were 18 ft (5.49 m) long and had a 6.5 × 36 in. (165 × 914 mm) deck. The stem was 9 × 29.5 in. (229 × 749 mm) in half of the specimen then transitioned to a cross section of 12.63 × 41.25 in. (321 × 1048 mm) through horizontal and vertical tapering after midspan. The clear span of the Mosier specimens was selected to produce similar shear and moment demands (M/V ratio corresponded reasonably to that of the controlling rating truck CTP3) at the critical section. To further simulate conditions of the existing bridge similar concrete strengths and reinforcing steel strengths were selected to have similar strengths and development lengths.

Three specimens were constructed: Mosier 1, the as-built condition, Mosier 2, the NSM titanium applied after first failing the reinforcing steel anchorages, and Mosier 3, the NSM titanium applied with the reinforcing steel anchorages fully intact. The concrete used in the specimens was provided by a local ready mix supplier and designed to provide relatively low strength to reflect the low prescribed concrete strength used in the original design. The concrete material properties are shown in Table 1.

2.2 Reinforcing steel details

Reinforcing steel configurations were determined from the 1953 as-built drawings. The drawings show the reinforcing steel was square cross section bars of intermediate grade steel (40 ksi). To provide similar strength and development lengths as the in situ conditions, smaller diameter bars of modern ASTM A706-09 Grade 60 (Grade 420) (ASTM, 2009a) steel were selected. Mill

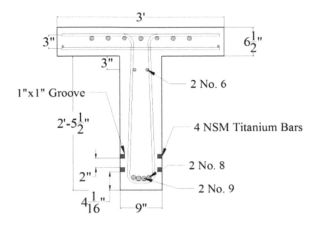

Figure 4.   View of cross section of specimen (shown with NSM titanium alloy bars) at midspan.

certifications from local rebar fabricators were reviewed to select bars from heats that produced the tensile force closest to those in the respective in situ bars.

All specimens contained similar longitudinal reinforcing steel layout and materials. Longitudinal reinforcing steel in the web consisted of two #7 (22M) and two #8 (25M) bars on the one half and two #9 (29M) bars through the other half. Two #6 (19M) steel reinforcing bars were placed in the upper stem on the right side. All longitudinal reinforcing steel bars were adequately anchored past the support locations. To resist negative moment, the Mosier specimens had two #7 (22M) and five #8 (25M) steel reinforcing bars in the flange. Double-legged, open #4 (13M) stirrups were spaced 12 in. (305 mm) on center throughout midspan in Mosier 1 and Mosier 2. Mosier 3 decreased the stirrup spacing before the NSM titanium strengthening initiated. The cross section at midspan is shown in Figure 4, the elevation views and plan dimensions for the specimens are shown in Figure 5. The reinforcing steel material properties are shown in Table 2.

2.3   NSM titanium details and installation

The design intent of NSM material strengthening procedure was to provide adequate flexural capacity in the Mosier Bridge girders so the bridge could remain open without load restriction and without shoring. To accomplish this, NSM titanium alloy bars were used. The titanium used for strengthening the specimens deserves special attention, as this is not a material commonly used in civil engineering practice. Titanium is nonmagnetic, has a low coefficient of thermal expansion of around 4.78 μin/in °F (8.6 μm/m °C) at operating temperatures, and is impervious to commonly encountered sources of environmental deterioration which result in excellent long-term durability. The material unit weight is 276 lb/ft^3 (4419 kg/m^3), about half that of steel. The nominal modulus of elasticity is 15,500 ksi (106,800 MPa), also about half that of steel. In this study, 0.625 in. (16 mm) diameter, round, titanium alloy bars were used in a NSM installation by placing them into concrete grooves cut into the concrete cover. NSM groove depth and spacing requirements were determined by ACI 440.2R-08 (ACI, 2008). The groove depth and width was 15/16 in. (23.8 mm) using a #5 (16M) NSM bar. The clear distance between grooves was approximately 3 in. (76.2 mm). The elevation of the NSM titanium was determined by optimizing the effectiveness of the titanium (providing a large lever arm) and being spaced over the #7 (22M) reinforcing steel bar.

Epoxy was used to bond the titanium alloy bars into the concrete grooves. The epoxy used in this test program was a commercially available general purpose gel epoxy adhesive widely used for bonding to concrete. The manufacturer reported epoxy material properties include tensile strength of 4 ksi (27.6 MPa), elongation at break of 1.0%, compressive yield strength of 12.5 ksi (86.2 MPa), and 2-day cure bond strength >2 ksi (13.8 MPa).

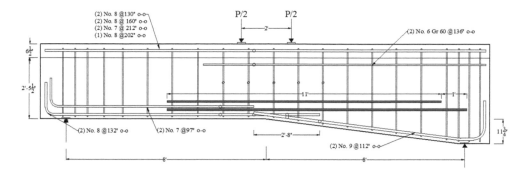

Figure 5a. Elevation view specimen Mosier 2 (Mosier 1 is same but without NSM titanium).

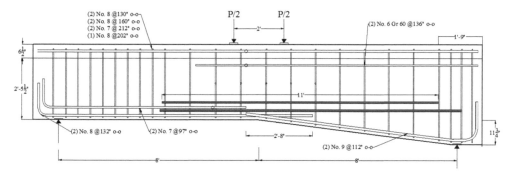

Figure 5b. Elevation view specimen Mosier 3.

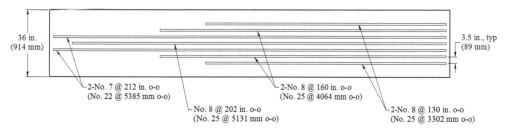

Figure 5c. Plan view of Mosier specimens showing flexural reinforcing steel in deck (all similar).

Table 2. Reinforcing steel bar properties for all Mosier specimens.

Material	Bar Diameter (in.) [mm]	Bar Area (in.^{2}) [mm^2]	Grade (ksi) [MPa]	Yield stress f_y (ksi) [MPa]	Ultimate stress f_u (ksi) [MPa]
Transverse #4 [13M]	0.500 [12.7]	0.20 [12.7]	40 [280]	50.2 [346]	79.6 [549]
Longitudinal #6 [19M]	0.750 [19.1]	0.44 [248]	60 [420]	63.0 [434]	106.3 [733]
Longitudinal #7 [22M]	0.875 [22.2]	0.60 [387]	60 [420]	65.3 [450]	104.6 [721]
Longitudinal #8 [25M]	1.000 [25.4]	0.79 [509]	60 [420]	63.6 [438]	112.1 [773]
Longitudinal #9 [29M]	1.128 [28.7]	1.00 [645]	60 [420]	62.6 [432]	102.0 [703]

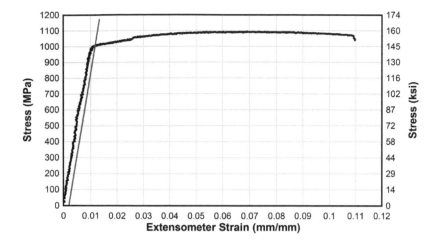

Figure 6.   Uniaxial tension stress-strain response of titanium alloy bar with special surface treatment.

The titanium bars were fabricated with a unique surface treatment to enhance bond between the bar and epoxy. The titanium material is an alloy with 6% aluminum and 4% vanadium, (so called Ti-6Al-4V) that meets ASTM B348 (ASTM, 2013), although these specifications do not adequately describe the tight tolerance material properties for the material selected in this research project. The annealed titanium alloy used in this study is produced to very tight tolerances on dimensions and material properties and is considered aircraft fastener grade. The as-received titanium bars (with the surface treatment) were tested in uniaxial tension using a 2 in. (50 mm) gage length extensometer according to ASTM E8 (ASTM, 2009b). A typical stress-strain plot from one of the samples is shown in Figure 6. As seen here, the material does not exhibit a well-defined yield plateau and the yield stress is determined from the 0.05% strain offset. The average material properties from the Ti-6Al-4V tension samples were: yield stress of 145.2 ksi (1000 MPa) with 1.01% coefficient of variation (COV); ultimate stress of 158.1 ksi (10,090 MPa) with 0.88% COV; elongation of 11.3% with 2.66% COV; yield/tensile of 0.92 with COV of 0.17%; and modulus of elasticity of 15,130 ksi (105,000 MPa) with COV of 0.55%. The cross sectional area of the bars was determined by weighing long bars ($\sim$14 ft (4.3 m)) of known length and dividing by the unit weight. The average area was 0.2975 in^2 (192 mm^2) with COV of 0.18%. As seen in these statistics, the variability of the titanium alloy bars was very low compared to most conventional civil engineering materials.

Using a nominal yield stress of 145 ksi (1000 MPa), four NSM titanium bars were required to provide the required member strength. A 6 in. (152.4 mm) hooked development length was assumed for the titanium NSM materials. The extent of the NSM reinforcing was determined by the diaphragm present at the left end of the interior girder, approximately 5 ft (1.5 m) from the south end of the Mosier specimen. Therefore, the NSM titanium was terminated 2 in. (51 mm) before the diaphragm to allow for drilling clearances. The NSM titanium on the right side of the Mosier 2 and Mosier 3 specimens was terminated after exiting the flexural tension zone. The upper layer of NSM titanium bars was terminated prior to the lower layer to minimize stress concentrations. In summary, the strengthening required two NSM titanium bars at a length of 11 ft (3.35 m) out-to-out, and two NSM titanium bars at a length of 12 ft (3.65 m) out-to-out.

After initial curing, the Mosier 2 and Mosier 3 specimens were placed in a stable configuration for saw-cutting the NSM grooves. Three longitudinal passes were made with a concrete saw. A roto-hammer was used to chip the grooves to their intended depth, and then holes for the hook were drilled using a 3/4 in. (19 mm) diameter hammer drill bit. The corner between the NSM groove and circular hole was chiseled to allow for the bend radius of the hook. The NSM titanium was anchored using a 2 in. (51 mm) diameter 90° hook with a 6 in. (152 mm) tail. Due to the width of the stem, the 6 in. (152 mm) tail could not be used if the NSM titanium bars were terminated at the same locations on each side of the stem. Mosier 2 used a smaller length of hook in the thinner stem on the left side and the 6 in. (152 mm) hook length in the stem on the right side. Mosier 3 used 6 in. (152 mm) long hooks on all ends of the NSM titanium. To allow for the increased hook length in Mosier 3, the NSM titanium bars were offset 1 in. (25.4 mm) on each side of the stem so the hooks would not intersect. The NSM titanium hooks were fabricated in a bar bending machine around a 2 in. (51 mm) diameter pin. Prior to bending, the NSM titanium was heated to a maximum of 900°F with an acetylene torch for all Mosier 2 hooks. The NSM titanium hooks of Mosier 3 were heated to approximately 1250°F.

Installation of the NSM titanium for the Mosier specimens was unique. The dead load of the Mosier Bridge produced negative moment in the section, therefore, the Mosier specimens must experience negative moment prior to and during the NSM titanium installation. After the specimen was at the specified load to produce negative moment, the applied load was held constant. The concrete gooves were cleaned and the epoxy was placed into the groove before the bar was installed. The bars were held flush within the groove at the haunch transition by a clamp at midspan. The groove was filled with epoxy and finished. Epoxy was cured for a minimum of seven days before unloading the dead load (negative moment) and applying the actuator load to simulate truck induced loading (positive moment).

To fulfill the designer assumption for the Mosier 2 specimen, the anchorage of the reinforcing steel at midspan was failed prior to installing the NSM titanium. An applied load brought specimen Mosier 2 to failure of the anchorage at the critical section. The cracked and debonded concrete was removed from the anchorage region of the intersecting reinforcing steel bars at midspan. Then the cracks around the anchorage were epoxy injected. Once the epoxy injection was cured, specimen Mosier 2 followed the similar testing protocols of Mosier 1 and Mosier 3.

The design intent of specimen Mosier 3 was to test the NSM titanium strengthened specimen with all reinforcing bars fully embedded in the concrete. Additionally, the NSM titanium was confined at the haunch transition at midspan with two steel plates. The confinement plate resists the outward force induced on the NSM titanium bars from the haunch transition geometry. The outward force on the NSM titanium was approximated at 1 kip (4.89 kN). The dimensions of the steel confining plates were $1/2 \times 3 \times 15$ in. ($13 \times 76 \times 381$ mm). The plate was attached with two 1/2 in. (13 mm) diameter steel bolts. A hole was drilled through the stem above the NSM titanium groove at midspan to accommodate the steel bolt.

3 EXPERIMENTAL AND ANALYTICAL RESULTS

Three specimens were tested to failure in the structural engineering laboratory at Oregon State University. Mosier 1 was loaded in 10 kip (44.5 kN) increments, then unloaded, and then proceeded to the next load step until eventual failure. Mosier 2 and Mosier 3, the NSM titanium specimens, had a larger predicted capacity and was loaded in 25 kip (111 kN) increments, unloaded to 5 kips (22.2 kN), and then proceeded to the next load step until failure. The overall response of Mosier 1 is shown in Figure 7 along with the analytically predicted strengths and demands. As seen here, the specimen representing the as-built conditions (Mosier 1) exhibited low strength and no ductility. The failure occurred in the poorly detailed splice region at the haunch transition and the cracking conditions looked similar to those observed in the field. The specimen strength was only marginally above the computed factored demands.

The NSM titanium strengthened Mosier specimens (Mosier 2 and 3) achieved much greater capacity than Mosier 1, as seen in Figure 8. The NSM strengthened specimens experienced ductile flexural failures and displayed distributed cracking and widespread signs of distress prior to failure. The applied actuator load, applied shear, V_{APP}, dead load shear, V_{DL}, total shear, V_{EXP}, midspan displacement, and observed failure crack angle are reported in Table 3. The reported midspan displacement corresponded to the peak load. The total shear is the applied shear from the actuator

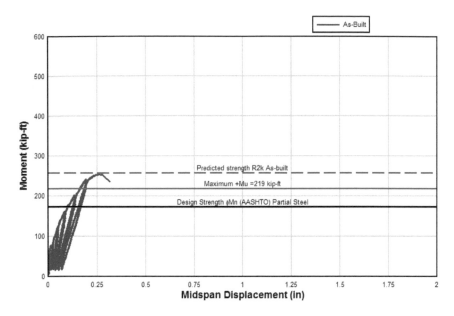

Figure 7.   Mosier 1 predicted capacity and demand shown with experimental response.

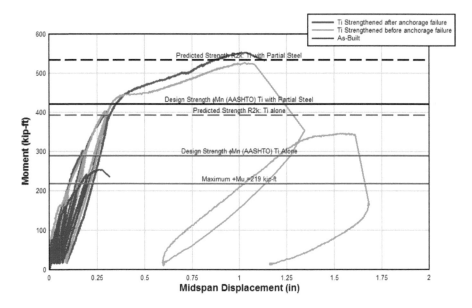

Figure 8.   All Mosier specimens with predicted capacity and demand shown with experimental responses.

plus the dead load shear. Dead load shear was calculated from the weight of concrete acting across the failure plane.

Mosier 2 and Mosier 3 specimens showed increased load capacity of 74.7 kip (332 kN) and 67.8 kip (302 kN), respectively, compared to Mosier 1. The deformation of the strengthened specimens increased by 0.762 in. (19.3 mm) and 0.752 in. (19.1 mm) for Mosier 2 and Mosier 3, respectively, compared to Mosier 1. Mosier 2 had a slightly lower stiffness than Mosier 3 because initial cracking of the concrete and slip of the cutoff reinforcing steel bars occurred prior to installation of the NSM titanium alloy bars. Mosier 3 contained NSM titanium alloy bars bonded to integral concrete that

Table 3. Summary of specimen capacity and midspan displacement.

Specimen	Applied Load kips [MN]	Applied Shear (V_{APP}) kips [kN]	DL Shear (V_{DL}) kips [kN]	Total Shear (V_{EXP}) kips [kN]	Midspan Disp. in. [cm]	Failure Crack Angle deg.
Mosier 1	63.7 [0.283]	31.9 [141.6]	0.27 [1.2]	31.1 [142.8]	0.258 [6.55]	68
Mosier 2	138.4 [0.615]	69.2 [307.8]	0 [0]	69.2 [307.8]	1.02 [25.9]	90
Mosier 3	131.5 [0.585]	65.8 [292.5]	0 [0]	65.8 [292.5]	1.01 [25.7]	90

enabled stiffer response until an applied load of approximately 109 kips (484 kN), when significant slip of the internal reinforcing steel occurred within the splice region that damaged the concrete to which the NSM titanium alloy bars were bonded. The damaged bond of the NSM titanium bars contributed to the slight decrease in ultimate capacity compared to Mosier 2.

Mosier 3 exhibited a significant amount of reserve capacity after achieving the peak load. This reserve capacity is seen as the last load cycle upon which Mosier 3 achieved a load of 88 kips (391 kN), even after bond failure along the length of the NSM titanium bars. The reserve capacity indicates that the mechanical anchorages on the titanium alloy bars could sustain loads above the factored demands even after bond rupture along the length that occurred at peak load.

The moment capacity of specimen Mosier 1 was computed using AASHTO-LRFD and the analysis program Response 2000 (R2K) (Bentz, 2000) and is shown relative to the experimentally observed responses in Figure 7. In the analyses, partial contribution of the poorly detailed reinforcing steel at midspan was used based on the available development length relative to the computed development length using the ACI 318 formulation (ACI, 2011). The R2K predicted flexural capacity of Mosier 1 was within 3 kip-ft (4 kN-m) of the observed value and shows that the partially developed steel should be included to best predict the available strength.

The NSM titanium specimen strengths were also predicted using AASHTO and R2K and are shown in Figure 8 relative to the experimental results. The solid lines show the AASHTO predictions with and without the partial contribution of the internal reinforcing steel at midspan. The dashed lines represent the R2K flexural capacities with and without partial contribution of reinforcing steel at midspan. As seen in the figure, improved predictions for member strength were achieved when the partially developed internal steel was used in the analyses.

The most conservative prediction of the NSM strengthened member capacity (AASHTO design moment capacity relying only on the NSM titanium and assuming the internal steel provides no resistance) exceeded the factored moment demand at the critical section. R2K reasonably predicted the capacity of the NSM strengthened specimens assuming fully anchored NSM titanium alloy bars.

After completion of the experimental tests and analysis, the strengthening approach was applied to the bridge. The final cost of the NSM titanium alloy strengthening scheme was 30% less expensive than the NSM carbon fiber alternative. After installation of the NSM titanium, the shoring was removed and the bridge returned to service without load restrictions. The Mosier Bridge is now the first reinforced concrete bridge in the world with titanium alloy reinforcing.

4 SUMMARY AND CONCLUSIONS

An in-service RCDG bridge built in 1953 exhibited wide cracks with vertical offset at a location containing a poorly detailed splice of the flexural reinforcing steel. The subsequent rating showed the girders did not have adequate capacity to carry truck loads without significant weight restriction. To allow continued operation of the bridge, it was shored until it could be strengthened. The bridge was strengthened with NSM titanium alloy bars. The methods and materials for the strengthening scheme were demonstrated in the laboratory on three (3) full-size specimens that mimicked the in situ details, materials, and load effects. Based on the experimental results, analysis, and field experience, the following conclusions are presented:

- The as-built girder details provided inadequate strength and no ductility. The bridge was operating at loads close to the member capacity and strengthening was necessary.
- The NSM titanium strengthened specimens increased the girder strength by a factor of 2 and were well above the required factored load effects using only 4 titanium alloy bars.
- The NSM titanium alloy bars were able to provide strength well above the required factored load effects even if the underlying flexural steel has failed due to anchorage slip at the splice location.
- The NSM titanium strengthened specimens increased the girder deformation capacity and exhibited significant distress before failure.
- Upon loss of bond along the NSM bars, the mechanical hooks used at the ends of the titanium alloy bars allowed the member to continue to carry load above the required factored load effects.
- The material and techniques were demonstrated on the actual Mosier Bridge and restored the bridge to full operation without shoring or posting at a cost 30% below alternatives.
- The high-strength, ductility, environmental durability, and ability to fabricate mechanical anchorages make titanium alloy reinforcement a promising material for strengthening civil infrastructure.

ACKNOWLEDGEMENTS

This research was funded by the Oregon Department of Transportation and Mr. Steven Soltesz served as the research coordinator. This support is gratefully acknowledged. Perryman Company, Houston, PA provided the BridgeAlloy™ titanium bars with the surface deformations developed specifically for this application, the assistance of Mr. Warren George in this endeavor was greatly appreciated. The findings and conclusions are those of the authors and do not necessarily reflect those of the project sponsors or the individuals or companies acknowledged.

REFERENCES

American Association of State Highway and Transportation Officials (AASHTO). 2013. "AASHTO-LRFD Bridge Design Specification," 6th edition with 2013 interims, AASHTO, Washington, D.C.
American Concrete Institute (ACI). 2011. "318-11/318R-11: Building Code Requirements for Structural Concrete and Commentary," ACI, Farmington Hills, Michigan.
American Concrete Institute (ACI). 2008. "Guide for the Design and Construction of Externally Bonded FRP Systems for Strengthening Concrete Structures," ACI, Farmington Hills, Michigan.
ASTM A706/A706M-09b. 2009a. "Standard Specification for Low-Alloy Steel Deformed and Plain Bars for Concrete Reinforcement," ASTM International, West Conshohocken, PA, 6 pp.
ASTM E8/E8M-09. 2009b. "Standard Test Methods for Tension Testing of Metallic Materials," ASTM International, West Conshohocken, PA, 27 pp.
ASTM B348/B348M-13. 2013. "Standard Specification for Titanium and Titanium Alloy Bars and Billets," ASTM International, West Conshohocken, PA, 8 pp.
Bentz, E.C. 2000. "Sectional Analysis of Reinforced Concrete Members," PhD Thesis, Department of Civil Engineering, University of Toronto, 310 pp.
Google Maps. 2014. [Mosier Bridge, Mosier, Oregon] [Street View].
Oregon Department of Transportation (ODOT). 2013. Bridge Section, Digital Photographs and Design Drawings.

Chapter 21

Fatigue investigation and retrofit of double-deck cantilevered truss I-95 Girard Point Bridge

Y.E. Zhou & M.R. Guzda
AECOM, Hunt Valley, MD, USA

ABSTRACT: This paper discusses a comprehensive investigation as well as retrofit design and construction for extensive fatigue cracks in the end connections of floor beams on a double-deck, cantilever-suspended steel truss bridge. The investigation involved 3D finite element analysis using global and local models, field measurement of strains and displacements due to live load and temperature, laboratory testing of steel samples for material properties and high-stress low-cycle fatigue characteristics, as well as development of an effective retrofit based on the analytical, experimental, and field testing results. It was concluded that the cracks were a result of distortion induced fatigue in the floor beam web due to interactive deformations of the global structural system under live load and temperature variations. The retrofit entailed removing the fatigue susceptible weld terminations and reinforcing local areas of the floor beam web for out-of-plane distortion. Construction of the fatigue retrofit was completed in July 2011. The repaired structure has performed satisfactorily since then indicating success of the retrofit.

1 INTRODUCTION

The Girard Point Bridge, carrying Interstate highway 95 (I-95) over the Schuylkill River, is located in Philadelphia, Pennsylvania. The bridge is an 18-span, double-deck, through-truss steel structure (Fig. 1(a)). The main spans are comprised of two 353-ft cantilevered anchor spans and a 700-ft center span including a 390-ft suspended portion. Construction of the truss was completed in 1973 and the entire bridge was opened to traffic in 1976. The upper deck carries the southbound traffic and the lower deck carries the northbound traffic of I-95. Fig. 2(a) and 2(b) depict the floor beam end connections of the upper deck and the lower deck, respectively.

(a)   (b)

Figure 1. Girard Point Bridge: (a) overview; (b) typical fatigue cracks around end of floor beam top flange with a flame cut opening below a knee brace.

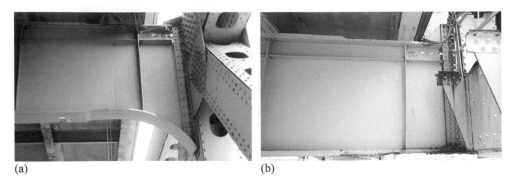

(a)   (b)

Figure 2.   End connections of: (a) upper deck FB 65 (PP 1); and (b) lower deck FB 82 (PP 18).

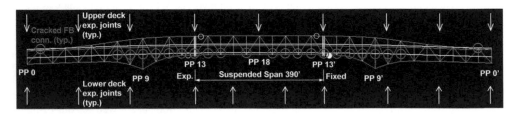

Figure 3.   Locations of cracked floor beam end connections, truss expansion joints, and deck expansion joints across three-span cantilevered-suspended truss unit.

Fatigue cracks were first reported in 1993 in some of the floor beam end connections in the three-span cantilever-suspended unit of the bridge. The cracks occurred in the floor beam web and the triangular knee brace, and initiated from the horizontal web-to-flange connection welds at the end of the floor beam top flange (Fig. 1 (b)). The fatigue cracks were found to have grown in length and location overtime, and spread over nearly all the floor beam end connections of the same construction detail by the late 1990's. As shown in Fig. 3: on the lower deck, all but two of the 23 welded plate girder floor beams from panel point (PP) 7 through PP 7' had fatigue cracks in both end connections; on the upper deck, only two floor beams have the welded knee braces and both cracked in both end connections. Fig. 3 also shows locations of expansion joints of the upper and lower decks, where the floor system becomes discontinuous.

Common features of the cracked connections include: (1) termination of the floor beam top flange approximately 1 in. from the vertical connection angles with a rough, flame-cut opening in between, and (2) a triangular knee brace welded to the top flange (Fig. 2 and Fig. 1(b)). The floor beam end connections that did not experience cracking have different construction details: those on the lower deck (PP 0 thru PP 6, and PP 6' thru PP 0') are truss type floor beams made of built-up sections and both webs of the top chord (consisting of two channels) are connected to the main truss through clip angles; those on the upper deck have their top flange extended over the vertical connection angles and connected to the truss through a clip angle and four bolts. Another paper (Zhou et al. 2014) provides more details about the un-cracked connections.

2   PRE-RETROFIT FIELD TESTING – DETERMINE THE CAUSE

2.1   *Field instrumentation and testing*

In 2003, an investigation was initiated through field measurement of strains and deformations of the end connections of three floor beams during a load test and seven days of regular traffic. Locations of the three instrumented floor beams are shown in Fig. 4; the panel point numbering is

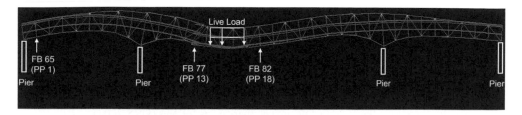

Figure 4.   Illustration of deformations and locations of pre-retrofit instrumentated floor beams.

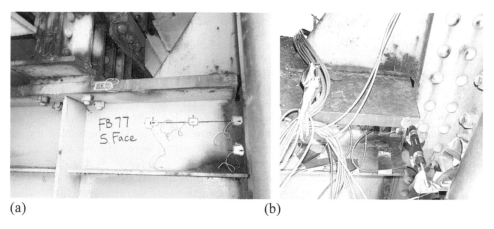

(a)   (b)

Figure 5.   Pre-retrofit instrumentation on lower deck FB 77 (PP 13): (a) strain gages on web (same positions on each face); (b) displacement transducer for out-of-plane movement.

also labeled in Fig. 3. The three floor beams (FB) included two on the lower deck (FB 77 at PP 13 and FB 82 at PP 18) and one on the upper deck (FB 65 at PP 1). Lower deck FB 77 is located at the expansion end of the suspended span under the pin-and-hanger members and is among those that had experienced the worst fatigue cracks. Lower deck FB 82 is located at the center of the suspended span and is one of the only two plate girder floor beams that experienced no fatigue cracks. Upper deck FB 65 is one of the two floor beams on the upper deck that have the connection detail as described previously and both experienced fatigue cracks.

At each instrumented floor beam connection, a number of strain gages were placed on both faces of the web at opposing positions to distinguish between in-plane and out-of-plane responses. In addition, a displacement transducer was installed to measure the longitudinal displacement of the top flange relative to the truss joint, which is the maximum out-of-plane deformation of the web at the top. Fig. 5 depicts the instrumentation on the lower deck FB 77.

A controlled load test was performed using test trucks of known weight crossing each of the three traffic lanes on the lower deck and the upper deck of the bridge. During each test run, the test deck was intermittently blocked to traffic while the other deck remained open. Strain and displacement range histograms, processed through the 'rainflow' algorithm for cycle counting, were collected along with time histories from all sensors due to normal traffic for approximately seven days after the load test. The rainflow processing was performed real-time using the internal software of Campbell Scientific CR9000 data acquisition system.

Results of the field testing indicated that the floor beams were subjected to a combination of in-plane bending due to vehicles directly above it and out-of-plane bending due to global deformations of the structural system under vehicles across both decks. The global structural deformations (Fig. 4) result in longitudinal forces at the stringer-to-floor beam connections and bearings due to incompatible deformations between the cantilever-suspended truss and the segmentally continuous bridge deck. These longitudinal forces are not considered in conventional bridge design but can be significant when expansion bearings and joints become dysfunctional or experience high resistance due to aging.

It was determined from the field testing that out-of-plane distortion of the floor beam web in the gap area between the end of the top flange and the vertical connection angles was the cause for the fatigue cracking and growth. The fatigue strength of the web gap detail was reduced by the poor workmanship of the flame cut (Fig. 1(b)).

Specific findings from the field testing are summarized as follows (URS 2003).

2.2   *Findings from lower deck FB 77 (PP 13)*

- The global structural response causes out-of-plane bending of the floor beam web about the double angle connection and corresponds to vehicles crossing the 389-ft suspended span on both the upper and lower decks; the local behavior causes the vertical in-plane bending of the floor beam and corresponds only to vehicles directly above it and within the two immediate adjacent deck panels of 39-ft each (Fig. 6).
- The seven-day strain/displacement histograms indicated that normal traffic had significant events with maximum live load effects higher than the sum of all six single-truck test runs from the load test. However, the occurrences of these events were very low and thus their contribution to fatigue damage may be insignificant.
- Strains in the floor beam web due to out-of-plane distortion are lower than (approximately 30% of) those due to in-plane bending for the same test truck crossing. However, the seven-day measurements of normal traffic indicated significantly higher out-of-plane distortion strains than the in-plane bending strains due to the collective effects of loads from both decks on the out-of-plane behavior.
- The relative longitudinal displacement between FB 77's top flange and the connected truss joint due to a single truck crossing was measured varying between 0.003 and 0.0075 in. depending on the truck's lateral position and the deck it was on. The seven-day histograms indicated drastically decreasing cycle counts for increasing displacement ranges. The maximum displacements were recorded twice in the 0.100–0.101 in. bin.

2.3   *Findings from lower deck FB 82 (PP 18)*

- The end connection is subjected to primarily in-plane stresses with little out-of-plane distortion. The likely reason is that FB 82 is located at the middle of the center span which is also the point of symmetry of the structure in the longitudinal direction.
- Longitudinal relative displacements between the floor beam top flange and the truss joint were measured around 0.0012 in. due to single truck crossings. This is very small compared with those measured at FB 77 and likely due to the flexure of the stringers on both sides of FB 82. The measured strains were also very small compared with FB 77, which explained why FB 82 did not experience fatigue cracking.

2.4   *Findings from upper deck FB 65 (PP 1)*

- Measured strains in the web due to out-of-plane distortion were significantly higher than those due to in-plane bending.
- Measured strains and displacements due to out-of-plane web distortion were similar for test truck crossings on the upper and the lower deck, suggesting that vehicles in all lanes on both decks contribute to the distortion induced stresses at the floor beam end.

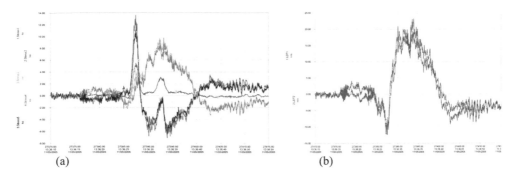

(a)   (b)

Figure 6.   Measured strain/displacement responses of lower deck FB 77 (PP 13) to traffic: (a) horizontal strains on both web faces, upstream end; (b) longitudinal out-of-plane displacements, both ends.

Figure 7.   "Dog-bone" cutout retrofit with strain gages on lower deck FB 77 (PP 13).

3   INITIAL RETROFIT – THE "DOG-BONE" CUTOUT

Based on the findings from the field testing, a retrofit scheme was developed to remove the cracked weld detail and enlarge the horizontal floor beam web gap (between the termination of top flange and the connection angles to the truss) to better accommodate the longitudinal distortion. As shown in Fig. 7, the retrofit cutout encompassed the entire knee brace, a portion of the floor beam top flange and web, as well as a portion of the vertical back-to-back connection angles, resulting in a "dog-bone" shape in the floor beam web. A finite element analysis was performed to investigate stress fields in the floor beam end connection for various cutout geometries (URS 2004; Russo et al. 2006). Field measurements on two different retrofit prototypes were also made to verify and refine the cutout detail (Mahmoud and Connor 2005). In 2005, the "dog-bone" retrofit was applied to both end connections of all 23 plate girder type floor beams on the lower deck (PP 7 thru PP 7') and two floor beams on the upper deck (PP 1 and PP 1').

The retrofit, though it removed weld terminations and flame-cut edges in the highly stressed areas and reduced stress concentrations, would make the floor beam end connections more flexible and thus subject to larger longitudinal displacements relative to the truss joints. Additionally, the removal of the knee brace welded to the top flange of the floor beam reduced its end section for in-plane bending and for shear. While the in-plane bending capacity was found adequate after the cutout, patch plates were installed on both sides of the web to supplement the shear capacity of the floor beam.

4 POST-RETROFIT FIELD TESTING – SHORT-TERM

In 2005, post-retrofit field testing was performed on both ends of lower deck FB 77 (PP 13) located at the expansion end of the suspended span. This floor beam was also instrumented before the retrofit and should provide the best indication on changes in the global structural behavior resulting from the "dog-bone" cutouts.

Strain gages were installed along the edge of the cutout on both sides of the floor beam web for stress effects of both in-plane bending and out-of-plane distortion (Fig. 7); a displacement transducer was also used at each end for longitudinal displacement of the floor beam top flange relative to the truss joint. Strain and displacement measurements were collected for nearly five days under normal traffic. The test results indicated higher stresses and deformations in the modified floor beam web area compared with the pre-retrofit condition under both live and temperature loads.

The maximum stress ranges measured from the five-day period are: 20.7 ksi in-plane bending stress due to a heavy truck crossing on the lower deck; 11.2 ksi out-of-plane bending stress due to a heavy truck crossing on the lower deck; 13.9 ksi out-of-plane bending stress due to a heavy truck crossing on the upper deck; and 49 ksi out-of-plane bending stress due to a 25°F daily temperature variation. Although the in-plane and out-of-plane bending stresses caused by the same vehicle are not directly superimposable due to their different durations and peak times (Fig. 6), vehicles on both decks contribute to the out-of-plane bending stress.

It should be noted that after the retrofit the floor beam web end has no connections of any kind in the highly stressed areas. Thus its fatigue resistance should be Category "A" for base steel, which has a constant-amplitude-fatigue-threshold (CAFT) of 24 ksi (AASHTO 2007). Although field measurements suggested most live load stresses below the 24 ksi CAFT, the maximum local stresses due to temperature seemed very likely to exceed the 50 ksi nominal yield strength of the web steel. Therefore, further investigation on the resistance of the web steel to high-stress low cycle fatigue was necessary.

5 LABORATORY TESTING FOR STEEL PROPERTIES AND HIGH-STRESS LOW-CYCLE FATIGUE

A review of original construction documents indicated that the 3/8 in. thick steel plates used for the floor beam webs and the knee braces should be ASTM A572-67 Grade 50 steel (High-Strength Low-Alloy Structural Columbium-Vanadium Steel). Laboratory tests were made using steel samples removed from the bridge during the 2005 rehabilitation.

5.1 *ASTM tests for steel properties*

Five tensile specimens were fabricated from four lower deck floor beam knee brace plates (samples). Standard ASTM tensile tests were performed at Lehigh Testing Laboratories Incorporated (LTL) in New Castle, Delaware. Three specimens indicated 55 ksi yield and 77 ksi tensile strengths for one knee brace sample and 43 ksi yield and 73 ksi tensile strengths for another sample. Two additional knee brace samples had results of 57 ksi yield, 81 ksi tensile strength, and 43 ksi yield, 69 ksi tensile strength.

Chemical analysis of the two knee brace samples indicated insignificant difference, within the variation limits specified by ASTM. Rockwell Hardness test of the two knee brace samples indicated similar hardness (79 HRBW for one and 74 HRBW for another). It was unclear why the four knee brace samples had a significant difference in yield strength (55 ksi vs. 43 ksi).

5.2 *Cyclic reverse bending test of plate specimens*

To determine the resistance of web steel to distortion induced high-stress low-cycle fatigue, cyclic reverse bending tests of the knee brace samples was performed in the laboratory of Villanova University (Gross and Yost 2009).

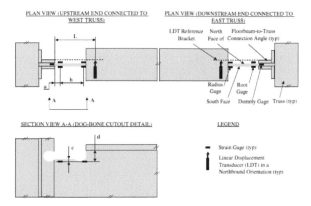

Figure 8.   Layout of strain and displacement sensors on lower deck FB 76, FB 77 and FB78.

Six plate specimens of dimensions 12 × 1.5 × 0.375 in. were tested: C1, C2, C3 and C4 had yield strength of 57 ksi; D1 and D2 had yield strength of 43 ksi. Two specimens were clamped to a test frame in tandem such that both were loaded simultaneously in reverse-curvature bending through a deflection controlled MTS test machine at their common clamped end. The clear length of each specimen between the clamped ends was 7 in. (similar to the "dog-bone" cutout) and the amplitudes of applied deflections were approximately 0.15 in. in each direction (also similar to field measurements) or what was necessary to produce yield-level stresses at the ends. Each specimen had strain gages on both surfaces at each end next to the clamps to control and monitor the maximum stresses before and during the cyclic testing.

Measurements from the strain gages indicated combined bending and axial behavior of the plate specimens as well as variation of strains among the ends (well exceeding the yield level at some locations). The cyclic stresses experienced by the plate specimens in the lab test were more severe than the actual stress conditions measured by the field test.

Test results show that of the six specimens tested (12 plate ends in total), seven plate ends cracked between 20,000 and 70,000 stress cycles, and the remaining five plate ends did not crack after 80,000 cycles. Visual examination of the fracture surfaces of the plate samples indicated signs of non-ductile behavior.

6   POST-RETROFIT FIELD TESTING – LONG-TERM

Long-term field testing was performed to measure stresses and deformations of the concerned floor beam end connections due to temperature and live load. The field testing focused on the vicinity of the expansion end of the suspended span (PP 13). Instrumented locations included both end connections of lower deck FB 76, FB 77 and FB 78 (Fig. 8), upper deck FB 75, as well as expansion bearings of the truss false chords and select expansion bearings of the stringers. A total of 39 sensors were employed including strain gages, displacement transducers, and thermocouples. "Dummy gages" were placed in zero stress areas to account for temperature effects.

Measurements were collected over a period of six months (from November 2008 to May 2009) for the effects of daily and seasonal temperature changes. The data acquisition system consisted of a Campbell Scientific CR9000X unit, a wireless modem, marine batteries, and solar panels.

Fig. 9 shows field measured strain and displacement responses of FB 77 to temperature during a 10-day period from December 21 to 31, 2008 (with a temperature rise of approximately 50°F). The vertical dashed red lines represent a relative "zero" state of bending stresses in the floor beam web measured by the strain gages. The horizontal dashed red line represents the web position corresponding to this "zero" stress state of web out-of-plane bending. The field measurements

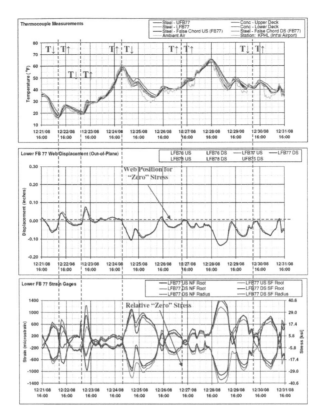

Figure 9.   Field measured strain/displacement responses of lower deck FB 77 to temperature.

show that each time the strain measurements reach the "zero" state the web displacement also returns to its "zero" position. However, the corresponding temperatures at these "zero" positions may or may not be the same, indicating a complicated process of thermal movements with static and dynamic frictions.

Major findings from the long-term field testing are summarized as follows:

- Measured movements of truss expansion joints (pin in slotted hole) at the lower deck false chords (at FB 77) were approximately 70% of their theoretical thermal expansions.
- Measured movements of the lower deck stringer expansion bearings (bronze shoe on steel bearing plate) at FB 77 were approximately 60% of the theoretical thermal expansions.
- Measured strains and displacements of the webs of lower deck FB 76, FB 77 and FB 78 have indicated significant out-of-plane distortion between the truss and the exterior stringer due to temperature changes. The levels of maximum elastic stresses corresponding to the measured strains due to temperature alone can well exceed the 50 ksi specified yield strength of the web steel.
- In the "dog-bone" cutout area of the floor beam web, field strain measurements over the six-month monitoring period indicated less than 60 cycles of stress range exceeding 40 ksi, and on the order of 80 cycles of stress range between 10 and 40 ksi.
- The field measurements have indicated a consistent and linear correlation between the strains and out-of-plane displacements of the floor beam web at its end connection to the truss. This indicates generally elastic out-of-plane bending behavior of the web plate despite the likelihood of localized steel yielding.

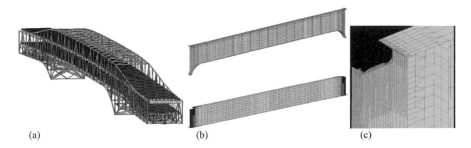

(a)   (b)   (c)

Figure 10. Finite element models: (a) overview of global model; (b) superelements for upper floor beam
(top) and lower floor beam (bottom); (c) end details.

- The correlation between the measured web strains and air temperature changes is rather erratic
  and usually unrepeatable. This may be attributed primarily to the erratic nature of the expansion
  bearings of the stringers and truss false chords at the FB 77 locations. The temperature differ-
  entials among different components of the double-deck, steel-concrete structure may also have
  secondary effects.

7 FINITE ELEMENT ANALYSIS

Comprehensive 3-D finite element analysis (FEA) including a global model and sub-models was
performed for the three-span, cantilever-suspended truss unit using the computer software GT-
STRUDL. The objective was to correlate local stresses in the floor beam ends with the behavior
of the global structural system due to temperature and live load. The global model included all
the truss members, both reinforced concrete decks, and all the expansion joints/bearings in the
truss and floor systems (Fig. 10(a)). Sub-models, named "Superelements" in GT-STRUDL, were
developed that accurately represent all the welded plate girder floor beams (Fig. 10(b)&(c)). The
Superelements consisted of shell elements with 12 boundary nodes (two at the ends, nine at the
stringer connections, and one at the lateral bracing connection where appropriate), through which
they were connected to the global model. Stiffness matrices and force vectors were developed for
the Superelements through these 12 boundary nodes and then incorporated into the global model.

Stiffness analyses of the global model were performed for several types of load (camber, dead
load, live load, and thermal load) as well as various combinations of them. For each load case,
nodal displacements were obtained from the global model at the boundary nodes and subsequently
applied to the Superelement of each floor beam to compute the local stresses and deformations at
its end connections.

The FEA found that conditions of truss expansion joints at the false chords and stringer expansion
bearings at both ends of the suspended span are critical to the out-of-plane bending, or distortion, of
floor beam connections. If all of these expansion bearings behave "as-designed" (free of restraint),
out-of-plane bending of the floor beam web would be negligible.

In order to identify contributing sources and better define the scope of retrofit, ten different cases
of expansion bearing conditions were formulated. These cases varied from Case 1: "as-designed"
condition at all expansion bearing locations; Case 2: "frozen" (fully restrained) condition at all
expansion bearing locations; to various combinations in between (Case 3 through Case 9). Case 10
included elastic releases at the expansion joints and bearings, where the stiffness properties were
developed through matching the FEA results with field measurements.

Fig. 11 shows a comparison between the results of all ten cases of FEA and field measured
stresses and displacements of lower deck FB 77 due to a 25°F temperature rise (normalized where
necessary). The comparison indicates that the vast majority of field measured web stresses and
displacements fall within the variation range of the ten FEA cases.

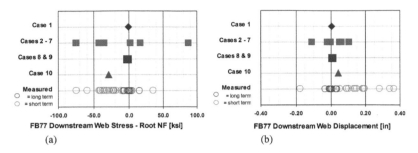

Figure 11.   Comparison between FEA and field measurements for 25°F temperature rise – lower deck FB 77 (PP 13): (a) web stresses; (b) out-of-plane displacements.

8   FINAL RETROFIT

8.1   *Floor beam web reinforcement and expansion bearing rehabilitation*

Based on the results of the investigation, final retrofit was proposed that involved strengthening the end connections of select floor beams and rehabilitating select expansion bearings of the truss and deck systems. The goal of the retrofit was twofold: to reduce the longitudinal forces at the floor beam ends caused by deformation interactions of the global structural system due to temperature and live load; and to strengthen the floor beam ends to resist possible distortions due to friction at the expansion joints and bearings.

The scope of the final fatigue retrofit included the following major items (Fig. 12):

- Install new 3/8 in. doubler plates of Grade 70 high-performance steel (HPS) to both sides of the web and connection angles at both ends of all the plate girder floor beams on the lower deck and two floor beams on the upper deck.
- At both ends of four lower deck floor beams, install tie plates between the top flange of the floor beam and the connection angles to the truss to strengthen the connection in supporting dead load during the retrofit construction.
- Replace all stringer expansion bearings with permanently self-lubricated bronze bearings over the floor beams of both decks at both ends of the suspended span.
- Rehabilitate truss expansion bearings in the false chords of the suspended span.
- Lubricate all corroded expansion bearings between the stringers and the floor beams.
- Replace existing open finger-dam expansion deck joints with closed modular dams at both ends of the suspended span.

8.2   *Retrofit design*

The detailed design of the retrofit, including dimensions and thickness of various reinforcing plates as well as connection bolts, was checked in accordance with the AASHTO LRFD Bridge Design Specifications (AASHTO 2007). The finite element models were used to perform analysis for stress fields under various design loads.

Temperature effects are a combination of frictional forces and thermal forces. Based on field measurements, it was determined that a change in temperature up to 15°F would likely not cause slippage of aged expansion bearings thereby imposing forces on the floor beam due to the frictional resistance of the bearings. Above 15°F slippage should occur at expansion bearings at both ends of the suspended span and thermal forces in the structure are only those resulting from the frozen expansion bearings beyond these. PennDOT specifies a temperature rise of 42°F and a temperature fall of 78°F for bridge design. Therefore the thermal loads are: $\Delta T_{rise} = (42° - 15°) = 27°F$, and $\Delta T_{fall} = (78° - 15°) = 63°F$.

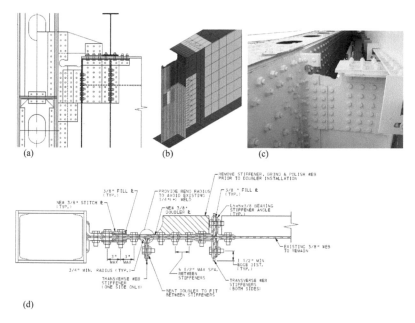

(a)   (b)   (c)

(d)

Figure 12.   Plating retrofit of floor beam end connections: (a) elevation, lower deck; (b) plate thickness contours, lower deck; (c) completed retrofit, upper deck; (d) horizontal cross section, lower deck.

The following LRFD load combinations were used to evaluate the retrofit.

LRFD STRENGTH I

For temperature rise (TR):

STR I TR = 1.25(DL) + 1.00(Camber) + 1.75(LL + I) + 1.00(FR + 15) + 0.50(TU + 27)

For temperature fall (TF):

STR I TF = 1.25(DL) + 1.00(Camber) + 1.75(LL + I) + 1.00(FR-15) + 0.50(TU-63)

LRFD SERVICE I

Similar to the above equations except that all load factors are 1.00, where, DL = effects of total dead load, Camber = locked-in forces due to camber; LL + I = effects of LRFD live load plus impact; FR ± 15 = forces due to $\Delta T = \pm 15°F$ with all expansion bearings "frozen" throughout structure; TU + 27 = forces due to $\Delta T_{rise} = +27°F$ with "as-designed" expansion bearings of truss and stringers at FB 77 & FB 87 and "frozen" for all other stringer bearings; TU-63 = forces due to $\Delta T_{fall} = -63°F$ with expansion bearing conditions similar to TU + 27.

8.3   *Retrofit construction*

Construction of the fatigue retrofit was completed in 2011 as part of a comprehensive bridge rehabilitation and preservation project that included painting, deck overlay, drainage system, and repair of corroded steel elements across the structure.

9   SUMMARY AND CONCLUSIONS

(1) Field testing and FEA indicated that the floor beam end connections are subjected to combined in-plane and out-of-plane responses due to live load, and primarily out-of-plane response due to temperature. The out-of-plane distortion of the web was the cause for the extensive fatigue cracking.

(2) Field measurements indicated significant out-of-plane distortion of the lower deck floor beam web at the expansion end of the suspended span due to temperature changes. The maximum local stresses in the floor beam web in the "dog bone" cutout area could well exceed the yield strength of web steel. Despite the high localized stresses, field measured floor beam web strain to out-of-plane displacement correlation exhibited essentially linear elastic behavior at the floor beam end connection.

(3) Cyclic reverse bending test of steel plate specimens was performed in the laboratory with the maximum local stresses at both ends well exceeding the steel yield strength in each direction. Test results suggested possible brittle cracking after 20,000 cycles.

(4) Field measurements of six months suggested less than 60 cycles of stress range exceeding 40 ksi and 80 cycles of stress range between 10 and 40 ksi at the floor beam end connection. This is equivalent to approximately 120 stress cycles per year near or in excess of the yield strength of web steel.

(5) The FEA considered ten different cases of truss and stringer expansion bearing conditions varying between "as-designed" and "frozen", or with an elastic release. A comparison between the FEA and field measurements indicated that the vast majority of field measured web stresses and displacements fall within the variation range of the ten FEM cases. The FEA also indicated that the functionality of expansion bearings of the truss and the stingers at both ends of the suspended span is critical to the magnitude of distortion induced stresses at the floor beam end connections.

(6) Final retrofit included strengthening the end connections of all lower deck plate girder type floor beams and two upper deck floor beams, and rehabilitating expansion bearings of the truss and stringers at both ends of the suspended span.

(7) Construction of the fatigue retrofit was completed in 2011. No signs of distress or dysfunction have been reported in the subsequent inspections.

ACKNOWLEDGEMENT

The authors would like to acknowledge other participants who contributed to the project: Mr. John Lang, Mr. Jason Beecher, Mr. Rob Cunningham, Mr. Brian Guzas, Mr. Mehul Dave, Dr. Piotr Podhorecki, Dr. Frank Russo, and Dr. Rob Connor. Also acknowledged are PennDoT staff: Mr. Thomas Macioce, Ms. Kristin Langer, Mr. Henry Berman, and Mr. Peter Berg.

REFERENCES

AASHTO (2007), *LRFD Bridge Design Specifications*, 3rd Edition, American Association of State Highway and Transportation Officials, Washington, D.C.

Gross, S.P. and Yost, J.R. (2009), Bending Fatigue Testing of Thin Rectangular Steel Strips, Final Research Report, Dept. of Civil and Environmental Engineering, Villanova University, Pennsylvania.

Mahmoud, H.N. and Connor, R.J. (2005), Field Monitoring Prototype Retrofits of Floorbeam Connections on the I-95 Girard Point Bridge, ATLSS Report No. 05-01, January 2005, Bethlehem, Pennsylvania.

Russo, F.M., Beecher, J.B. and Lang, J.W. (2006), Diagnosis and Retrofit of Floorbeam to Truss Connections in the Double Deck Girard Point Bridge, First International Conference on Fatigue and Fracture in the Infrastructure, Philadelphia, Pennsylvania.

URS (2003), Girard Point Bridge Summary Report of Strain/Displacement Measurements (FB 65, FB 77 and FB 82), Hunt Valley, Maryland.

URS (2004), Girard Point Bridge TS&L Bridge Rehabilitation Report, King of Prussia, Pennsylvania.

URS (2005), Girard Point Bridge Summary Report of Strain/Displacement Measurements of Retrofitted Connections of Lower Deck (NBR) Floor Beam 77, Hunt Valley, Maryland.

URS (2009), Girard Point Bridge Temperature/Fatigue Study of Floor Beam Connections, Fort Washington, Pennsylvania.

Zhou, Y.E., Beecher, J.B., Guzda, M.R. and Cunningham, D.R. (2014), Investigation and Retrofit of Distortion Induced Fatigue Cracks in a Double-Deck Cantilever-Suspended Steel Truss Bridge, *Journal of Structural Engineering*, ASCE.

Chapter 22

Flexural fatigue performance of ECC link slabs for bridge deck applications

K.M.A. Hossain, S. Ghatrehsamani & M.S. Anwar
Ryerson University, Toronto, Ontario, Canada

ABSTRACT: The leaking expansion joints are a major source of multi-span bridge deteriorations in Canada and North America. Expansion joints can be replaced by flexible link slabs made of Engineered Cementitious Composite (ECC) forming a joint free bridge. ECC is a special type of high performance fiber reinforced cementitious composite with high strain hardening characteristic and multiple micro-cracking behavior under tension and flexure. The locally available aggregates and supplementary cementitious material (SCM) have been used to produce sustainable and cost effective ECC mix for the link slab application. The use of flexible ECC link slab in joint free bridge deck has been an emerging technology. Limited research has been conducted on the fatigue performance of such ECC link slabs. This paper presents the results of experimental investigation on ECC link slabs subjected to flexural fatigue loading at stress levels of 40% and 55% for 400000 cycles. The structural performance of ECC link slabs is compared with their self-consolidating concrete (SCC) counterparts based on load-deformation/moment-rotation responses, residual strength, strain developments, cracking patterns, ductility index and energy absorption. The ECC link slabs showed superior performance even at higher fatigue stress level compared to their SCC counterparts exhibiting higher post-fatigue residual strength, energy absorbing capacity and ductility with overall lower fatigue damage evolution.

1 INTRODUCTION

Bridge rehabilitation is becoming a major issue for transportation authorities due to decreasing budget allocation. Mechanical expansion joints are major source of bridge deterioration in North America (Au et al. 2013). The possible approach for improving the durability performance of the bridge structures is the elimination of the mechanical expansion joints. Besides high installation cost, these expansion joints need constant maintenance due to low durability performance. The leakage of chloride-containment and water through these expansion joints lead to corrosion of the steel girders and deterioration of the substructures (Kim et al. 2004). Mechanical expansion joints are major source of bridge deterioration in North America (Au et al. 2013).

Different design methodologies have been proposed by scholars and transportation authorities for substituting these mechanical expansion joints (Au et al. 2013; Alampalli &Yannotti 1998; Gilani 2001; Zia et al. 1995). The first approach is to use the haunched deck for developing a continuous deck system over the piers while keeping the girders as simply supported span. The second method is the construction of integral concept with continuity of the girders and the third approach is the implementation of the link slabs on the deck for developing a joint-free bridge deck with simply supporting spans as shown Figure 1. The joint free-bridge deck is the most economical and efficient rehabilitation strategy compared to other proposed methods (Alampalli & Yannotti 1998; Hossain & Anwar 2014).

The link slab is defined as a section of the deck connecting the two adjacent simple-span girders. The total length of the link slab consists of the debonding and transition zone as it is illustrated in Figure 2. The length of debonding zone (where all the shear connectors are removed and debonding mechanism is placed on top of the steel girder flange) is limited to 5% of each adjacent bridge

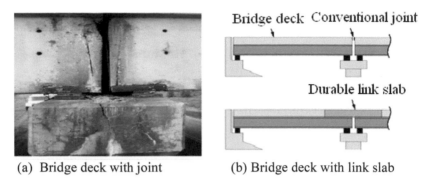

(a) Bridge deck with joint   (b) Bridge deck with link slab

Figure 1.   Joint free bridge deck with link slab.

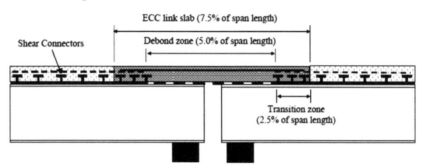

Figure 2.   Schematic of link slab showing components.

span. For reducing the stress concentration on the debonding zone, transition zone is introduced with additional 2.5% increase in length and 50% additional shear stud connectors than required by AASHTO (2012) design code (Caner & Zia, 1998; Qian et al. 2009).

Engineered cementitious composite (ECC) is a special type of high performance fiber reinforced concrete material that offers significant potential for resolving the durability performance of reinforced concrete structures (Li & Kanda 1998; Li et al. 2002; Li 2003). The high strain hardening characteristic and multiple micro-cracking behavior under tension and flexure while minimizing the amount of the reinforcing fibers (less than 2% by volume) makes ECC an ideal material for the link slab application (Sahmaran et al. 2010; Fischer & Li 2003; Hossain & Anwar 2014). Over the last few years, the locally available aggregates and supplementary cementing materials (SCM) have been incorporated into the traditional ECC mixes to produce green, sustainable and cost effective mixes for structural applications (Ozbay et al. 2014; Hossain & Anwar 2014; Sherir et al. 2013; Sahmaran et al. 2009; Sherir 2012). The use of ECC link slab has been a new emerging technology and over the years few research studies have been conducted on the behavior under static monotonic loading. However, very limited research studies have been conducted until to date on the flexural fatigue performance of link slabs made with green ECC mixtures (Sherir 2012; Sherir et al. 2013; Suthiwarapirak et al. 2004). The flexure fatigue performance of flexible ECC link slabs under repetitive fatigue loading is very important for their application in the construction of sustainable joint-free bridge construction with enhanced durability and service life.

This paper presents the results of experimental investigation on green ECC link slabs (made with high volume class CI fly ash and local mortar sand) subjected to flexural fatigue loading applied at variable fatigue stress levels and cycles. The structural performance of ECC link slabs is compared to their self-consolidating concrete (SCC) counterparts based on load-deformation/moment-rotation response, strain developments, cracking patterns/crack width, ductility and energy absorbing capacity.

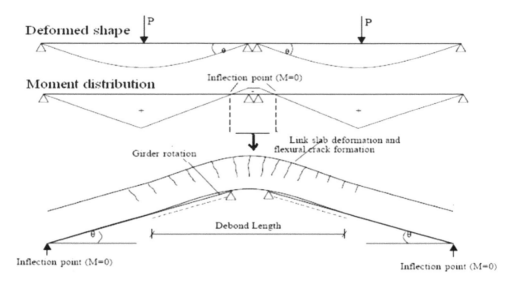

Figure 3. Two span bridge deformation with focus on link slab section (Kim et al. 2004).

2 EXPERIMENTAL INVESTIGATION

An extensive experimental research has been planned for studying the structural performance of ECC link slab subjected to flexural fatigue loading. A green ECC mixture made with locally available mortar sand (in place of expensive silica sand) and class CI fly ash at 50% replacement level by weight of cement was selected for the construction of link slab specimens. Flexure fatigue performance of ECC link slabs was evaluated by conducting tests at different fatigue stress levels of 40 and 55% at constant 400000 fatigue cycles. For a comparative analysis, both ECC and conventional SCC link slabs were tested. Flexural monotonic tests to failure were conducted on link slab specimens before fatigue tests to determine the loads for various stress levels of fatigue testing. Post-fatigue flexural monotonic testing (to failure) was performed after fatigue tests for determining the residual strength/stiffness, load-deformation/moment-rotation response, strain development, cracking characteristics and failure modes of the link slabs.

2.1 *Link slab configuration, test specimens, material properties and casting*

The typical deformed shape and moment distribution due to the applied loading for a two span bridge structure with a link slab is schematically represented in Figure 3. Flexural crack formations were expected at the top of the link slab due to the development of tensile stresses. The link slab section was designed between the points of inflections. Based on the stiffness of the link slab, the location of the inflection points varies from 0 to 20% of the adjacent span's length (Kim et al. 2004; Caner & Zia 1998).

The testing was focused on the link slab section within the inflection points. The location of inflection points (length of the link slab) based on the stiffness of the section (concrete and the steel rebar) was determined as 6.5% of the adjacent span's length. Overall, the total length of the links slab for the 1/4th scale model was 810 mm. The length for the debonding zone based on the proposed design guidelines by Caner and Zia (1998), was set as 5% of simple spans; a total length of 330 mm. The length of the transition zone with additional 50% shear stud connector (total of 16 studs) was 300 mm (150 mm at each side) (Qian et al., 2009). The cross-sectional dimensions for the 1/4th link slab models were 175 mm in width and 60 mm in depth. A minimum longitudinal steel reinforcing ratio of 1.1% (three 6 mm bars) was provided for maintaining a low structural stiffness of the link

Table 1. Link slab specimens configuration subjected to monotonic and fatigue loading.

	Geometric properties of link slab					Testing parameters	
Designation	Length (mm)	Debond zone length 2.5% (mm)	Transition zone length 5% (mm)	Concrete in transition zone	Concrete in debond zone	Number of Cycle	Fatigue stress level (%)
LS-ECC-control	810	330	150	SCC	ECC	0	0
LS-ECC-40-400000	810	330	150	SCC	ECC	400000	40 ± 20
LS-ECC-55-400000	810	330	150	SCC	ECC	400000	55 ± 20
LS-SCC-control	810	330	150	SCC	SCC	0	0
LS-SCC-40-400000	810	330	150	SCC	SCC	400000	40 ± 5

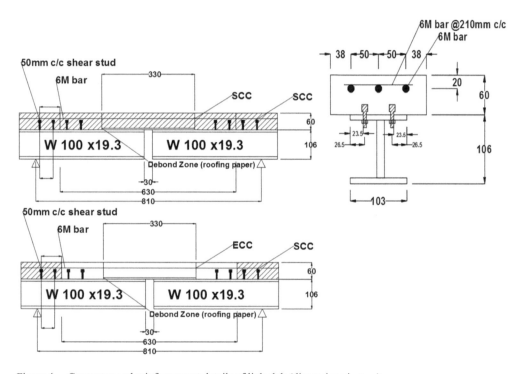

Figure 4. Geometry and reinforcement details of link slab (dimensions in mm).

slab. Additional transverse reinforcements at 210 mm c/c spacing were provided using 6 mm bars. The summary of the specimen's configurations, reinforcement detailing and concrete mixtures are presented in Table 1 and Figure 4. A W100x19 steel structural section in accordance to Canadian Institute of Steel Construction (CISC 2010) was used as supporting adjacent girders for the link slab specimens. The reinforcing steel bars had a mean yield and ultimate strength of 407 MPa, 550 MPa and 224 GPa, respectively.

Two concrete mixtures namely ECC and SCC were used for the deck construction of the scaled down link slab specimens. The green ECC mixture (developed at Ryerson University) incorporated local mortar sand aggregate (instead of traditional micro-silica sand) and class CI fly ash as per ASTM C618 standard (at 50% replacement of cement by weight). The mix design of the developed ECC mixture is presented in Table 2 in terms of weight of the cement content.

Table 2. ECC mix design proportions by weight of cement.

Specimen	Water	Cement	Fly Ash CI	Mortar Sand	Fiber Content	HRWRA
ECC-CI-1.2-MS	0.585	1	1.2	0.798	0.045	0.0095

The cement used for the ECC mixture was type GU Portland cement. Mortar sand was used with maximum aggregate size of 1.18 mm for maintaining the strain hardening characteristic of the ECC mixture. The provided PVA fibers were 39 μm in diameter and 8 mm in length with a tensile stress capacity of 1600 MPa. For improvement of ECC's workability, the polycarboxylic-ether type high-range water-reducing admixture (HRWRA) with 30% of solid content was used. A commercial ready-mix SCC mixture having 10 mm maximum aggregate size, sand, Portland cement, silica fume and air-entraining admixtures was used

Similar to the general field practice, the bridge deck portion at the ends were cast initially with flowable SCC mix, and left for 24 hours for setting. Then the link slab segment was cast with ECC or SCC after initial 24 hours. At least three control specimens for each types of concrete in forms of cylinders, beams and cubes were also casted at the same time. All the specimens were cured for 28 days using wet burlaps in the laboratory conditions with a relative humidity and temperature of $50 \pm 2.5\%$ and $24 \pm 2°C$, respectively. The 28 days mean compressive and flexural strength of the ECC mix were 55.5 MPa and 11.1 MPa, respectively while for SCC mix, they were 36.4 MPa and 4.1 MPa, respectively.

2.2 *Test set-up, instrumentation and testing*

Flexural fatigue test (under four point loading) was performed to investigate the cyclic behavior of 1/4th scale ECC link slab specimens in compare to the control SCC specimen at the age of 28 days. The static four point monotonic loading test to failure of specimen was also performed (prior to flexural fatigue testing) under displacement control condition at a loading rate of 0.005 mm/s.

The flexural fatigue tests were performed by applying different mean fatigue stress levels of 40% and 55% of the ultimate load (determined from pre-fatigue static loading) at fixed cycle of 400000 at 4 Hz (4 cycles per second). The specimens were pre-monotonically loaded to the mean fatigue stress levels of 40% and 55% under load control condition, at a rate of 0.5 kN/min as a ramp to fatigue loading. Then fatigue loading was applied with an upper and lower limit of 20% of the applied mean stress level.

The testing was performed using a closed-loop controlled servo-hydraulic system. During the fatigue flexural tests, the deflection evolution at various points including mid-span was recorded by LVDT's using data acquisition system as shown in Fig. 5a. Data acquisition system was also used for recording the concrete and rebar strain (installed at mid span as shown in Fig. 5a) as well as girder end rotation (using inclinometer) developments. The crack propagation was visually observed and recorded for every 10000 fatigue cycles. Figure 5 illustrates the instrumentations and test set-up for the flexural fatigue testing of the link slabs.

After fatigue flexural test, static flexure test was performed on the exhausted fatigued link slab specimens to determine the residual strength, ductility, end-rotation and strain development. Similar approach was adopted to test companion non/pre-fatigued link slab specimen tested under static monotonic loading to failure.

3 RESULTS AND DISCUSSION: FATIGUE BEHAVIOUR AT DIFFERENT STRESS LEVELS FOR CONSTANT FATIGUE NUMBER OF CYCLES

The link slabs were loaded to 40% and 55% of ultimate load before commencing flexure fatigue tests up to 400000 cycles. During fatigue loading, 40% and 55% stresses were selected as the mean

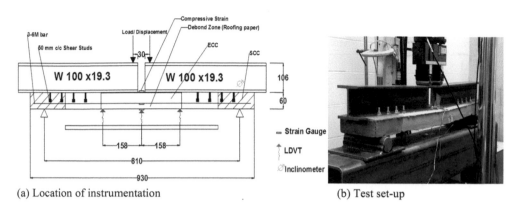

(a) Location of instrumentation

(b) Test set-up

Figure 5.   Test set-up and instrumentation of 1/4th scale link slab specimens (dimensions in mm).

Table 3.   Mid-span deflection evolution.

Designation	Fatigue stress level (%)	Difference between first and last value of mid-span deflection (mm)
LS-ECC-40-400000	40 ± 20	2.39
LS-ECC-55-400000	55 ± 20	1.72
LS-SCC-40-400000	40 ± 20	0
LS-SCC-40-400000	40 ± 5	0.42

stress level and were varied with stress peak amplitude of ±20% from mean (that means stress ranges were 40% ± 20% and 55% ± 20%, respectively). ECC link slabs allowed fatigue tests with 20% amplitude. However, SCC link slab with 20% fatigue amplitude failed at the beginning of the fatigue testing prior to completion of the first cycle. Due to the weak response of SCC link slab, the peak amplitude was reduced to 5% of the stress level as shown in Table 3.

3.1   *Deflection evolution, strain development and crack characterization*

The mid-span deflection evolutions and strain developments for the ECC and SCC link slabs subjected to different fatigue stress levels are presented in Figures 6–7 and Table 3.

It is observed from Figure 6a that the mid-span deflection for the ECC link slabs (at 40% ± 20% and 55% ± 20% stress level) increased drastically during fatigue loading due to formation of larger deflection band widths in compare with the SCC link slab (at 40% ± 5 fatigue stress level) with no band width development. While the increase in deflection for the SCC link slab at 40% ± 5 stress level was only 0.42 mm, the increase for both ECC link slabs at 40% ± 20 and 55% ± 20 stress level were 2.39 mm and 1.79 mm, respectively (Table 3). The large deflection band width for the ECC link slabs demonstrate a large bending capacity and strain hardening characteristic even at the high fatigue stress range (±20% of ECC's ultimate loading). This shows that after formation of cracks under repetitive fatigue loading at high stress range of ±20%, the bending capacity of the ECC is enhanced; therefore ECC controls the fatigue performance. SCC link slab at the same stress range of ±20% failed before completing a cycle.

From the mid-span deflection evolution curves (Figure 6b), it is observed that all link slab specimens (ECC and SCC) deformed more at the initial fatigue cycles. The ECC link slab at 40% ± 20 fatigue stress level deforms from 2.31 mm to 4.10 mm and the SCC link slab at 40% ± 5%

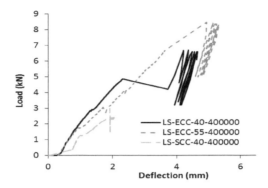

Figure 6(a).   Load-deflection response.

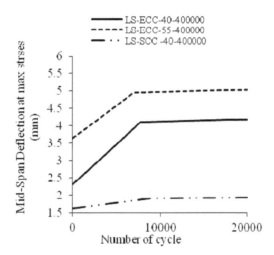

Figure 6(b).   Deflection evolution up to 20000 cycle.

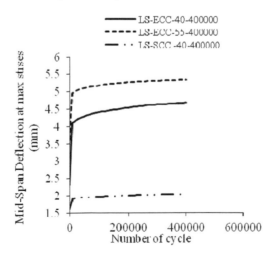

Figure 6(c).   Deflection evolution up to 400000 cycle.

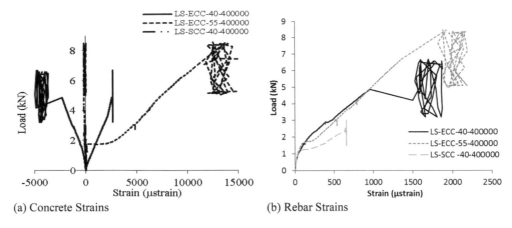

(a) Concrete Strains

(b) Rebar Strains

Figure 6.   Strain development for link slabs at different fatigue stress level.

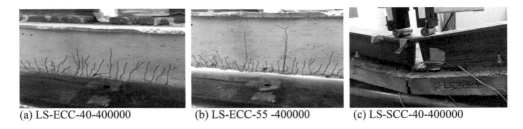

(a) LS-ECC-40-400000   (b) LS-ECC-55 -400000   (c) LS-SCC-40-400000

Figure 7.   Crack formation and propagation in CC link slabs tested at different fatigue stress level.

fatigue stress level deformed from 1.62 to 1.91 mm at the initial 10000 fatigue cycle (Figure 6b). The high deflection evolution at the initial fatigue cycle was caused by the stress jump from the mean stress (attained by the monotonic loading) to high fatigue stress ranges (±20% for the ECC and ±5% for the SCC link slabs).

It is also noted from Table 3 and Figure 6a that larger deflection band width was developed for the ECC link slab at 40% ± 20 fatigue stress level in compare with the ECC specimen at 55% ± 20 stress level. The change in mid-span deflection for the ECC link slab at 40% ± 20 was 2.39 mm whereas the change for the ECC link slab at 55% ± 20 stress level was 1.72 mm only (Table 3). The high bending capacity of the ECC link slabs under flexural fatigue loading was attributed to the high interfacial bond among the PVA fibers and the matrix, where fibers were ruptured at its ultimate strength rather than a pull-out (Suthiwarapirak et al. 2004). In contrast, the fatigue behavior of the SCC link slab was dependent to the failure of the reinforcing bars (Zanuy et al. 2011). For comparative purpose, a minimum amount of reinforcing bars were provided, which led to a small increase in deflection of 0.42 mm for the SCC link slab (Table 3).

As it is evidenced from Figure 7(a), no concrete tensile strain band width was formed for the SCC link slab subjected to 40% ± 5% stress level. In fact, SCC concrete tensile strain was much lower at the beginning of fatigue performance because of stress release due to crack formation. Figure 7(a) also shows lower concrete tensile strain development at the initial fatigue cycle (at 10000 cycle). The rebar strain development was increased from 529.5 to 640 micro-strain for the first 10000 fatigue cycle. The increase in rebar strain for the SCC link slab at 40% ± 5% may be associated with the transfer of load due to crack formation to the steel reinforcement. Beyond the first 10000 fatigue cycle, rebar strain development remained constant since the applied stress range was very low (±5% of the mean stress level). The strain development characteristics justifies that the flexural fatigue resistance of SCC link slab was derived predominantly from the rebar

Table 4.   Crack characterization of link slabs at different fatigue stress level (at the end of test).

Designation	Fatigue stress level (%)	Number of cracks formed	Crack widths (μm)
LS-ECC-40-400000	40 ± 20%	17	100
LS-ECC-55-400000	55 ± 20%	16 micro-cracks and	150
		2 major cracks	560
LS-SCC-40-400000	40 ± 5%	1 Major crack	1000

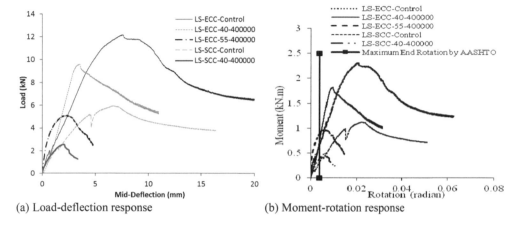

(a) Load-deflection response   (b) Moment-rotation response

Figure 8.   Load-deflection and moment-rotation responses.

contribution. In contrast ECC link slab at higher fatigue stress range (±20%) developed larger concrete tensile strain band width during the fatigue loading (Figure 7a) – which demonstrates its strain hardening characteristic that allows more bending and higher strain development compared to its SCC counterpart. Although the rebar strain was increased for the ECC link slab at 40% ± 20% stress level specifically for the first 10000 cycles, but it did not reach the yield strain (2000 micro-strain) even at the end of the 400000 fatigue cycles (Figure 7b). This is an indication of damage in ECC due to micro-crack formation during fatigue loading and gradual transfer of load to the steel rebar causing an increase in rebar strain. The higher increase in rebar strain for the ECC link slab was caused by higher fatigue stress range of ±20% compared to ±5% of SCC link slab. The strain development signifies that ECC plays a significant role in enhancing flexural fatigue resistance and in mitigating damage.

The ECC tensile strain development for the link slab tested at higher fatigue stress level of (55% ± 20%) was greater compared to specimen tested at 40% ± 20% as indicated by the formation of larger tensile strain band width (Figure 7a). However, developed ECC tensile strain was still lower compared to its tensile strain capacity 0.05. The rebar strain was close to the yield strain at the beginning of the fatigue loading and reached the yield strain at 147000 fatigue cycle. The increased rebar strain at 147000 fatigue cycle justifies that stresses were transferred to the reinforcing bars due to formation of micro-cracks along with PVA fiber rupturing bridging the crack. The rebar tensile strain for the ECC link slab tested at 55% ± 20% fluctuated at yield strain beyond 147000 fatigue cycle showing an increasing trend. Overall, the large width of strain band development (both steel and concrete) for ECC link slabs compared to their SCC counterparts is an indication of ECC's superior performance in terms of inducing greater flexibility, ductility and energy absorbing capacity. The inferior performance is attributed to SCC's weak tension stiffing capacity under fatigue loading (Giordano & Mancini 2009).

The cracking propagation was visually observed for every 10000 cycles. The number of the cracks, widths and cracking patterns are presented in Table 4 and Figure 8.

Table 5.   Summary of the load-deflection and moment-rotation responses of the fatigued link slabs.

Designation	Fatigue cycle	Fatigue stress level (%)	Ultimate load (kN)	Ultimate deflection (mm)	Rotation at ultimate (Radian)	Moment at ultimate (kN.m)
LS-ECC-Control	0	0	12.16	7.59	0.0205	2.31
LS-ECC-40-400000	400000	40 ± 20	9.60	3.48	0.0094	1.82
LS-ECC-55-400000	400000	55 ± 20	5.12	2.31	0.0062	0.97
LS-SCC-Control	0	0	5.93	6.74	0.0222	1.27
LS-SCC-40-400000	400000	40 ± 5	2.57	2.01	0.0063	0.48

Flexural cracks were developed on the tension face at mid span of the link slabs. The first cracking appeared during pre-monotonic loading for all ECC link slabs followed by initiation of additional micro-cracks during flexural fatigue loading. In contrast, the first cracking for the SCC link slab at 40% stress level appeared at the first fatigue cycle, even though the peak stress amplitude was reduced to 5%. Few micro-cracks, in addition to a major crack were formed in the mid-span of link slab until the flexural strength of the concrete was degraded at 80000 cycle. As the number of the cycles increased for the SCC link slabs, the crack width grew wider and formed one major crack (Figure 8c). As Zanuy et al. (2011) described, the reduction of bond strength among the reinforcing bars and the concrete after repeated fatigue loading increased the SCC crack width to 1 mm (Table 4). It should be noted that crack widths for all ECC link slab remained below 560 μm after 400000 fatigue cycle due to fiber bridging phenomena (Table 4 and Figure 8a–b).

The high fatigue stress level developed higher number of the cracks and higher crack width (Table 4 and Figure 8). Micro-cracks were formed initially for the link slab subjected to 55% stress level along with two additional major cracks after the PVA fiber was ruptured at 147000 cycle (Figure 8b). The crack widths grew wider until it reached 560 μm due to the presence of the reinforcing bar (Table 4). In contrast, the crack widths for the ECC link slab at 40% stress level remained below 100 μm.

3.2   Post fatigue monotonic behavior of link slabs at different fatigue stress levels

The flexural monotonic loading was applied immediately to the exhausted fatigued link slab specimens (tested at different stress level for 400000 cycles) to determine the post-fatigue residual strength, deflection capacity, strain characteristics, rotational stiffness and ductility.

3.2.1   Load-deflection and moment-rotation responses

The summary of the residual fatigue strength, mid-span deflection, rotation and moment resistance are presented in table 5, while load-deflection and moment-rotation responses are presented in Figure 9.

The ultimate load and mid-span deflection of the fatigued link slabs were reduced significantly compared to the controlled specimens. The ultimate load and the mid-span deflection of the SCC link slab was reduced by 57% and 70%, respectively (Table 5). This shows that SCC exhibited more damage compared to the ECC link slabs even though its peak fatigue stress amplitude was limited to 5%. The ECC link slab at the same mean stress level but higher peak of amplitude exhibited greater residual ultimate load and bending capacity (ductility) due to its multiple cracking characteristic and strain hardening behavior that is attained by the PVA fibers (Suthiwarapirak et al. 2004). From figure 9 it is evidenced that after fatigue performance, the residual support end-rotations of the damage link slabs are satisfying the maximum AASHTO (AASHTO 2012) serviceability limit of 0.00375 radian. However the SCC link slab exhibited a smaller factor of safety, since the end-rotation of 0.00375 was attained at 80% of the ultimate loading, near the failure point of the specimen. In contrast the ECC link slab at same mean stress level but higher peak stress amplitude demonstrated the residual allowable rotation at only 40% of the ultimate loading.

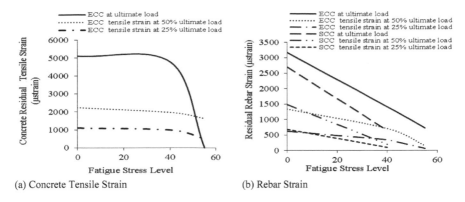

(a) Concrete Tensile Strain   (b) Rebar Strain

Figure 9.   Residual strain development for the link slabs at different fatigue stress level.

Table 6.   Residual strain developments, energy absorbing capacity and ductility index of link slabs.

Designation	Fatigue cycle	Fatigue stress level (%)	Tensile strain at ultimate load (μstrain)	Rebar strain at ultimate load (μstrain)	Compressive strain at ultimate load (μstrain)	Energy (Joules)	Ductility index
LS-ECC-Control	0	0	5106	3172	−1481	93.125	37.2
LS-ECC-40-400000	400000	40 ± 20%	4777	1368	−4855	31.372	16.9
LS-ECC-55-400000	400000	55 ± 20%	0	724	−445	13.293	2.2
LS-SCC-Control	0	0	0	2723	−783	35.62	13.4
LS-SCC-40-400000	400000	40 ± 5%	0	650	−329	4.035	2.5

The residual strength and bending capacity for the ECC link slab was reduced significantly at higher mean stress level. The steep slope of the post-peak branch of the load-deflection response in Figure 9 indicates the brittle failure of the ECC link slab at 55% mean fatigue stress level. This characteristic demonstrates the loss of the ECC tensile strain hardening, and proves that PVA fibers were ruptured to their ultimate strength during the fatigue loading. Further, the small increase in residual ultimate load and deflection at the post-monotonic loading was due to the provided minimum steel rebar. In contrast the strain hardening characteristic of the PVA fibers within the ECC link slab at 40% stress level were maintained since the post-peak failure slope of the load-deflection response was similar to its non-fatigued controlled specimen.

The reduced ductility response of the fatigued link slabs at higher fatigue stress level resulted in a lower moment capacity and residual support end-rotations as it is evidenced by LS-ECC-55-400000. At failure the residual end-rotation still satisfied the AASHTO allowable maximum end-rotation but at the lower factor of safety. This indicates that the steel rebar that carried all the bending stresses also reached its yielding stage. In contrast, the ECC link slab at 40% fatigue stress level exhibited the allowable end-rotation at higher factor of safety.

3.2.2 *Strain developments, energy absorbing capacity and ductility index*
The summary of residual strain developments, energy absorbing capacity and ductility index of the fatigued link slabs compare to control (non-fatigued) specimens are represented in Table 6.

Figure 10 compares the residual concrete tensile and rebar strain developments for various loading cases of 25%, 50% and 100% of the ultimate loading at different fatigue stress levels. As expected, the residual strain development for the fatigued link slab specimen was significantly reduced compared to the non-fatigued controlled link slabs (Figure 10 and Table 6). At the same

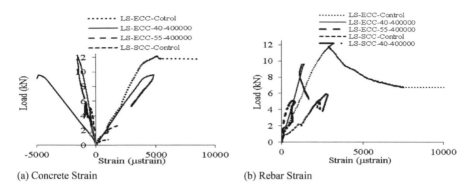

(a) Concrete Strain   (b) Rebar Strain

Figure 10.   Residual strain for the link slabs at different fatigue stress level.

fatigue mean stress level of 40%, the ECC link slabs exhibited a very high tensile residual strain of 4777 compared to the SCC link slab with no strain development. The loss of concrete tensile strain due to the crack formation for the SCC link slab (resulted as a tension stiffening at the end of the first cycle) caused the bond reduction between the concrete and the rebar (Chen et al. 2011; ACI 1992). The loss of concrete tensile strain and transferred stresses to the rebar caused the significant plastic deformation for the rebar. For the same reinforcing ratio, the residual rebar strain was significantly lower compared to the ECC link slabs, even though the peak stress amplitude for the ECC was 15% higher than SCC link slabs (Figure 11).

The small residual rebar strain indicates failure due to the significant plastic deformation during the fatigue loading – therefore the continuation of the post-monotonic test for the SCC link slab at the post-peak failure, would have resulted due to the fracture of the reinforcing bar (Zanuy et al. 2011). In contrast, the high residual rebar strain for ECC link slab demonstrated the active presence of concrete tensile strain during the fatigue loading. Although the concrete tensile strain was reduced compared to the non-fatigued ECC link slab, but the loss of both concrete and rebar strain was not significant. The high residual strain development of the fatigued ECC link slabs proves ECC's stiffness remains unchanged during the fatigue loading (Li 2003).

The ECC link slab subjected to 55% fatigue mean stress level exhibited a similar residual strain development like SCC. The small residual rebar strain and negligible post peak failure strain development with no residual ECC tensile strain formation indicates the failure of the link slab at 55% fatigue loading. The high stress fatigued loading ruptured the PVA fiber at early fatigue cycle which caused the stress transfer to the reinforcing bar. The residual concrete tensile and rebar strain development for the ECC link slab at 40% mean fatigue stress level were was 6% and 57% lower than the non-fatigued control specimens. The fatigued link slabs have lower residual energy absorbing capacity and ductility index, than the non-fatigued control specimens (Table 6). At the higher peak stress amplitude of 20% and constant mean fatigue stress level, the ECC link slab demonstrated 8 times higher energy absorbing capacity and ductility index in compare to the SCC link slab. This attribute represents the strain hardening and fiber bridging characteristics of ECC matrix attained through the PVA fibers (Nawy 2008).

4   CONCLUSIONS

This paper describes the fatigue flexural performance of link slabs made with engineered cementitious composite (ECC) compared to their self-consolidating concrete (SCC) counterparts. The following conclusions are drawn from the study:

- ECC link slabs showed superior structural performance subjected to high fatigue stress levels (maximum 60% and 75%) compared to their SCC counterparts subjected to comparatively low fatigue stress level of 45% up to 400000 cycles. For ECC link slabs, the ultimate strength, rebar and concrete strain developments, ductility and energy absorbing capacity were reduced only by 20% of the non-fatigued specimen whereas for the SCC link slab even with lower fatigue stress level (maximum 45%), they were reduced by 60%. Flexible ECC links slab will develop reduced negative moment for allowable girder rotation prescribed by codes and can be designed accordingly.

• While the polyvinyl alcohol (PVA) fibers were mainly responsible for resisting the high fatigue stress levels for the ECC link slabs, the steel rebar resisted the fatigue loading for the SCC link slab. Therefore with low reinforcing ratio, the SCC link slab demonstrated a poor performance under fatigue loading. The maximum fatigue stress level for the ECC link slab should be limited to 65% of its ultimate strength for strength and serviceability.

• At the higher peak stress amplitude of 20% and constant mean fatigue stress level, the ECC link slab demonstrated eight times higher energy absorbing capacity and ductility index compared to their SCC counterparts. The study confirmed the viability of constructing link slabs using fly ash and local mortar sand based green ECC for building cost-effective sustainable joint-free bridge decks with satisfactory structural performance.

REFERENCES

Alampalli, S. & Yannotti, A.P. 1998. In-service performance of integral bridges and joint less decks. *Transportation Research Record* 1624(98): 1–7.

ACI Committee 215. 1992. Considerations for design of concrete structures subjected to fatigue loading (ACI 215R-92). American Concrete Institute, Detroit.

ASTM C 618. 2012. Standard specifications for coal fly ash and raw or calcined natural pozzolan for use in concrete. American Society for Testing and Materials, West Conshohocken, PA, USA.

AASHTO. 2012. *AASHTO LRFD Bridge Design Specifications, 6th Edition*, American Association of State and Highway Transportation Officials (AASHTO), Washington DC, USA.

Au, A., Lam, C., Au, J. & Tharmabala, B. 2013. Eliminating deck joints using debonded link slab: research and field test in Ontario. *Journal of Bridge Engineering* 18(8): 768–778.

Caner, A. & Zia, P. 1998. Behavior and design of link slabs with jointless bridge deck. *PCI J.:* 68–80.

Chen, H. J., Liu, T. H., Tang, C. W. & Wen, P. T. 2011. Influence of high-cycle fatigue on the tension stiffening behavior of flexural reinforced lightweight aggregate concrete beams. *Structural Engineering and Mechanics Journal* 40(4): 1–20.

CISC, 2010. *Handbook of Steel Construction, 10th edition*, Ontario, Canada: CISC.

Fischer, G. & Li, V.C. 2003. Intrinsic response control of moment resisting frames utilizing advanced composite material and structural elements. *ACI Structural Journal* 100(1): 166–176.

Gilani, A. 2001. Link slabs for simply supported bridges incorporating engineered cementitious composites. MDOT SPR-54181. *Structural Research Unit, MDOT*, USA.

Giordano, L. & Mancini, G. 2009. Fatigue behavior simulation of bridge deck repaired with self compacting concrete. *Journal of Advanced Concrete Technology* 7(3): 415–424.

Hossain, K.M.A. & Anwar, M.S. 2014. Strength and deformation characteristic of ECC link slab in joint-free bridge decks, Istanbul Bridge Conference, Istanbul Bridge Conf., August 11–13, Turkey.

Hossain, K.M.A. & Anwar, M.S. 2014. Properties of Green Engineered cementitious composites incorporating volcanic materials, *Structural Faults & Repair,* 8th–10th July, Imperial College, London, UK.

Kim, Y.Y., Fischer, G. & Li, V.C. 2004. Performance of bridge deck link slabs designed with ductile ECC. *ACI Structural Journal* 1(6): 792–801.

Li, V.C. & Kanda, T. 1998. Engineered cementitious composites for structural applications. *Journal of Materials in Civil Engineering* 10(2): 66–69.

Li, V. C., Wang, S., Ogawa, A. & Saito, T. 2002. Interface tailoring for strain-hardening PVA-ECC. *ACI Materials Journal* 99(5): 463–472.

Li, V.C. 2003. Engineered cementitious composites (ECC) – A review of the material and its applications. *Journal of Advanced Concrete Technology* 1(3): 215–230.

Li, V.C., Lepech, M.D. & Li, M. 2005. Field demonstration of durable link slabs for jointless bridge decks based on the strain-hardening cementitious composites, Michigan Department of Transportation Report No. RC-1471, pp. 1–147.

Nawy, E. 2008. *Concrete Construction and Engineering Handbook*, 2nd Edition, New York, CRC Press.

Ozbay, E.M., Lachemi, M., Karahan, O., Hossain, K.M.A. & Atis, C.D. 2014. Investigation of the properties of ECC incorporating high volumes of fly ash and Metakaolin. *ACI Materials J.* 109(5): 565–571.

Qian, S., Lepech, M.D., Kim, Y.Y. & Li, V.C. 2009. Introduction of transition zone design for bridge deck link slabs using ductile concrete. *ACI Structural journal* 106(1): 96–105.

Sahmaran, M., Lachemi, M., Hossain, K.M. & Li, V.C. 2009. Influence of aggregate type and size on ductility and mechanical properties of ECCs. *ACI Material Journal* 106(3): 308–316.

Sahmaran, M., Lachemi, M., Hossain, K.M.A., Ranade, R. & Li, V.C. 2010. Internal Curing of ECC's for prevention of Early Age Autogenous Shrinkage Cracking. *Cement & Concrete Res.* 39(10): 893–901.

Sherir, M. 2012. *Fracture energy, fatigue and creep properties of ECC's incorporating fly Ash/slag with different aggregates.* MASc Thesis, Ryerson University, Toronto Canada.

Sherir, M.A.A., Hossain, K.M.A. & Lachemi, M. 2013. Behaviour of Engineered Cementitious Composites Under Fatigue Loading, *3rd Specialty Conference on Material Engineering & Applied Mechanics*, Montréal, Québec, May 29 to June 1.

Suthiwarapirak, P., Matsumoto, T. & Kanda, T. 2004. Multiple cracking and fiber bridging characteristic of ECCs under fatigue flexure. *Journal of Material in Civil Engineering* 16(5): 433–443.

Zanuy, C., Maya, L.F., Albajar, L. & Fuente, P.D. 2011. Transverse fatigue behavior of lightly reinforced concrete bridge decks. *Journal of Engineering Structure* 33(10): 2839–2849.

Zia, P., Caner, A. & El-Safte, A. 1995. Jointless bridge decks research project 23241-94-4. *Center for Transportation Engineering Studies*, North Carolina State, pp. 1–117.

Bridge bearings

Chapter 23

Modern bearings for key bridges – special functions & type selection

A. Kutumbale & G. Moor
Mageba USA, New York, NY, USA

ABSTRACT: With recent advancements in bridge design technology, bridge bearings are required to address significant further challenges in addition to their primary functions of resisting loads and accommodating movements and rotations. This paper presents developments in the design of bridge bearings, focusing on innovative solutions such as uplift-restraining bearings subjected to fatigue loading, temporary locking of bearings to resist construction loads, bearings with adjustable height and easily replaceable bearings. Case studies include key bridges in the United States such as the Ohio River Bridge (Kentucky/Indiana), the St. Croix Bridge (Minnesota/Wisconsin) and the Bayonne Bridge (New York/New Jersey).

1   INTRODUCTION

Bridge bearings play a critical role in the structures in which they are installed – in particular, by accommodating superstructure rotations, by allowing movements or transferring horizontal forces, and perhaps most significantly, by transferring vertical forces between superstructure and substructure. Such vertical forces are not always downwards; as uplift forces may also need to be resisted by bearings in certain circumstances. Such resultant uplift forces, where they arise at all, are generally very infrequent, occurring only in exceptional circumstances. A typical solution for resisting such infrequent uplift forces is shown in Figure 1.

In recent years, with advancements in bridge design and the advent of new construction methods, bridge designers have been requiring bearings to perform further functions in addition to those listed above. These include resisting frequent uplift forces (with impacts, durability and even dynamic loading and fatigue becoming significant), providing dynamic restraint (temporary locking), and accommodating bumper contact between bridge elements such as deck and pylon. As well as performing such auxiliary functions, bearings are also increasingly being expected to offer greater

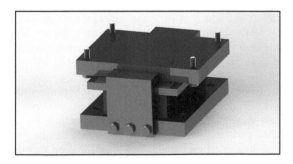

Figure 1.   A sliding elastomeric bearing, with uplift clamps preventing upward forces (Revere Beach Bridge, Massachusetts).

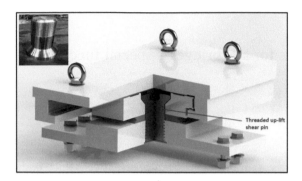

Figure 2.   A sliding disc bearing with simple threaded pin connection providing internal uplift restraint (Bayonne Bridge, New York/New Jersey).

durability and life-cycle performance, resulting in demands for the development of innovative replacement techniques and the use of improved bearing materials. Some of these aspects of bearing design and supply are described below, with reference to case studies from applications on recent landmark bridge construction projects in the United States.

2   BEARING DESIGN FOR FREQUENT UPLIFT FORCES

Uplift forces can arise in bridge bearings for a variety of reasons. For example, bridges that are designed with severe skews or unbalanced continuous spans, or subjected to very strong winds or significant dynamic forces (e.g. railway bridges) may require their bearings to resist uplift forces. Traditionally, uplift forces have been resisted by concrete counterweight over the bearings, or by changing continuous spans to simple spans. However, if the bearings can be designed to resist uplift forces, then modifications to the bridge structure can be avoided, resulting in significant cost savings.

In designing bridge bearings to resist uplift forces, the frequency of the uplift force is a crucial factor.

Where uplift forces will occur only infrequently, during exceptional circumstances, relatively simple solutions such as uplift clamps at the bearings sides can suffice – for example, as shown in Figure 1. Alternatively, where the bearing type allows (unlike, for example, in the case of pot bearings), the uplift resistance can be provided by an internal pin or similar, on the bearing's central vertical axis. An example of such a solution is shown in Figure 2, where a disc bearing, which generally features a central pin as standard to connect the bearing's upper and lower connection plates, has been designed to resist infrequent uplift forces.

For cases where uplift forces may occur in service at frequent intervals, the design of bearing requires greater deliberation, as frequent uplift introduces elements such as impacts, stress reversal (tension/compression cycles) and fatigue. The solution shown in Figure 2, for example, may suffer wear of threads over time under the action of frequent load reversals because the central pin must leave enough play for the central steel plate to rotate about a horizontal axis as the superstructure rotates.

To address this, it may be possible to provide an additional concave sliding interface to accommodate rotations. This can be done by replacing the simple internal pin with an uplift restraint, designed to accommodate sliding rotations, as shown in Figure 3. This can be designed, as illustrated, with two individual parts making the new "pin", which can be pre-tensioned together to ensure that the concave sliding interface remains in a state of compression, even under the action of the expected uplift forces, and the change in load direction only causes an increase or decrease in tension in the threads as opposed to stress resversal. Hammering of (and damage to) the sliding material is thus avoided, and the durability of the threaded connection is increased.

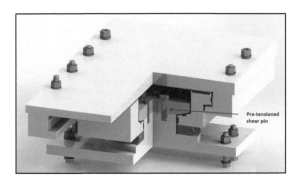

Figure 3.   A sliding disc bearing with pre-tensioned internal uplift mechanism.

Such connections should be thoroughly tested for fatigue before it can be assumed that they will perform satisfactorily over the entire life of bearing. It is important that the durability of the selected solution be properly evaluated before use, because replacement of an uplift bearing generally requires provision of temporary uplift restraints as well as lifting of the bridge deck and traffic diversions, and should therefore be avoided where possible.

Further examples of uplift-resisting bearings for special applications are presented by Spuler et al. (2013).

3   BUMPER BEARINGS – HEIGHT ADJUSTABILITY AND REPLACEMENT OF SLIDING MATERIAL

In the case of cable stay and extradosed bridges, bearings are often installed vertically between the bridge's deck and its pylons to provide lateral and longitudinal restraint. In addition to resisting high compressive loads and movements, such "bumper bearings" must also withstand hammering – a phenomenon which generally does not occur in horizontally-installed bearings where the minimum dead load is sufficient to keep the bearing in compression at all times.

Additionally, the horizontal gap between the deck and pylon varies over time, making bearing installation very challenging, given a constant bearing thickness. This also is generally not a problem in the case of horizontally installed bearings, since pedestal heights can be easily adjusted for concrete connections and filler plates may be used for steel connections.

Any gap between the sliding material (e.g. PTFE) and mating surface (generally stainless steel) will lead to hammering, which will directly impact on the life of the sliding material. Although modern sliding materials such as UHMWPE (ultra high molecular weight polyethylene), which offer higher compressive strength and greater durability than PTFE, are suitable for use in bumper bearings, hammering will still reduce their durability. The adverse effect of hammering on sliding material depends directly on the size of gap between the sliding material and the mating surface, as greater gaps during rotations lead to eccentric loading on the sliding material.

3.1   *Bumper bearings of the Ohio River Bridge (Downtown Crossing) – Kentucky & Indiana*

The main structure of the Ohio River Bridge (Downtown Crossing) is a cable stayed bridge, connecting Kentucky and Indiana across the Ohio River (Figure 4). Of the 20 bearings (10 elastomeric & 10 disc) required for the main span, 14 were bumper bearings (four providing longitudinal restraint and ten providing transverse restraint). The type selection and design of these bumper bearings posed several challenges, due to the limited plan area and height available, the high movement and load requirements, and the varying gap width between deck and pylon.

Figure 4.   The new Ohio River Bridge (Downtown Crossing) connecting Kentucky & Indiana.

Figure 5.   Design of an elastomeric bearing with replaceable sliding material and shim plate.

Per the bridge engineer's initial estimate, the accumulated sliding path on the bridge is expected to be 900 miles over 50 years. This amount of movement is exceptionally high for any bridge type, and requires the bearings to be designed with replaceability in mind, considering that PTFE, the most commonly used sliding material, suffers serious deterioration after a sliding distance of just 14 miles.

Due to the exceptional demands, the bearings supplied for this bridge are equipped with a high-grade UHMWPE sliding material known as Robo-Slide. This has been tested to a sliding distance of over 30 miles without any wear, proving far better durability than PTFE. Additionally, the compressive strength of Robo-Slide is twice that of PTFE, making it more suitable for high compressive stresses and hammering forces.

Nevertheless, for very high accumulated sliding paths, bearings must still be expected to be replaced during the bridge's life. For obvious reasons, replacement of bumper bearings can be a very cumbersome process. With this in mind, the Ohio River Bridge bearings were designed in such a way that the sliding material of all bearings can be replaced without the need to replace the entire bearing assembly. This was made possible by the provision of an additional replaceable plate (with sliding material embedded in it), bolted to the bearing plate (see Figures 5 and 6). Once the need to replace the sliding material is established, the replaceable plate can be unbolted and removed from the bearing, considering that bumper bearings have a gap between sliding material and mating surface. A new sheet of sliding material can be inserted in the replaceable plate, which can then be easily bolted back in place. The replaceable plate was also provided with lifting lugs, making the replacement process yet easier.

Another issue that arises with bumper bearings is the available gap for bearing installation and variation of this gap in service. As mentioned above, after installation of the bearing, the gap must remain as small as possible in order to avoid eccentric loading and hammering on the bearing's

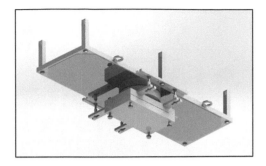

Figure 6.   Design of a disc bearing with replaceable sliding material and shim plate.

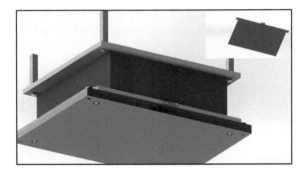

Figure 7.   Rendering of an elastomeric bearing with shim plates.

Figure 8.   Disc bearing equipped with shim plate and adjustable presetting lugs (*Ohio River Bridge, Kentucky & Indiana*).

sliding material. The Ohio River Bridge (Downtown Crossing) bearings were each equipped with a sleeve-shaped shim plate, which can be used to adjust the height of bearing as required. Figures 7 & 8 below show the shim plate details. Removing a shim plate in the loaded condition is not required, as the bearing is designed to resist compressive stress at the maximum anticipated load. However, in cases where the gap between sliding material and stainless steel increases (over ¼″ in this case), a shim can be placed between the bearing plate and masonry plate to fill the gap.

Another interesting feature of the disc bearings of the Ohio River Bridge (Figure 8) is that they are provided with presetting slots, which allow the preset of each bearing to be adjusted on site while the bearing is still held together by transport fixings. Keeping all the parts of a bumper bearing together while in vertical position can be difficult, but these slotted lugs achieve this while

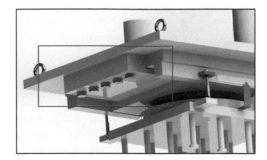

Figure 9.   An expansion disc bearing, equipped with longitudinal stoppers to serve as a fixed bearing.

enabling the presetting of the bearing to be changed as required by the bridge deck's position at the time of installation by releasing the lock nuts and sliding the top assembly to the desired position with respect to the bottom assembly.

4   TEMPORARY LOCKING OF BEARINGS

Segmental bridge construction is a common way of building bridges, involving the use of gantry cranes to install bridge deck segments in a sequential order. After placing of a segment on its bearings, the moving gantry crane exerts horizontal loads on the bearings, which may be of the fixed or sliding (expansion) type. For fixed bearings, this is not an issue as the additional forces resulting from crane movements can be accounted for in the design of bearings. However, sliding bearings that allow sliding in the direction of crane movement cannot resist the resulting forces. Since these forces are active only during construction, these bearings are required to function as fixed bearings during bridge construction and as sliding bearings after construction is complete. In certain cases, a bearing is required to be of the free sliding type during the construction phase, but fixed after completion of construction, when the bridge is in service.

In order to accommodate such requirements, mageba bearings can be equipped with additional locking stoppers which can be added or removed as appropriate to prevent or allow sliding movement.

4.1   *Bearings of the St. Croix Bridge – Minnesota & Wisconsin*

The new St. Croix Bridge connects the states of Minnesota and Wisconsin across the St. Croix River. The extradosed bridge has 68 bearings supporting the main and approach structures, designed for vertical load capacities of up to 6,500 kips (29,000 kN) in service, or 9,700 kips (43,000 kN) in extreme events.

33 of the bridge's bearings are required to be locked during construction, but free to slide in one or both directions when the bridge is in service. The design of the bearings required that the load exerted by the gantry crane (2% of the vertical dead load in this case) be accounted for in the design of the bearings' shear resisting mechanisms. To accommodate this, external add-on locking stoppers were designed that can be added to the bearing by bolting to the top part, preventing the bottom part from sliding relative to the upper part, as illustrated in Figures 9 and 10.

5   BEARING TRANSPORTATION FIXINGS

After a bearing has been fabricated, it can be subjected to rough handling and potentially damaged at various instances during transportation and installation, and may require disassembly and

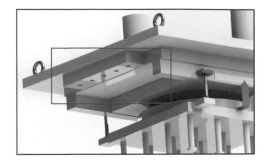

Figure 10.   An expansion disc bearing with stoppers removed to enable it to slide as required.

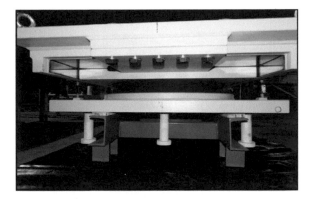

Figure 11.   A disc bearing as fabricated, equipped with longitudinal stoppers (*Bayonne Bridge, New York & New Jersey*).

reassembly for testing or other purposes. Simple elastomeric bearings without separate connection/sliding plates generally do not need banding or transportation fixings, as they do not consist of independent components. However, most bearing types, such as disc, pot and spherical bearings, are made up of individual components which can fall apart if this is not prevented, causing complications for installation and increasing the risk of damage. Such bearings must be delivered from the factory to the construction site in such a way that the entire bearing assembly stays together during transportation and installation. In the case of large bearings, the transportation fixings must be able to support the entire weight of the bearing and to withstand the dynamic forces exerted during transport. Ideally, such fixation systems should facilitate quick and easy removal and replacement as required.

In addressing these needs, mageba provided a modular arrangement of threaded rods with coupler nuts, as shown in Figures 12 and 13, which have proven their value as a solution that can be applied to any bearing type at only minimal additional cost. The speed with which these fixings can be closed and opened results in accelerating bearing testing, handling and inspection. Additionally, since this solution does not require any cutting or welding on site, it does not present any risks for the bearing's core components or its corrosion protection.

6   CONCLUSIONS

A bridge's bearings, being among its most critical components, can serve several purposes in addition to their core function of resisting loads, allowing movements and accommodating rotations. Modifications to bridge bearings generally do not affect other bridge components, as they are largely

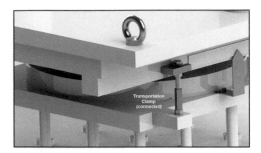

Figure 12.   A disc bearing with the transport fixation in locked position.

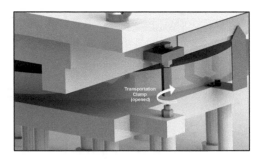

Figure 13.   A disc bearing with the transport fixation in released position.

independent of the bridge structure in terms of design and fabrication – except, of course, in relation to the space available, etc. However, modifications to the bearing design, such as enabling them to durably withstand frequent uplift loads, can significantly reduce bridge construction costs – for example, by avoiding the need for additional bracing or counterweights. Such modifications can also save costs associated with bearing handling, installation or replacement, or avoid the need to change bridge construction methods or to use alternative devices to avoid loading on bearings during construction.

The costs associated with such changes to a bearing's design are generally insignificant compared to the total cost of bridge construction, so the development and application of innovative bearing solutions, which can considerably reduce a bridge's overall life-cycle costs, should always be considered.

REFERENCE

Spuler, T., Moor, G. & O'Suilleabhain, C. 2013. Uplift bearings – Selection and design considerations. *Proc. 7th New York City Bridge Conference*, New York City.

Chapter 24

High load multirotational bearings for an extradosed bridge

R.J. Watson & J.C. Conklin
R.J. Watson, Inc., Amherst, NY, USA

ABSTRACT: The Pearl Harbor Bridge known locally as the Q Bridge carries I-95 traffic over the Quinnipiac River in New Haven, Connecticut. The original plate girder structure was built in 1958 and was designed to accommodate 40,000 vehicles per day. When that total approached 140,000 vehicles per day the Connecticut Department of Transportation decided a new structure was needed. The new twin $635 million cable stayed extradosed bridges are now nearly complete and feature some high load multirotational bearings that were designed for a vertical capacity in excess of 44,400 kN (10,000 kips) which makes them some of the largest bridge bearings ever fabricated in the world. This paper covers the issues surrounding the design, manufacture and testing of these devices. Additionally the testing conducted at the University of California at San Diego's SRMD facility on these bearings will be discussed in detail.

1 INTRODUCTION

The I-95 Corridor in the Greater New Haven Connecticut area is one of the most heavily travelled sections of roadway in the country carrying traffic between New York and Boston. Originally constructed in the 1950's, the bridge over the confluence of the Quinnipiac and Mill Rivers carries traffic volumes in excess of 140,000 vehicles per day, which is more than 3 times the 40,000 vehicles per day it was designed for.

The Connecticut Department of Transportation (CDOT) is improving traffic operations along this section of I-95 with their New Haven Harbor Crossing Corridor Improvement Program, which features roadway improvements along this 12 km (7.2 miles) of I-95. CDOT and the Federal Highway Administration have worked closely with the South Central Regional Council of Governments along with other federal and local agencies to develop an acceptable strategy to address the transportation needs in this corridor.

One of the key features of this program is a new signature bridge crossing the New Haven Harbor known as the Pearl Harbor Memorial Bridge, Figure 1. The Q Bridge as it is known locally will be a ten lane extradosed bridge. The Extradosed System is a hybrid design that combines the benefits of a post tensioned concrete girder bridge with that of a cable stayed design. An Extradosed Bridge was chosen because it is cost effective in this span range plus the reduced tower height was needed due to zoning requirements. One of the benefits of an extradosed bridge is longer main span, which has allowed for the old Q Bridge foundation to remain in place. The northbound 157 m (515 foot) main span, which is now open to traffic is the first extradosed bridge built in the USA. This $635 million project is scheduled to be completed in 2016 (Merritt et al., 2012).

The low bid contractor for the main span was a joint venture of Walsh Construction and PCL Constructors. The new Pearl Harbor Bridge designed by URS Corporation is being built in 3 stages. The first half of the new bridge was built alongside and to the South of the existing bridge.

This span carries the northbound lanes of I-95 and was opened to traffic in June, 2012. The northbound structure now carriers 3 lanes of traffic in each direction while the existing bridge is demolished and the remaining half of the new span is built. Once complete, the southbound lanes

Figure 1. Q Bridge rendering.

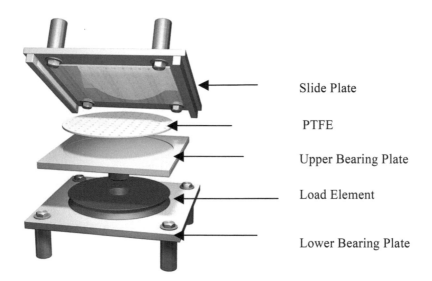

Figure 2. Disk bearing schematic.

will be shifted to the second span and the bridge will be opened to 5 lanes in each direction (Cho, 2012).

A complex structure such as the Extradosed Q Bridge requires a high quality bearing device design capable of accommodating very large vertical loads along with the required rotations and displacements. High load multirotational bearings (HLMRB) were considered the best candidate for this bridge due to their ability to distribute very large loads while simultaneously accommodating rotations in any directions.

After a thorough search of available HLMRB the engineers at URS selected the disk bearing for this demanding project based on their long history of trouble free performance, Figure 2.

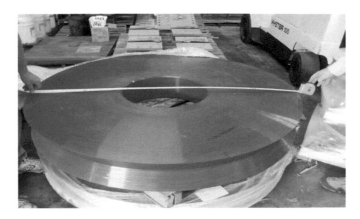

Figure 3.   Polyurethane load element for the Q Bridge.

Disk bearings were developed in the late 1960's as a cost effective means to safely transmit the loads, rotations, and translations of a bridge superstructure to the substructure (Watson, 1986). The primary component of the disk bearing is the load and rotational element, which is comprised of a polyether urethane elastomer, Figure 3.

Due to this material's high compressive strength, there is no need for confinement of the elastomer, which acts much like a conventional elastomeric bearing. This immediately eliminates the sealing ring problem, inherent with many pot bearings. The urethane material used in the disk has outstanding weathering properties and remains stable from −70 to +121 degrees centigrade (−94 degrees F to +250 degrees F). Therefore under normal atmospheric conditions there is no problem with the rotational element softening or crystallizing during temperature extremes. The unconfined disk accommodates rotation by the differential deflection of the elastomeric element.

The horizontal loads of the structure are transmitted through a shear resisting mechanism (SRM). This ball and socket type connection allows the free rotation of the superstructure up to .04 radians or 2.3 degrees in the standard design but inhibits shear from being applied to the rotational element (Watson, 2002). The standard mechanism accommodates a horizontal force of 10% of the vertical capacity of the bearing device. However, this can be easily modified for higher horizontal force conditions that are commonplace with today's longer spans, curved girders and increased awareness of potential seismic activity.

Now in service for well over 40 years, disk bearings have an outstanding track record on bridges all over the world. One of the reasons for this success is the simplicity in design, which also allows for ease of inspection and maintenance free performance. Some notable installations of disk bearings are on the Pasco-Kennewick Bridge in Washington, Figure 4.

Other installations include the Penobscot Narrows Bridge in Maine (Figure 5), the I-35W Bridge in Minnesota (Figure 6), (Cho & Thilmany, 2007) the Hoover Dam Bypass (DeHaven et al., 2009) (Figure 7) and the Manhattan Bridge (Figure 8).

2   Q BRIDGE BEARING DESIGN

There were 20 high load multirotational bearings required for the main span on the Q Bridge. One of the other reasons for specifying disk bearings is that the high vertical loads were such that the plan area of confined elastomer bearings was getting too large for the pedestals to accommodate this type of bearing design. Confined elastomer bearings are limited to 24 MPA (3500 psi) allowable pressure while the disk bearing allowable load according to AASHTO is 35 MPA (5000 psi) which significantly reduces the plan area (Duel Purposes, 2008).

Figure 4.   Pasco-Kennewick Bridge.

Figure 5.   Penobscot Narrows Bridge.

Figure 6.   I-35W Bridge Minneapolis.

The largest disk bearings required for the Q Bridge are located at Tower 2 where two 47,500 kN (10,700 kip) vertical load bearings were supplied. These are the largest capacity disk bearings ever manufactured for a bridge and have a disk diameter 1486 mm (58.5 inches). The base plates were 2550 mm (8.4 feet) 3330 mm × (10.9 feet) and weighed nearly 19 metric tons (21 tons), Figure 9 & Figure 10.

Tower 3 required two bearings nearly as large with a capacity of 42,700 kN (9600 kips). Also at Towers 2 and 3 were eight bearings with vertical capacities of 23,100–25,700 kN (5200–5800 kips) which are large bearings as well. Piers 1 and 4 utilized smaller disk bearings ranging from 3850–5050 kN (866–1135 kips) in vertical capacity.

Figure 7.   Hoover Dam bypass.

Figure 8.   Manhattan Bridge.

Figure 9.   10,700 kip (47,500) disktron bearing.

One of the features of the disk bearing is their ability to handle rotations in excess of the design level. The compressive strength of the polyether urethane is well beyond the levels it is subjected to under the current AASHTO Specifications. This is especially important during construction when loading levels are unpredictable and the possibility of over rotation is high. Service rotations are typically in the 2% range. However, when girders or segments are set during construction these rotations can be way in excess of this value. The polyurethane load element has the ability to handle loads and rotations well beyond design levels. The disk element is virtually indestructible and performs well under extreme conditions.

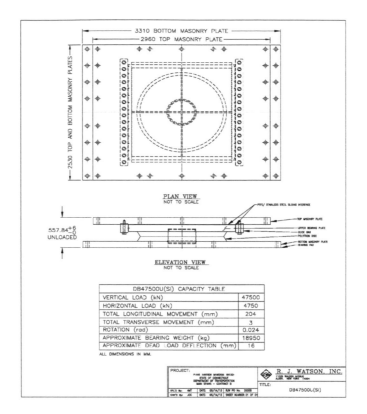

Figure 10.   Q Bridge bearing details.

3   TESTING

Testing is normally a challenge however testing a 19 metric tons (21 ton) bearing is a logistical nightmare. Very few facilities in the world have the capacity to test such a device. One such place is the CALTRANS SRMD Test Facility located at the University of California at San Diego (Van Hampton, 2007).

The Connecticut DOT High Load Multirotational Bearing (HLMRB) Specification calls for proof load testing which requires a 12 hour pre-load along with a rotation test and 100 movement/50 cycle slide testing of the expansion bearings. The testing on the large disk bearings was carried out over a 3 day period in San Diego back in October of 2010. One fixed and one expansion bearing were tested and performed flawlessly as expected, Figure 11.

Another challenging aspect of the Connecticut DOT bearing specification is the long-term deterioration test required for HLMRB. The test is to be conducted on a representative sample of the bearings and consists of 5000 cycles of rotation at full load followed by 1000 cycles of sliding for expansion bearings. For the rotation test one of the load plates in the test fixture has an extension on which acts as a lever. A load is applied to this level to simulate rotation. Again the disk bearings performed well in this rigorous test program.

It should be noted that long term deterioration testing is complex and time consuming. Manufacturers should work together with bridge authorities in advance of contract lettings to conduct this type of testing so that it does not delay project schedules.

It is also important to point out that while disk bearings perform well under long term deterioration testing, there is evidence of wear and abrasion on the PTFE and polyurethane materials following

Figure 11.   Test frame at the UCSD SRMD facility.

Figure 12.   Disktron bearing installation at the Q Bridge.

the tests (Watson, 2010). If long term deterioration is conducted on production bearings the owner should specify that these materials be replaced prior to installation.

Following the successful completion of the extensive test program the disk bearings were installed on the main span of the Q Bridge, Figure 12.

4   CONCLUSIONS

Disk Bearings have been proven to be a valuable tool for engineers looking to accommodate loads, movements and rotations on bridges of all types. Disk bearings are typically specified when these load combinations prove too complex for conventional bearings. Due to the excessive loading criteria on extradosed bridges disk bearings offer a low profile and compact design compared to other types of HLMRB. The ease of installation and problem free design of disk bearings have made it a popular choice of contractors as well.

Extensive testing and field applications in excess of 40 years have shown that the disk bearing is a long term maintenance free device. The disk bearings supplied for the Q Bridge will successfully accommodate the extremely high loads, movements and rotations for the life of the structure.

REFERENCES

Cho, A.T. 2012. "Extradosed Crossing Gives Extra Boost in New Haven," *ENR*.
Cho, A.T. & Thilmany, J. 2007. Monumental Milestone, *ENR*.
DeHaven, T.A., Western, K.L., Watson, R.J. 2009. "Selection and Design of the High Load Multirotational Bearings for the St. Anthony Falls Bridge in Minnesota", New York City Bridge Conference.
Duel Purposes, 2008. Bridge Design & Engineering, Issue No. 51 Second Quarter.
Merritt Jr, R., Bonzon, W.S., Dunham, J.S. 2012. "Pearl Harbor Memorial Bridge", Aspire.
Van Hampton, T. 2007. Minneapolis Bridge Rebuild Draws Fire, *ENR*.
Watson, R.J. 2010. "Applications of Disk Bearings on Large Structures", IABSE Symposium on Large Structure.
Watson, R.J. 2002. "Performance of Disk Bearing on Short and Medium Span Bridges", 6th International Conference on Short and Medium Span Bridges, Vancouver, BC.
Watson, R.J. 1986. "Some Considerations Involved at Points of Bearing in Contemporary Bridges and Structures" ACI SP-94-18.

Chapter 25

Seismic isolation of highway bridges: Effective performance of LRBs at low temperatures

C. Mendez Galindo
Mageba Mexico, Mexico City, Mexico

G. Moor
Mageba USA, New York, NY, USA

B. Bailles
Mageba International, New York, NY, USA

ABSTRACT: Curved highway bridges are widely used in modern highway systems, often being the most viable option at complicated interchanges or other locations where geometric restrictions apply. Among the great variety of seismic isolation systems available, the lead rubber bearing (LRB), in particular, has found wide application in highway bridge structures. However, conventional LRBs, which are manufactured from standard natural rubber and lead, display a significant vulnerability to low temperatures. This paper describes the challenge faced in the seismic isolation using LRBs of a curved highway viaduct where low temperatures must be considered in the design. Specifically, the LRBs must be able to withstand temperatures as low as $-30°C$ for up to 72 hours, while displaying acceptable variations in their effective stiffness. This extreme condition required the development of a new rubber mixture, and the optimization of the general design of the isolators.

1 INTRODUCTION

Increasing awareness of the threats posed by seismic events to critical transport infrastructure has led to the need to seismically retrofit highway viaducts and other bridges to improve their ability to withstand a strong earthquake. Continually evolving technology and the improving evaluation and design abilities of practitioners have also contributed to the need for such solutions – as have, of course, increasingly stringent national design standards. In recent years, curved highway bridges (Figure 1) have become more widely used, as the most viable option at complicated interchanges or river crossings. Curved structures are more prone to seismic damage than straight ones, and may sustain severe seismic damage owing to rotation of the superstructure or displacement toward the outside of the curve line due to the complex vibrations that arise during strong earthquake ground motions.

2 SEISMIC ISOLATION OF STRUCTURES

A bridge's bearings have historically been among its most vulnerable components with respect to seismic damage. Steel bearings in particular have performed poorly and have been damaged by relatively minor seismic shaking (Ruiz, 2005). So a strategy of seismically isolating a bridge's superstructure, by replacing these vulnerable bearings with specially designed protection devices, has much to offer.

Figure 1.   Construction of a curved highway viaduct.

Seismic isolation systems provide an attractive alternative to conventional earthquake resistance design, and have the potential for significantly reducing seismic risk without compromising safety, reliability, and economy of bridge structures (Pan et al., 2005). Furthermore, with the adoption of new performance-based design criteria, seismic isolation technologies will be the choice of more structural engineers because they offer economical alternatives to traditional earthquake protection measures (Mendez et al., 2008).

Seismic isolators provide the structure with enough flexibility so the natural period of the structure differentiates as much as possible from the natural period of the earthquake, as shown in Figure 2. This prevents the occurrence of resonance, which could lead to severe damage or even collapse of the structures. An effective seismic isolation system should provide effective performance under all service loads, vertical and horizontal. Additionally, it should provide enough horizontal flexibility in order to reach the target natural period for the isolated structure. Another important requirement of an effective isolation system is ensuring re-centering capabilities, even after a severe earthquake, so that no residual displacements could disrupt the serviceability of the structure. Finally, it should also provide an adequate level of energy dissipation, mainly through high ratios of damping (Figure 2), in order to control the displacements that otherwise could damage other structural elements.

2.1   *Application in bridges*

Bridges are ideal candidates for the adoption of base isolation technology due to the relative ease of installation, inspection and maintenance of isolation devices. Although seismic isolation is an effective technology for improving the seismic performance of a bridge, there are certain limitations on its use. As shown in Figure 2, seismic isolation improves the performance of a bridge under earthquake loading partially by increasing the fundamental vibration period. Thus, the vibration period of a bridge is moved away from the high-energy seismic ground period and seismic energy transfer to the structure is minimized. Therefore, the use of seismic isolation on soft or weak soil, where high period ground motion is dominant, reduces the benefits offered by the technology (Turkington et al., 1989). The seismic isolation system has a relatively high vibration period compared to a conventional structure. Due to the principle of dynamic resonance, a larger difference between the dynamic vibration frequencies of the isolation system and the superstructure results in a minimized seismic energy transfer to the superstructure. Therefore, seismic isolation is most effective in relatively rigid structural systems and will provide limited benefits for highly flexible bridges.

Another consideration is related to the large deformations that may occur in seismic base-isolation bearings during a major seismic event, which causes large displacements in a deck (Pan et al., 2005). This may result in an increased possibility of collision between deck and abutments. Damping is crucial to minimize the seismic energy flow to the superstructure and to limit the horizontal displacements of the bearings (Mendez et al., 2008).

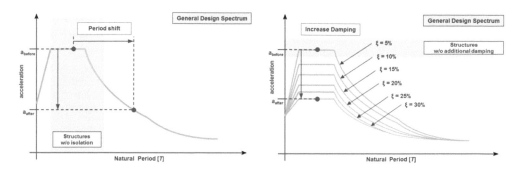

Figure 2.   Reduction of acceleration by seismic isolation (left) and by additional damping (right).

Figure 3.   Cut-out view of a multi-directional LRB, showing the lead core at its center.

3   LEAD RUBBER BEARINGS (LRB)

Among the great variety of seismic isolation systems, lead rubber bearings (LRB) have found wide application in bridge structures (Moehle and Eberhard, 1999). This is due to their simplicity and the combined isolation and energy dissipation functions in a single compact unit. Using hydraulic jacks, the superstructure of a bridge that requires seismic retrofitting can typically be lifted to remove the original bearings, easily replacing them with suitable LRB bearings.

LRBs consist of alternate layers of natural rubber (NR) and steel reinforcement plates of limited thickness, and a central lead core (Figure 3). They are fabricated with the rubber vulcanized directly to the steel plates, including the top and bottom connection plates, and can be supplied with separate anchor plates, facilitating future replacement.

LRBs limit the energy transferred from the ground to the structure in order to protect it. The rubber/steel laminated isolator is designed to carry the weight of the structure and make the post-yield elasticity available. The rubber provides the isolation and the re-centering. The lead core deforms plastically under shear deformations at a predetermined flow stress, while dissipating energy through heat with hysteretic damping of up to 30%.

In practice, bridges that have been seismically isolated using LRB bearings have been proven to perform effectively, reducing the bridge seismic response during earthquake shaking. For instance, the Thjorsa River Bridge in Iceland survived two major earthquakes, of moment magnitudes (Mw) 6.6 and 6.5, without serious damage and was open for traffic immediately after the earthquakes as reported in (Bessason and Haflidason, 2004).

LRB bearings of seismically isolated bridges, due to their inherent flexibility, can be subjected to large shear deformations in the event of large earthquake ground motions. According to exper-imental test results, LRB bearings experience significant hardening behavior beyond certain high shear strain levels due to geometric effects (Turkington et al., 1989).

3.1   *LRB analytical model*

LRB bearings have been represented using a number of analytical models, from the relatively simple equivalent linear model composed of the effective stiffness and equivalent damping ratio

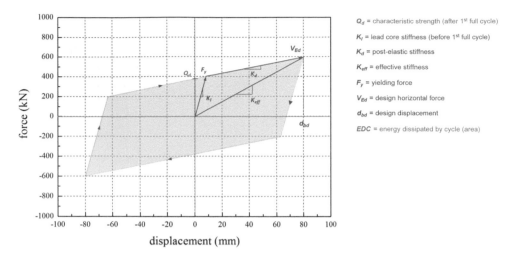

Figure 4. Analytical model of an LRB elastomeric isolator.

formulated by (Huang and Chiou, 1996) to the sophisticated finite element formulation developed by (Salomon et al., 1999). However, the most extensively adopted model for dynamic analysis of seismically isolated structures is the bilinear idealization for the force-displacement hysteretic loop (Ali et al., 1995). Due to its simplicity and accuracy in identifying the force-displacement relationship of the isolation devices, LRB bearing supports can be represented by the bilinear force-displacement hysteresis loop given in Figure 4. The principal parameters that characterize the model are the pre-yield stiffness K_l, corresponding to the combined stiffness of the rubber bearing and the lead plug, the stiffness of the rubber K_d and the yield force of the lead plug Q_d. The value of Q_d is influenced primarily by the characteristics of the lead plug, but it is important to take into account that in areas of cold temperatures, the use of natural rubber will result in significant increases in force values.

4   TESTING OF LRB SEISMIC ISOLATORS

Prototype testing is frequently required by contracts for the supply of LRB seismic isolators, due to the fact that applications tend to be unique in various ways, considering both the structure and the seismic characteristics of the region where it is located. An example of such testing is included in the case study below.

5   CASE STUDY: SEISMIC ISOLATION OF NEW HIGHWAY VIADUCT IN QUEBEC

In general, highway bridges in Quebec have not being designed to withstand high seismic demands. However, even though Quebec is not as seismically active as other areas, a certain risk of earthquake damage exists. A recent example of how seismic engineering is now being more widely applied in the design and construction of new structures in Quebec is the seismic isolation of a curved highway viaduct, serving the city of Levis as part of the A20/A73 Interchange.

The new viaduct was constructed adjacent to an existing structure in order to increase highway capacity. LRBs were selected to support the bridge superstructure in normal service and to protect the structure during an earthquake by isolating it from the destructive movements of the ground beneath. The LRBs thus ensure the constant serviceability of the structure, even after the occurrence of a strong earthquake, facilitating the passage of emergency vehicles and contributing to the safety

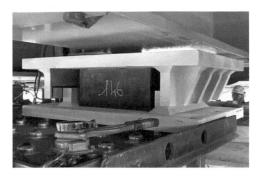

Figure 5.   Lead rubber bearings installed in the bridge – guided (left) and multi-directional (right).

of the population. The viaduct is a six-span superstructure with steel girders, with spans of between 40 m and 60 m and a total length of over 300 m. With a horizontal radius of 270 m, it has a prominent curve which heightens the risk of serious damage during an earthquake and thus increases the need for its deck to be seismically isolated from its supports. The end spans of the deck are supported by conventional pot bearings (at the abutments and on the first pier at each end), while the three internal piers support the deck via LRBs.

5.1   *Design of the LRBs*

Each of these internal piers has six LRBs, one supporting each of the deck's main longitudinal girders. The LRBs at each side of these piers, supporting the outer girders, are *multi-directional* (facilitating horizontal movements in all directions insofar as these are permitted by deformation of the elastomeric pad and its lead core). The remaining LRBs, supporting the internal girders, are *guided*, with steel fittings preventing all transverse movements. An LRB of each type is shown in Figure 5. Each LRB has a vertical load capacity of approximately 3,000 kN – primarily to serve its primary purpose of supporting the deck under normal service conditions. Due to the structure's location, the LRBs were designed for temperatures as high as 40°C (104°F) and as low as −30°C (−22°F). In addition to these severe temperature conditions, the LRBs also had to be designed to fulfill the following requirements:

- Facilitate movements of up to 95 mm in the longitudinal direction;
- In the case of guided bearings, restrict movements in the transverse direction;
- Provide damping of up to 24%;
- Dissipate hysteretic energy up to 40 kNm per cycle;
- Ensure re-centering following an earthquake;
- Increase the period of the deck of the bridge to more than 2 seconds;
- Transmit horizontal loads of up to 410 kN at a typical ambient temperature of 20°C (68°F); and
- Transmit horizontal loads of up to 600 kN at a low temperature of −30°C (−22°F)

These demands presented a significant challenge for design and manufacture – especially in relation to low temperature performance. The bearings were designed to provide optimal performance at 20°C and to minimize variations in dynamic characteristics at very low temperatures. Considering the sensitivity of rubber to low temperatures, this was very difficult to achieve. However, after a detailed analysis of the effects of temperature on the rubber and the lead, and evaluation of the overall performance of the devices during extensive full-scale testing, it was possible to develop an optimal solution according to Canadian Highway Bridge Design Code CAN/CSA-S6. This solution included design of a new rubber mixture – based on an extensive development program which included testing of a number of rubber samples – and resulted in an optimized LRB design considering all conditions.

Table 1.   Testing protocol required for room temperature performance.

Test No.	Test Name	Specification	Main DOF [–]	Amplitude [mm]	Cycle duration [sec]	Compression load [kN]	Horizontal load [kN]	Cycles [–]
1	Thermal/Service	AASHTO 13.2.2.1 CSA 4.10.11.2 (c)(i)	L	±60	20	1,715	±190	20
2	Wind and Braking: Pre-seismic 1/2	AASHTO 13.2.2.2	L	7	20	1,715	±26	20
	Wind and Braking: Pre-seismic 2/2		V	0	60		0	0
3	Seismic	AASHTO 13.2.2.3	L	±95	20	1,715	300	3
		CSA 4.10.11.2 (c)(ii)	L	±24	20		75	3
			L	±48	20		150	3
			L	±71	20		225	3
			L	±95	20		300	3
			L	±119	20		375	3
4	Seismic verification	CSA 4.10.11.2 (c)(iii)	L	±95	60	1,715	293	10
5	Wind and Braking: Post-Seismic 1/2	AASHTO 13.2.2.4	L	7	20	1,715	±26	3
	Wind and Braking: Post-Seismic 2/2		V	0	60		0	0
6	Stability 1/3	CSA 4.10.11.2 (d)	L	105	60	1,072	325	loading ramp
	Stability 2/3		L	105	60	2,155	325	loading ramp
	Stability 3/3		V	0	60	4,677	0	0

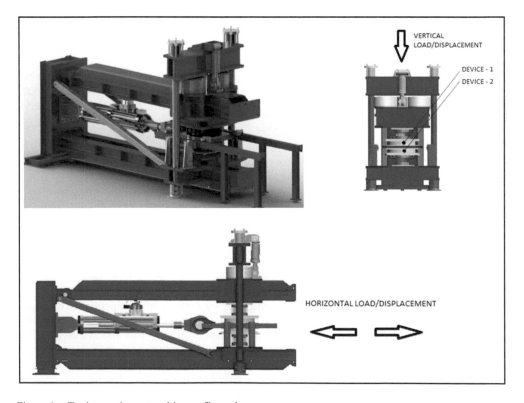

Figure 6.   Testing equipment and its configuration.

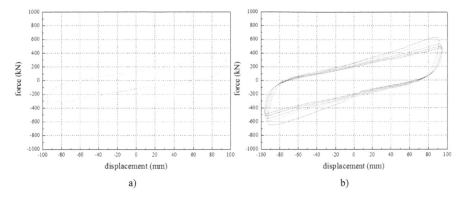

Figure 7.   Test results at a) room temperature of 20°C (68°F) and b) low temperature of −30°C (−22°F) after 72 hours of exposure.

5.2   *Prototype testing of LRBs*

Prototype testing was carried out in accordance with the isolator supply contract, to verify the performance of the LRBs in accordance with their design and the project specifications. The testing included evaluation of the dynamic performance of each device in terms of effective stiffness, damping, energy dissipated per cycle and other parameters such as displacements and forces. The testing protocol for room temperature testing is shown in Table 1. Similar testing was required at the specified very low temperature.

The test equipment and its configuration, which allows the simultaneous testing of two isolators, is shown in Figure 6. The steel frame holding the isolators was designed to counter the thrust forces that are created during testing of seismic isolation devices. The maximum horizontal load depended on the characteristics of the servo actuators installed, and a nominal value of 1400 kN was considered. The maximum vertical load of 10000 kN was provided by two actuators, each 5000 kN.

The project required consideration of both the AASHTO Guide Specifications for Seismic Isolation Design (AASHTO GSSID) and the Canadian Highway Bridge Design Code (CAN/CSA-S6-06). While AASHTO GSSID requirements are well known and applied, the application of CAN/CSA-S6-06 requirements presented an additional challenge. This code specifies in Section 4.10.11 the main requirements for the testing of seismic isolation devices.

The specimens each had a plan dimension of 500×500 mm and a total height of 284 mm, and were designed for a total design displacement of 95 mm and a test maximum vertical load of 4,677 kN. The samples were subjected to 23 different tests, most of them including dynamic conditions, and with frequency and amplitude varying from one test to the next. For all dynamic testing, a vertical load of 1,715 kN was applied to each of the samples.

The testing protocol presented in Table 1 fulfills all specified requirements, incorporating necessary adjustments as required by the project engineer. The following special considerations were taken into account for the prototypes testing:

1. Room Temperature Tests (with isolators conditioned at the temperature of 20 ± 5°C for 48 hours prior to testing):
   a. 5 fully reversed sinusoidal cycles at amplitude of 95 mm and peak velocity of 200 mm/s (frequency of 0.333 Hz).
   b. 3 fully reversed sinusoidal cycles at amplitude of 95, 24, 48, 72, 95 and 119 mm and frequency of 0.333 Hz.
2. Low Temperature Tests (with isolators conditioned at the temperature of −30°C for 72 hours prior to testing):
   • 5 fully reversed sinusoidal cycles at amplitude of 95 mm and peak velocity of 200 mm/s (frequency of 0.333 Hz).

Table 2.   Average results of the last three cycles of the prototype testing, at room and low temperatures.

Parameter	Unit	Room temperature 20°C (68°F)	Low temperature −30°C (−22°F)
Displacement	mm	95	95
Horizontal force	kN	302	589
Post-elastic stiffness	kN/mm	1.91	3.88
Effective stiffness	kN/mm	3.17	6.33
Characteristic strength	kN	120	220
Energy dissipated per cycle	kN-m	43.12	88.06
Damping	%	24	25.6

5.3   *Low temperature results*

The extensive testing carried out on the two specimens provided a large amount of data. Here, only the key performance at room temperature, and a comparison with the performance at low temperature, are presented. Figure 7 shows the main hysteretic responses at room temperature (a) and low temperature (b).

The results in Table 2 demonstrate that the key dynamic parameters such as effective stiffness, horizontal force, post-elastic stiffness and characteristic strength increase by a factor of about two at very low temperatures. However, considering the severe variation of temperature and the strong dependence of rubber's behavior on temperature, these results verified well the effectiveness of these specially developed LRBs at low temperatures, as well as compliance with the project specifications.

6   CONCLUSIONS

Lead rubber bearings (LRB), which are widely used to seismically isolate highway bridge structures, display a significant vulnerability to low temperatures (e.g. −30°C) unless designed and fabricated for such conditions. In particular, their design should ensure that they display only minor variations in their effective stiffness at such temperatures. As in the case study presented, this may require the development of a new rubber mixture, the optimization of the general design of the isolators, and verification of low-temperature performance by means of extensive full-scale prototype testing.

REFERENCES

Bessason, B., and Haflidason, E.: Recorded and numerical strong motion response of a base-isolated bridge, *Earthquake Spectra*, **Vol. 20**, No. 2, pp. 309–332, 2004.

Huang, J. S., and Chiou, J. M.: An equivalent linear model of lead-rubber seismic isolation bearings, *Engineering Structures*, **Vol. 18**, No. 7, pp. 528–536, 1996.

Mendez Galindo, C., Hayashikawa, T., and Ruiz, J. D.: Seismic damage due to curvature effect on curved highway viaducts, *Proceedings of the 14th World Conference on Earthquake Engineering*, IAEE, Beijing, China, October 12–18, 2008.

Moehle, J. P., and Eberhard, M. O.: Chapter 34: Earthquake damage to bridges. In: Chen, W. F., and Duan, editors. *Bridge Engineering Handbook*, Boca Raton, CRC Press, 1999.

Ruiz, J. D.: Seismic performance of isolated curved highway viaducts equipped with unseating prevention cable restrainers, *Doctoral Dissertation,* Graduate School of Engineering, Hokkaido University, Japan. December 2005.

Salomon, O., Oller, S., and Barbat, A.: Finite element analysis of base isolated buildings subjected to earthquake loads, *International Journal for Numerical Methods in Engineering*, **Vol. 46**, pp. 1741–1761, 1999.

Turkington, D. H., Carr, A. J., Cooke, N., and Moss, P.J.: Seismic design of bridges on lead-rubber bearings, Journal of Structural Engineering, ASCE, **Vol. 115**, No. 12, pp. 3000–3016, 1989.

Bridge history & aesthetics

Chapter 26

Charles Ellet, Jr., the pioneer American suspension bridge builder

K. Gandhi
Gandhi Engineering, Inc., New York, NY, USA

ABSTRACT: Charles Ellet, Jr. (1810–1862) was a multi-talented engineer who was far ahead of his time and who made important contributions in the fields of long span suspension bridge-building; river training and flood controls in western rivers; transportation planning and economics; canal and railroad building; and demonstrating merits of iron-clad steam rams in naval warfare. Ellet built the first permanent wire suspension bridge in the U.S. over the Schuylkill River in 1842, first suspension bridge across the Niagara Gorge in 1848, and the first suspension bridge with a span over 1,000 feet at Wheeling, Virginia in 1849. This paper highlights Ellet's contributions in building and promoting suspension bridges in the U.S.

1 WHO WAS CHARLES ELLET, JR.

According to Gambrel (1931) Charles Ellet, Jr. (Figure 1) was born at Penn's Manor, Bucks County, PA, sixth of the fourteen children of Charles Ellet, a Quaker farmer, and Mary, daughter of the high sheriff of Philadelphia. Ellet did not get proper guidance from his eccentric litigious father, who opposed Ellet's determination to become an engineer.

Ellet left home at 17, working as a rodman on the Susquehanna survey. Then, in 1828, he joined Chesapeake & Ohio Canal Co. in Maryland as an unpaid assistant in the field and office, and finally as an assistant engineer for a salary of $800 per year. He learned some mathematics and French on his own with little formal instruction.

In March 1830, with his mother's financial help, he went to France. There he was able to schedule a meeting with General Lafayette and the American Ambassador to France. With their references, he was able to attend the Ecole Polytechnique. There he learned the theory of designing wire suspension bridges. He decided to travel by foot through southern France and Switzerland. On this trip he wrote of inspecting many suspension bridges on the Rhone, Loire, Garonne, and Seine. He

Figure 1.   Charles Ellet, Jr. (Stuart 1871).

made notes on these bridges and observed firsthand a suspension bridge being constructed across the Loire, and witnessed the manner in which the wire cables for these bridges were manufactured (Lewis 1968).

On his travels through Switzerland, he most likely visited what Tyrrell (1911) claims is the first wire suspension bridge in Europe. The bridge at Geneva over the river Fosse was constructed in 1823 with two equal spans of 132'-6" by Colonel Dufour. The author, after reading the development of wire suspension bridges in France by Peters (1987), believes that it was Marc Seguin and his four younger brothers who built the Geneva Bridge in 1823. According to Peters (1987), "Dufour drew up a measured cross-section through the double moat on the reverse of the sketch Seguin had sent to Pictet. The sketch showed the whole extent of the proposed bridge site. Dufour carefully noted the differences in elevation of the abutment emplacements which Seguin had lacked".

He returned from France to the U.S. in early 1832. In 1834, he proposed a suspension bridge across the Potomac, and later he proposed suspension bridges across the Mississippi River at St. Louis, and across the Connecticut River at Middletown in 1848.

In 1842 he built the first permanent wire suspension bridge in the U.S. over the Schuylkill River at Fairmount in PA. He built the first pedestrian bridge over the Niagara Gorge in 1848, and the longest suspension bridge in the world with a span of 1,010 ft. at Wheeling, Virginia over the Ohio River in 1849.

His plan for controlling flood and improving navigation in western rivers such as the Mississippi and the Ohio by impounding surplus waters in an upland reservoir was more than 75 years ahead of his time as can be judged by the fact that Ellet's reports were reissued in 1927-28 for the Flood Control Committee of the 70th Congress.

Visiting Europe during the Crimean War (1855), Ellet urged Russia to employ "ram-boats" for the removal of the blockade of the Port of Sebastopol (Lewis 1968). Returning to the U.S., he urged his ram-boat scheme to the successive secretaries of the navy, and circulated his pamphlet "Coast and Harbor Defences" in 1855. His scheme was ignored until 1862 when the boat Merrimac demonstrated the efficacy of the ram. Two weeks later he was commissioned a colonel, and he was asked by the Secretary of the Navy to prepare a ram fleet to clear the Mississippi. He sustained a bullet wound and died as his boat touched shore at Cairo, Illinois on June 21, 1862.

Ellet was a prolific writer and had 46 publications to his credit. Besides the flood control, he also wrote several articles on transportation economics and rate making.

In his book on Charles Ellet, Jr., Lewis (1968) has summarized Ellet's personality as follows: "One notable characteristic of Ellet's personality throughout life was his almost superhuman drive in the initial pursuit of an idea or a task, and then often he would become weary and drop it. Undoubtedly greater permanent accomplishments would have been his had he directed his efforts towards fewer endeavors. Ellet was destined always to be the inventor, and as is frequently the case, was forced to witness others achieve the recognition for his innovations."

2   CHAIN AND WIRE SUSPENSION BRIDGES IN THE U.S. BEFORE ELLET

According to a report prepared by Austin N. Hungerford of San Francisco and published in a Bulletin of the American Iron & Steel Association (Engineering Record 1904) one of the early chain bridges was "built by James Finley in 1801 across Jacob's Creek in Western Pennsylvania according to a system on which Finley received a patent several years later.

The same article provides the circumstances under which the first wire suspension bridge for pedestrians was built in the U.S., and most likely in the world. Joseph White and Erskine Hazard built a rolling mill and a wire factory at the Falls of Schuylkill River. Robert Kennedy and Conrad Carpenter agreed to build a bridge at this location under an act passed on February 22, 1809 using the Finley patent which would permit White and Hazard to levy tolls. They built a 3-span bridge in 1809 with two equal spans of 153 ft. in length. In January of 1811, the bridge failed under the weight of a large drove of cattle. Finley investigated the bridge failure and determined that an "ill-judged clip or coupling piece broke, with which two parts of the chain were joined together".

Frustrated with the failure, White and Hazard decided to build a wire suspension bridge across the Schuylkill River at Fairmount for foot passengers only, and limited 8 persons on the bridge at any time. They fastened suspension wires at one of the top windows of the mill, stretched them across the river and tied them to some large trees on the other side. They provided steps to descend down to the ground from the bridge. According to Schuyler (1931) the cable used were made of six 3/8 inch wires. The bridge had a single span 408 ft. and a passageway of only 18 inches. The bridge collapsed under a weight of snow and ice in 1816, the same year it was built.

In January 1817 an act was passed by the Legislature which authorized the Schuylkill Falls Bridge Co. to sell all its corporate rights to certain persons "who will undertake to erect a permanent bridge" at this point. The third bridge was designed by Louis Wernwag, built by Isaac Nathans, and opened in December 1817. This was a 340 ft. span covered wooden arch truss bridge and became known as the "Colossus".

3   FIRST U.S. WIRE SUSPENSION BRIDGE OVER SCHUYLKILL RIVER

In 1839, Ellet published a pamphlet titled, "A Popular Notice of Wire Suspension Bridges". In this article Ellet described particular advantages of wire cables, and covered the history and development of wire suspension bridges in the U.S., France, and England, and indicated the potential of bridging the Mississippi near its confluence with the Missouri, the Ohio River, and the Niagara River below the Falls using the wire suspension bridges. One of Ellet's motives for publishing the article on wire suspension bridges was to make the people of Philadelphia aware of the advantages of a new bridge system.

The real opportunity for Ellet to build the first wire suspension bridge in the U.S. came when the Collossus Bridge over the Schuylkill River burned down on September 1, 1838. The cornerstone of that bridge was laid with Masonic ceremonies on April 28, 1817.

Ellet submitted his plan for a wire suspension bridge to the county and learned that five other plans were also submitted (Lewis 1968). In July 1839, he learned from his mother in Philadelphia when Ellet was visiting his brothers in Illinois, that his plans had been selected by the Commissioners under the Free Bridge Act of 1839 for the replacement of the "Colossus". However, the City of Philadelphia decided to construct the bridge in the Spring of 1841.

According to Steinman (1945) the contract was awarded to a local contractor, Andrew Young, to build the bridge according to Ellet's plans. He sought help from Roebling for building the bridge. However, the County Board rescinded Young's contract, and awarded it to Ellet in June 1841 to erect a wire suspension bridge of his own design with a span of 358 ft.

Following the French practice for the Schuylkill Bridge, Ellet laid wires in separate strands side by side with iron bars fastened across them from which the suspenders were hung. Roebling preferred the system where the suspenders were hung from clamps surrounding the cables, which were generally in planes sloping at angle from the vertical, with systems of auxiliary stay cables radiating from the towers to successive panel points of the floor system (Tyrrell 1911). Roebling believed that by laying the wire in compact cables some of the integral resistance of solid bar could be obtained. Ellet's reasoning was that it was impracticable to combine the wires into a cable of large diameter, but the wires should be laid in cables of small diameter, adding to their number as additional strength was required (Steinman 1945).

The Schuylkill River Bridge (or Fairmount Bridge) (Figures 2 and 3) was opened to traffic in the Spring of 1842. The bridge was 26 ft. wide and included 18 ft. carriageways and two four-foot footwalks. It cost Ellet $53,000 to build the bridge whereas his quoted price was $50,000. He considered the loss well-justified as he received favorable publicity in newspapers and periodicals all over the country for building the first permanent suspension bridge in the U.S.

In 1853, Ellet wrote a letter to the county commissioners who were responsible for the care and management of the Fairmount Bridge, and requested them to examine the point of fastening which was hidden from sight, and offered his services. When Ellet received no response, he directed public attention to the issue of inspection of the Fairmount Bridge through the columns of the

Figure 2.   Line Diagram of the Schuylkill River Bridge (Stuart 1871).

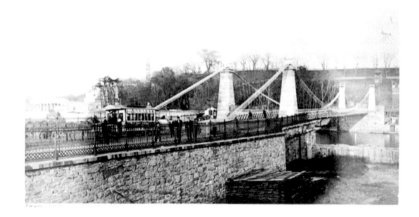

Figure 3.   Schuylkill River Bridge at Fairmount (Sayenga 1983).

Philadelphia "Ledger". For greater safety, he suggested to add a new cable on each side of the bridge, and attach it to an independent anchorage. Scientific American (1855) endorsed this idea whole heartedly noting the failures of many suspension bridges in different parts of the country in the preceding six months. Because of increased traffic and vehicles with heavier loads, the bridge was replaced in 1874 with a double-deck truss span.

4   NIAGARA SUSPENSION BRIDGE

In 1845, Charles B. Stewart, a prominent civil and military engineer and who later became the first state engineer and surveyor of New York State, and eventually engineer-in-chief of the U.S. Navy (Steinman 1945), invited leading American and European engineers to build a bridge over the Niagara River between the falls and "Whirlpool". Only four engineers, namely Ellet, Roebling, Samuel Keefer, and Edward W. Serrell responded saying that the project was feasible. Each of them later constructed a suspension bridge over the Niagara River (Stuart 1871).

In response to the inquiry from Stuart, Ellet sent a reply dated October 12 1845 from Philadelphia which is reproduced below (Stuart, 1871):

> *"In the case which you have presented, I can, however, say this much with all confidence: A bridge may be built across the Niagara below the Falls, which will be entirely secure, and in all respects filled for railroad uses. It will be safe for the passage of locomotive engines and freight*

trains, and adapted to any purpose for which it is likely designed, and properly put together; there are no safer bridges than those on the suspension principle, if built understandingly, and none more dangerous if constructed with an imperfect knowledge of the principles of their equilibrium. To build a bridge at Niagara has long been a favorite scheme of mine. Some twelve years ago I went to inspect the location, with a view to satisfy myself of its practicability, and I have never lost sight of the project since. I do not know in the whole circle of professional schemes a single project which it would gratify me so much to conduct to completion."

When two companies, one in the U.S., and the other in Canada with the power to connect with each other were formed in 1846, and when sufficient money was raised by selling stock to build the bridge, Stuart invited plans and estimates from prospective bidders. Ellet submitted his response from Philadelphia dated February 13, 1847 which is reproduced below (Stuart, 1871):

Charles B. Stewart,
 Commissioner of the Niagara Bridge Company
 Dear Sir, I promised to give you my views of the practicability and probable cost of the proposed bridge across Niagara river below the Falls. Immediately after inspecting the site, in eighteen hundred and forty-five, I gave the whole subject a careful investigation, and made a fair, but not extravagant, estimate of the cost of such a structure as I thought would be appropriate and of adequate strength.
 This estimate amounted to two hundred and twenty thousand dollars for a railroad bridge competent to sustain the weight of locomotive engines and heavy freight trains, and one hundred and ninety thousand dollars for one suitable for common travel, with a railway track in the centre, to be crossed by passenger and burthen cars drawn by horses.
 When I made my estimate, I had in view a work of the first order, and as I do not wish to be in any way connected with one of a lower grade, I cannot offer to reduce my proposition. But I will now repeat, that a secure, substantial and beautiful edifice, not one, however, equal to the claim of the locality – for nothing can match that – but a noble work of art, which will form a safe and sufficient connection between the great Canadian and the New York railways, and stand firm for ages, may be erected over the Niagara river for the latter sum named. If it should be built by me, or under my charge, it will cost about that sum, and I trust it will be worth the money.
 With my best wishes for the success of the enterprise in all its magnificence,
 I remain dear sir,
 Yours truly,
 Charles B. Ellet, Jr.,
 Civil Engineer

On November 9, 1847, the Director of the American and the Canadian Niagara Bridge Companies made a contract with Ellet for the construction of a railway and carriage bridge over the Niagara River, two miles below the falls for $190,000. The 28-foot width of the bridge would accommodate two footways each 4 ft. wide; two carriageways, each 7'-6" wide; and one railway track in the middle 5 ft. wide. The approximate length of the span was about 800 ft.

The ultimate strength of cables was 6,500 tons. The weight of each train was limited to 24 tons to be drawn by a locomotive with a maximum weight of 6 tons. The towers were to be made of stone. The bridge was to be tested for 200 tons, and had to be completed by May 1, 1849 (Stuart, 1871).

Ellet was ready to begin construction early in 1848. He needed to build a foot bridge prior to the construction of the main bridge. This required connecting the two sides (the U.S. and Canada) of the river by a cable. Ellet, using his ingenuity and flair for publicity, offered a prize of $5 to the first boy who would fly a kite to the opposite shore.

Ellet used the kite string to draw over large cords, and then hemp ropes, and finally he pulled the first wire cable across and suspended over the gorge (Lewis 1968).

Ellet sent a letter to Stuart (1871) describing the first basket ride over the Niagara (Figure 4) which is included here:

Niagara Falls, March 13th, 1848.
Charles B. Stuart.

Dear Sir, – I raised my first little wire cable on Saturday, and anchored it securely both in Canada and New York. To-day (Monday) I tightened it up, and suspended below it an iron basket which I had caused to be prepared for the purpose, and which is attached by pulleys playing along the top of the cable.

In this little machine I crossed over to Canada, exchanged salutations with our friends there, and returned again, all in fifteen minutes.

The wind was high and the weather cold, but yet the trip was a very interesting one to me – perched up as I was two hundred and forty feet above the Rapids, and viewing from the centre of the river one of the sublimest prospects which nature has prepared on this globe of ours.

My little machine did not work as smoothly as I wished, but in the course of this week I will have it so adjusted that anybody may cross in safety.

Truly yours,
Charles Ellet, Jr.

The details of Ellet's Niagara proposed Suspension Bridge are as follows (Scientific American 1848b):

No. of cables for bridge	16
No. of strands in each cable	600
No. of strands in the ferry cable	37
Diameter of the cable	$\frac{1}{2}$ inch
Height of stone tower	68'-1"
Height of wood tower for ferry	50'
Base of the stone tower	20' × 20'
Top of the stone tower	11' × 11'
Span of the bridge	800'
Total weight of the bridge	650 tons
Height from the water	230 ft.
Depth of the water under the bridge	380 ft.

Ellet built a 7'-6" wide light suspension footbridge hung on wooden towers as a service bridge to carry men and materials across the gorge. The last plank in the floor was laid on July 29, 1848 and railing was completed only one-third of way, Ellet drove over and back in a buggy drawn by a high spirited horse (Lewis 1968).

After the footbridge was completed, Scientific American (1848c) described the crossing of the bridge by pedestrians as follows: "Foot passengers now walk across from the dominions of Uncle Sam to the dominions of Queen Victoria for 25 cents. This is a great work, not only physically but morally. It will promote intercourse and good will among the republicans and royalists. Difference of opinion regarding governments should never make men enemies".

The cost of the footbridge was about $30,000 and toll collected by Ellet for 10 months was about $5,000. There arose a dispute between the Directors and Ellet about who keeps the toll. There ensued a litigation between the two parties, and a compromise was reached by which Ellet relinquished his contract; and his connection with the work was terminated on December 27, 1848 (Stuart 1871).

The Niagara Bridge Co. hired John Roebling to construct the bridge and after a seven year interval, using Ellet's footbridge as a scaffold, Roebling completed the railroad suspension bridge across the Niagara in 1855. Roebling's two level Niagara Railway Suspension Bridge is covered in detail by Gandhi (2006).

Figure 4.   Ellet's basket ride over Niagara River (Stuart 1871).

5   SUSPENSION BRIDGE OVER THE OHIO RIVER

In 1847 the Virginia legislature authorized the construction of a wire suspension bridge over the Ohio River between Wheeling, Virginia (now W. Virginia) and Zane's Island. Ellet informed the directors of the company formed to build the bridge that the directors should review the proposals submitted by others, and if they were not satisfied, then he (Ellet) would submit his proposals (Lewis 1968).

In July 1947, the directors informed Ellet that they had reviewed the plans submitted by Roebling, and they did not approve those plans. As a result Ellet was selected to build the bridge. The details of the bridge are given below (Scientific American 1848a):

Span between centers of towers	1,010 ft.
No. of strands of the wire (No. 10)	9,000
Minimum strength of each strand of wire 500 lbs.	
Height of bridge above the low water mark	87 ft.
Width of the bridge floor	24'-0"
Width of footway on each side	3'-6"
Width of the carriageway in the center	17'-0"
No. of cables	12
Length of each cable	1,350 ft.
Timber used	white pine and white oak
Estimated Construction cost of the bridge	$210,000

The 1,010 ft. span would make the Wheeling Bridge the longest suspension bridge in the world. The bridge was designed for carriages only, and not for carrying railroads. Construction of the bridge started in the summer of 1848, and was completed in December of 1849 (Figure 5). To signify the opening of the Bridge to the general public, Ellet crossed the bridge in a horse and buggy (Scientific American 1849a).

The bridge as built had 12 cables of iron with each cable four inches in diameter and having 550 strands of wire. The cables rested on iron rollers placed on the top of the towers, and the cables were anchored into heavy masonry of wingwalls at each end of the bridge (Figure 6). The length of the deck supported by the cables was 960 ft., and the weight of each linear foot of deck was 546 lbs. The dead load of the bridge including the weight of the 12 cables, bolts, casting, suspenders, etc. was 920 lbs. per linear foot or 441 tons (Scientific American 1949b). The actual cost of the Wheeling Bridge was $225,000 (Scientific American 1850a).

Figure 5.   Image of the Wheeling Bridge as it appeared in court documents (U.S. Supreme Court 1851).

Figure 6.   Tower and anchorages Wheeling Bridge (Gralian 2014).

6   LAWSUIT AGAINST WHEELING BRIDGE

6.1   *Lawsuit for obstruction to marine traffic*

In August of 1849, while the bridge was still under construction, a lawsuit was filed by the State of Pennsylvania on behalf of the steamboat interests of Pittsburgh against the Wheeling and Belmont Bridge Company. A motion was filed before the Supreme Court Justice Robert C. Grier for an injunction to stop the construction of the bridge. In the December term, the case was heard by the Supreme Court, and by an order of the Court, the case was referred to Chancellor Walworth, of New York, to act as special commissioner to gather the facts of the case (Lewis 1968). In July of 1850, Chancellor Walworth submitted to the Supreme Court that the bridge was a nuisance to boats with high funnels and chimneys (Scientific American 1850b). The lawyer for the bridge company, William W. Hubbel, responded that Chancellor Walworth had not yet heard all of the evidence nor the arguments, and had made no report on the subject (Scientific American 1850c).

In February of 1851, the Supreme Court intimated that a decision in the Wheeling Bridge case would not be rendered that term as the records from which to elicit a comprehensive brief were too voluminous. In addition, a copy of the testimony needed to be delivered into the hands of each of the judges, and as the evidence constituted a volume of 1,000 pages, it was preposterous to hope for an opinion until the subsequent term commencing in December of 1851 (Scientific American 1851a).

In reporting news from Alton Telegraph, Illinois on the Wheeling Bridge, Scientific American (1851b) noted that there were six railroads interested in crossing the Ohio at Wheeling, and there were six steamboats owned at Pittsburgh having long funnels who would not lower their funnels for the benefit of the passengers of these six railroads. The Supreme Court had to decide between the rights of the people of the U.S. and those of the boat owners. And, if the people could not have continuous railroads across the great rivers, that they should be informed of it in a timely manner.

6.2   *U.S. Supreme Court decision in favor of steamboat operators*

At the December 1851 term, Justice McLean delivered the majority opinion, with Chief Justice Taney and Justice Daniel dissenting. On March 1, 1852 the Court ruled the bridge to be an obstruction to the navigation of the Ohio River, and ordered that the bridge be elevated no less than 11 feet over the channel of the river, and "that unless this or some other plans shall be adopted, which shall relieve the navigation from obstruction on or before the first of February next (1853) the bridge must be abated" (Lewis 1968).

Ellet was successful in convincing the Supreme Court that a draw in the bridge would remove the obstruction to the boats with tall chimneys. After delivering an opinion against the bridge, the Court directed William J. McAlpine, one of the most well-known engineers of that period from New York City to report to the Court in May 1853 concerning the feasibility of a draw.

McAlpine presented eight different plans for modifying the existing suspension bridge, which he characterized as totally unfit for railroad purposes. All of these plans involved great expenses, the best costing no less than $156,243.50. He concluded that owing to the peculiar nature of the navigation and the principles upon which steamboats running on the Ohio River were built, that were was no doubt that the bridge as it was presented great obstructions (Scientific American 1852c).

The Supreme Court rendered its final decision in the case on May 27, 1852; and agreed that a draw in the bridge over the west channel of the river would adequately eliminate the obstruction to the navigation, if it were completed by February 1, 1853 (Lewis 1968).

While the Supreme Court was waiting for the report from McAlpine regarding the feasibility of a draw, Ellet (1852) sent a letter to Scientific American and presented six leading facts related to the case. The key points were that (1) Boats with chimneys less than 60 ft. in height could pass under the bridge at all times; (2) There were only 7 out of 270 more recently constructed boats with chimney heights varying from 70 to 80 ft. navigating the river, and all these seven boats had been built after the plans of the bridge were published two years before the construction of the bridge; and (3) The Supreme Court ruled for tearing down a bridge which cost more than $200,000 to build, all for the benefit of these seven boats.

The Chief Justice Taney in his dissent reasoned that the U.S. Courts had no jurisdiction over the matter, as the court has no law to guide them, and the jurisdiction exercised was without precedent. Scientific American (1852a) agreed with the Chief Justice and added that only Congress had the power over this case. The bridge was in the State of Virginia over an inland river, and Congress had made no laws for deciding such a case.

6.3   *Supreme Court injunction and Ellet's refusal to comply*

After the failure of the Wheeling Bridge in May 1854, Ellet was served by the Supreme Court on July 3, 1854 with an injunction, restraining him from rebuilding the bridge, except at a certain elevation (meaning at a higher elevation). Ellet declared that if he was not obstructed by the Court, he would have started the repairs and the bridge would have been open by July 1854 (Scientific American 1854). In the end, the Bridge Co. and Ellet ignored the injunction, and the repairs were completed by the middle of July for a cost of $37,000, and the bridge was opened to regular traffic.

In 1860, Roebling spent $55,000 more in repairing the bridge. As of 1968, the Wheeling Bridge was the oldest existing suspension bridge in America (Lewis 1968).

6.4   *Congress bypasses Supreme Court decision*

The citizens of Wheeling, Virginia appealed to congress to legalize the Wheeling Bridge, and both houses of Congress passed the bill by a large margin. The bill required the steamboats on the Ohio River to shorten their pipes. The Supreme Court had ordered the bridge to be taken down or alterations of a most expensive character to be made, such as building a draw bridge in the middle of a suspension bridge so that the steamboats with high chimneys could pass under the bridge without shortening the height of their chimneys (Scientific American 1852d). On August 31, 1852 President Millard Fillmore signed the bill into a law that declared the bridge a portion of a post road, and therefore not subject to the decree of the Supreme Court.

7   FAILURE OF THE WHEELING BRIDGE AND ROEBLING'S OBSERVATIONS

On May 17, 1854 the Wheeling Bridge collapsed due to high winds. Based on the eyewitness account of a reporter for the Wheeling "National Intelligencer" the failure mode was similar to that of the Tacoma Narrows Bridge before it collapsed in 1940 (Lewis, 1968). The floor was torn by the force of the wind into three sections: the eastern portion measured 500 ft.; the western 300 ft., leaving the central part about 200 ft. long. Ten out of twelve cables broke in succession from the anchorage; one cable composed of 150 wires broke in the center (Stuart 1871). At that time, Roebling was constructing the Niagara Railway Suspension Bridge, which was opened to traffic on March 18, 1855.

In his final report to the presidents and directors of the Niagara Falls Suspension and Niagara Fall International Bridge Companies, Roebling (1855) addressed the failure mechanism of the Wheeling Bridge, as follows: "Weight is a most essential condition, where stiffness is a great object, provided it is properly used in connection with other means. If relied upon alone, as was the case in the plan of the Wheeling Bridge, it may become the very means of its destruction. That Bridge was destroyed by momentum acquired by its own dead weight, when swayed up and down by the force of the wind". Roebling used stays in his design to prevent uncontrolled motion of his bridge.

8   ELLET AND ROEBLING

They were the two giants of the 19th century in terms of developing and perfecting the art of designing and building the longest suspension bridges in the world. Roebling single-handedly developed the wire rope industry by finding the use of wire ropes in mining, bridge-building, and other industries. They both started in the surveying industry and ended up building suspension bridges. The Roeblings became rich and survived as industrialists for over 100 years (Schuyler 1931) because of the single-minded pursuit of John Roebling in developing the wire rope industry, and the successive generations diversifying into relative industries and taking advantage of the strong foundation built by John Roebling.

Although Ellet had 13 brothers and sister, the Ellet family did not coalesce around Charles Ellet, Jr. who was impulsive and argumentative, and not a great team builder. He achieved national fame as a brilliant suspension bridge builder, but this fame did not translate into creation of wealth for him or his family because of his changing interests into other fields, notably, controlling flooding in western rivers, economics of railroad and transportation operations, and the ramboat scheme which ultimately led to not only his death but the death of his 19-year old son.

It is unfortunate that in each of the three head-to-head competitions for the construction of the Schuylkill, Niagara, and Wheeling Bridges with Roebling, Ellet won all three of these assignments but today the average person does not know his name and his achievements in the U.S. The name of Roebling is known all over the world among bridge engineers. Sayenga (1983) has presented and compared the live and times of both Ellet and Roebling in his interesting book.

9 CONCLUSIONS

Ellet was a brilliant thinker and thought about solving the problems of a young and growing nation, namely the U.S., at the national level. On some issues, such as flood control in western rivers, he was 75 years ahead of his time (Petterson, 1914), and his ideas were not considered seriously by the people in power due to his young age. It was he who developed the art of building wire suspension bridges in the U.S., and stimulated the thinking of Roebling in the same subject. If he had stayed with his first interest of building long wire suspension bridges for the rest of his life, his name would have been as famous as Roeblings today.

ACKNOWLEDGEMENTS

The author thanks Dr. Bojidar Yanev, P.E. of NYCDOT for his careful review of this paper and constructive suggestions; and Brenda Hill and Jaclyn Rabinowitz of Gandhi Engineering for their patience and expert assistance in preparing and editing of this paper.

REFERENCES

Ellet, Jr., Charles 1839. A Popular Notice of Wire Suspension Bridges. *American Railroad Journal and Mechanic's Magazine* VIII: 343–348.
Ellet, Jr., Charles 1852. The Wheeling Bridge Case. *Scientific American* 7(26): 204.
Engineering News 1905. A Note on Early American Suspension Bridges. 53(11): 269–271.
Engineering Record 1904. Early American Chain and Wire Bridges. 49(16): 496–497.
Gambrell, Herbert P. 1931. Charles Ellet. In Allen Johnson and Dumas Malone (ed.), *Dictionary of American Biography*. New York: Charles Scribner's Sons.
Gandhi, K. 2006. Roebling's Railway Suspension Bridge over Niagara Gorge. In Khaled M. Mahmoud (ed.), *5th International Cable-Supported Bridge Operators' Conference*, CRC Press 2006:
Gralian, James 2014. Wheeling Nailers vs. Cincinnati Cyclones (Preseason): The Bridge [blog]. Jerseys and Hockey Love [accessed 2015 Jun 12]. http://jerseysandhockeylove.com/blog/2014/10/13/wheeling-nailers-vs-cincinatti-cyclones-preseason-the-bridge.
Lewis, Gene D. 1968. *Charles Ellet Jr., The Engineer as Individualist, 1810–1862.* Urbana: University of Illinois Press.
Peters, Tom F. 1987. *Transition in Engineering – Guillaume Henri Dufour and the Early 19th Century Cable Suspension Bridges*. Boston: Birkhauser Verlag.
Petterson, H.A. 1914. Comparison of Systems of Flood Control. *Engineering Record* 69(20): 560.
Roebling, John A. 1855. *Final Report to the Presidents and Directors of the Niagara Falls Suspension and Niagara Fall International Bridge Companies*. Steam Press of Lee, Mann & Co., Buffalo, NY, 1855.
Sayenga, Donald 1983. *Ellet and Roebling.* York, PA: American Canal and Transportation Center.
Schuyler, Hamilton 1931. *The Roeblings: A Century of Engineers, Bridge-Builders and Industrialists 1831–1931*. Princeton: Princeton University Press.
Scientific American 1848a. The Bridge over the Ohio at Wheeling, Virginia. 3(31): 246.
Scientific American 1848b. Items of Niagara Suspension Bridge 3(43): 337.
Scientific American 1848c. The Niagara Bridge 3(47): 370.
Scientific American 1849a. Wheeling Suspension Bridge. 5(6): 41.
Scientific American 1849b. Wheeling Suspension Bridge. 5(7): 53.
Scientific American 1850a. Wire Suspension Bridge. 5(28): 218.
Scientific American 1850b. Wheeling Bridge. 5(44): 348.
Scientific American 1850c. The Wheeling Bridge. 5(45): 356.
Scientific American 1851a. The Wheeling Bridge Case. 6(21): 161.

Scientific American 1851b. The Wheeling Bridge. 7(9): 70.

Scientific American 1852a. The Wheeling Bridge – Steamboat Chimneys. 7(31): 245.

Scientific American 1852b. Wheeling Bridge – Explosions. 7(32): 250.

Scientific American 1852c. The Wheeling Bridge Case. 7(38): 403.

Scientific American 1852d. The Wheeling Bridge. 7(51): 403.

Scientific American 1854. The Wheeling Bridge. 9(45): 353.

Scientific American 1855. Fairmount Suspension Bridge. 10(45): 357.

Steinman, D.B. 1945. *The Builders of the Bridges, The Story of John Roebling and His Son*. New York: Harcourt, Brace & Co.

Stuart, Charles Beebe 1871. *Lives and Works of Civil and Military Engineers of America*. New York: D. Van Nostrand.

Tyrrell, Henry Grattan 1911. *History of Bridge Engineering*. Chicago: Published by the Author.

U.S. Supreme Court 1851. Pennsylvania v. Wheeling and Belmont Bridge Co., 54 U.S. 518.

Chapter 27

Lindenthal and his pursuit of a bridge across the Hudson River

K. Gandhi

Gandhi Engineering, Inc., New York, NY, USA

ABSTRACT: Gustav Lindenthal (1850–1935) at his death was referred to by some journals as "The Dean of American Bridge Builders". He was born in Bruun, Austria in 1850, and immigrated to the United States in 1874. He started his own consulting engineering firm in 1881. In 1888 he initiated the pursuit of building a major suspension bridge across the Hudson River connecting New Jersey with Manhattan, a pursuit which continued for the next 45 years. This paper examines the various schemes developed by Lindenthal, and the circumstances which prevented Lindenthal from achieving his lifelong dream of building a bridge across the Hudson River.

1 INTRODUCTION

The oldest written reference to bridging the Hudson River at New York is in the book "A Treatise on Bridge Architecture" by Thomas Pope, published in 1811. The author shows an arch with variable moment of inertia and labels it "View of T Pope's Flying Lever Bridge" (Pope 1811). Another scheme was proposed by Messrs. Anderson and Barr in 1884 which required placing piers in water to provide a clear span of 500 ft. (152.4 m) for navigation (Engineering News 1884a). This scheme was considered impractical and was ignored (Engineering News 1884b).

At a meeting of the American Society of Civil Engineers held in Kaeterskill, NY on January 4, 1888 Gustav Lindenthal read the paper "The North River Bridge Problem, with a Discussion on Long Span Bridges" (Engineering News 1888a). This was the first time that someone had seriously thought about building a bridge across the Hudson River. The basic details of Lindenthal's bridge are shown in Figures 1 and 2. The editorial in Engineering News stated that this is "the first definite description of a work which has at least a very fair chance of becoming the greatest of its kind on this continent, or in the world" (Engineering News 1888c). The details of the various elements of this suspension bridge such as foundations and masonry, piers for the towers, anchorages, towers, superstructure and loads, construction of details of cables, web system between cables, provision against tornados, and architectural features were covered in Engineering News (1888b, 1888d). The estimated construction cost was $15,000,000 and time of construction about $3\frac{1}{2}$ years.

After the publication of his paper, Lindenthal patented details of sheet metal cable cover to protect the cable from weather and changes of temperature; and the method of making attachments to the cable and of joining the cable itself in sections (Engineering Record 1888).

Max Am Ende (Ende, 1889a) criticized Lindenthal's scheme and suggested to build an arch bridge instead. Lindenthal (1889) responded by personally attacking Ende, which was unusual. Ende (1889b) responded by pointing out the unwarranted personal attack by Lindenthal.

2 FORMATION OF NORTH RIVER BRIDGE CO.

In the late 1880s the Pennsylvania Railroad (PennRR) wanted a direct rail entrance into New York City, and Lindenthal was retained by Mr. Samuel Rae of the PennRR to study the feasibility of such a move. The original scheme was to enter Lower Manhattan with a terminal at Washington Square. However, the cost seemed prohibitive for the PennRR to bear it alone. Lindenthal was asked by the

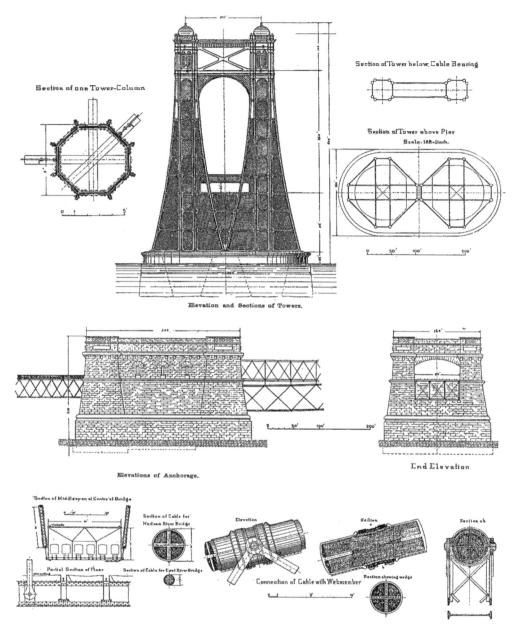

Figure 1.   Details of the railway suspension bridge across the Hudson River proposed by Lindenthal in 1888 (Engineering News, 1888a).

PennRR to find out if other railroads would participate in building the bridge across the Hudson and sharing the cost of construction.

The result was the formation in 1890 of a Corporation called the North River Bridge Co., and the incorporators included Jordan L. Mott, Jr., King McLanahan, James Andrews, Thomas F. Ryan, Garrett A. Hobart, F.W. Roebling, Charles J. Canda, Edward F.C. Young, Henry Flad, Gustav Lindenthal, A.G. Dickinson, John H. Miller, William Brookfield, Samuel Rea, William F. Shunk, Philip E. Chapin, and their associates (Engineering News 1890b).

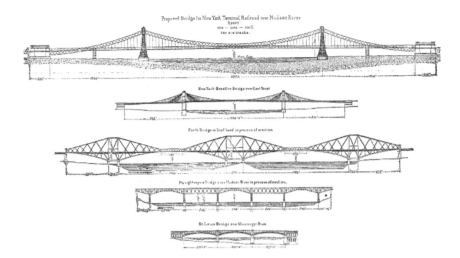

Figure 2.   The proposed elevation of the Hudson River bridge (top) compared to already constructed bridges (Engineering & Building Record 1888a).

3   APPROVAL OF NORTH RIVER BRIDGE BILL IN CONGRESS

The supporters of Gustav Lindenthal introduced bills in July 1888 in both houses of Congress to build a bridge with a single span of 2,830 ft. (862.6 m). A hearing was scheduled to determine possible injury to navigation, but it did not take place until 1890.

The North River Bridge Bill was before the Senate in 1890 and was looked upon favorably by the War Department. General Casey of the U.S. Engineers Corps suggested two minor amendments, one to use the word "towers" instead of "piers", and the second to provide a clearance under the bridge of 145 ft. (44.2 m) at the towers and 155 ft. (47.2 m) at the center of the span. The War Department also suggested a new section in the bill, requiring "the submission of plans satisfactory to the Secretary of War within one year after the passage of the bill, starting the construction one year after the approval of plans by the Secretary of War, and completing the bridge within 10 years of such approval; "otherwise the provisions of the act shall be null and void" (Engineering News 1890a).

The North River Bridge Bill was passed by both Houses of Congress by June 1890. This bill in express terms granted the right to make the necessary condemnation on both banks of the Hudson River without getting specific approvals from the States of New York and New Jersey (Engineering News 1890c). The plans for the North River Suspension Bridge designed by Lindenthal were approved by the Secretary of War in April of 1891 with a clear headway of 150 ft. (45.7 m) above high water (Engineering News 1891a).

Based on the annual report filed with the Interstate Commission, as of September 1891 Lindenthal's company, the North River Bridge Co., had nearly completed the preliminary work prerequisite to actual construction. The bridge had two levels with a capacity of 14 standard gage railway tracks. The limits of the bridge were between Bloomfield and Twelfth Streets in Hoboken, and 23rd Street and Tenth Avenue in New York. From this point in New York, a steel viaduct would carry the 14 tracks to a high level station on Sixth Avenue, extending from 25th to 28th Streets with a connection with the yards of the New York Central on 39th Street. The estimated total cost was $70,000,000 (Engineering News 1891b).

4   CONDEMNATION OF PROPERTY BY NORTH RIVER BRIDGE CO.

In March 1892, the North River Bridge Co. commenced proceedings to condemn property which was within its right-of-way under its charter from the Congress. A hearing took place in Trenton,

NJ on March 20, 1892. The property owner argued through his lawyer that the bridge company, in the absence of State legislation, had no authority for such a proceeding. The judge appointed a commission to appraise the property (Engineering News 1892c). As of December 1893 the matter was pending before the U.S. Supreme Court, which stopped all construction field activities of the North River Bridge Co. (Railroad Gazette 1893). The lawsuit was decided in favor of the North River Bridge Co., but the business depression of 1893 forced indefinite postponement of the Hudson River Bridge project (Frankland 1940).

However, this did not stop Lindenthal from promoting his bridge. He wrote two articles, "Bridging the Hudson at New York" (Lindenthal 1893) and "Will a bridge across the Hudson River Pay?" (Lindenthal 1895).

5   LINDENTHAL'S COMPETITION TO BUILD A PARALLEL BRIDGE

There was a rival bridge company known as the Consolidated New York & New Jersey Bridge Co., organized under a consolidation of old and new charters obtained from the States of New York and New Jersey, but as of the end of 1891 was still without authority from the U.S. Government to cross the river. This company obtained a charter from the New York Legislature, which was then tacked on to an old New Jersey Bridge charter. However, it did not obtain any permission from the U.S. Government to build a bridge across the Hudson River (Engineering News 1892a). On December 24, 1891 the New York & New Jersey Bridge Co. conducted the ground breaking ceremony without having approval from Congress. There was no coordination between the organizers in New Jersey and the invited guests from New York. The heavy rain made any ceremony and speech delivery almost impossible (Engineering News 1892b).

Mr. Thomas C. Clarke, the Chief Engineer of the New York & New Jersey Bridge Co. proposed a cantilever bridge with a center span of 2,200 ft. (670.6 m) and two flanking spans of 1,050 ft. (320 m) each. The piers and the masonry would be built for an eight track bridge, but the superstructure would be built only for four tracks to satisfy the needs for the near future rather than for the life of the bridge. On the New Jersey side, the west approach would be carried through Bergen Hill either by a tunnel or an open cut. On the New York side, the landing would be at about 71st Street, and the line carried down Eleventh Avenue to 42nd Street and across to Broadway with ample terminals. The plans were changed to build a suspension bridge when permission was denied to build a pier in water for a cantilever bridge.

There was equal (if not more) coverage in technical journals of the activities of the New York and New Jersey Bridge Co. about building of a parallel bridge to the north of Lindenthal's bridge. However, to limit the size of this paper, the author has decided to concentrate on Lindenthal and his bridge.

6   APPOINTMENT OF A BOARD OF FIVE EXPERT BRIDGE ENGINEERS BY THE PRESIDENT IN 1894

In January of 1894, the New York and New Jersey Bridge Bill was filed with the House Committee on Commerce in Washington, D.C. Instead of determining whether a pier should be permitted in water for a cantilever bridge with a minimum span of 2,000 ft. (609.6 m), the House Committee passed a resolution on June 7, 1894 that the President shall appoint a board of five competent, practical, disinterested, expert bridge engineers, of whom the Chief of Engineers of the United States Army shall be one, to decide what length of spans, not less than 2,000 ft. (609.6 m), is safe and practicable for the Hudson River Bridge between 59th and 69th Streets in New York City (Engineering News 1894a).

The engineers appointed to the Board by President Grover Cleveland included G. Bouscaren, W.H. Burr, Theodore Cooper, George S. Morison, and Major C.W. Raymond.

The Board submitted its Report in September 1894. The Board considered feasibility of constructing a cantilever bridge of (a) 2,000 ft. (609.6 m) span and (b) 3,130 ft. (954 m) span, and their cost of construction, including the cost of piers in the Hudson where the rock was more than 200 ft. (60.96 m) below the surface. The Board also considered the construction of a 3,130 ft. (954 m) span railway suspension bridge. (Engineering News 1894b).

The Board concluded that:

1. A single span from pierhead to pierhead, built on either the cantilever or suspension principle, would be safe.
2. The estimated cost of the 3,100 ft. (944.9 m) clear span cantilever was about twice that of the short span, making this option impracticable on financial grounds.
3. The 3,100 ft. (944.9 m) span railway suspension bridge was practicable. However, as the cost of a single span suspension bridge was almost one-third greater than that of the 2,000 ft. (609.6 m) cantilever, the Board was unable to say that such greater cost was enough to render the suspension bridge impracticable.

The Board also felt "that the contingency attending the construction of the deep river foundation of the cantilever bridge is enough to balance a part of the greater cost of the suspension bridge", meaning that the suspension scheme would meet all the requirements of a single span, and costwise would be competitive.

The Report of the Board of Engineers also included several Appendices consisting of (a) A letter from Charles Macdonald of the Union Bridge Co. who would build the cantilever bridge for the New York and New Jersey Bridge Co. (Engineering Record 1894a), (b) Theory of continuous stiffening truss with ends anchored down but not fixed in a vertical plane (Engineering Record, 1894b), (c) Letters from Wilhelm Hildenbrand (Engineering Record 1894c), (d) A modified plan and estimate from Hildenbrand (Engineering Record 1894d), (e) A letter from Gustav H. Schwab expressing views of New York State Chamber of Commerce (Engineering Record 1894e), and (f) A description of the proposed North River Bridge (Lindenthal 1894).

7   APPOINTMENT OF A BOARD BY THE SECRETARY OF WAR IN 1894

While the Hudson River Bridge was debated in Congress, the Secretary of War Daniel S. Lamont appointed on January 29, 1894 a Board consisting of Major C.W. Raymond, and Captains W.H. Bixby and Edward Burr, all of the Corps of Engineers, to investigate and report its "conclusions as to the maximum length of span practicable for suspension bridges, and consistent with an amount of traffic probably sufficient to warrant the expense of construction".

To the above instructions Brigadier General Thomas L. Casey, of the Chief of Engineers, added the further instructions to include in the investigation "strength of materials, loads, foundations, wind pressure, oscillation and bracing". Captain Bixby (Bixby 1895) later published a well-researched and scholarly paper on wind pressure in engineering construction related to the Hudson River Bridge.

8   MAXIMUM PRACTICABLE LENGTH FOR SUSPENSION BRIDGES

This Board reviewed the Report prepared by the board of five expert engineers appointed by President Cleveland before publishing its report in October 1894 (Engineering News 1894c,d,e). The most significant finding as far as the Hudson River Bridge was concerned was that it was not only possible to erect a single-span structure of 3,200 ft. (975.4 m) at that point, but it was practicable to build a suspension bridge of 4,335 ft. (1,321 m) span. And, cost-wise, a suspension bridge with a 3,200 ft. (975.4 m) span would be very competitive to a cantilever bridge of a 2,000 ft. (609.6 m) span.

The report prepared by the three engineers from the U.S. Army Engineers Corps forced the New York and New Jersey Bridge Co. to revise its plan from a 2,200 ft. (670.56 m) cantilever to a 3,000 ft. (914.4 m) span suspension bridge.

9 ACTIVITIES FROM 1895 TO 1912

The recession of 1893 forced almost all railroads to withdraw their support of the Hudson River Bridge project. The railroads were not willing to discuss their plans for expansion to New York with other railroads and the Hudson River Bridge project was practically abandoned by the railroads who would provide the funding for the bridge. The promoters were telling the public that traffic that would make the bridge pay could be secured from the beginning.

The charter of the North River Bridge Co. was to expire on July 11, 1895 unless something was done by that time showing the sincere purpose of the North River Bridge Co. to construct the work for which it obtained the powers. It was to show this intention and prevent the charter from lapsing that the North River Bridge Co. quietly commenced work upon the New Jersey anchorage.

The North River Bridge Co., at its annual meeting in November 1896, elected Mr. Jordan L. Mott as President, Mr. Samuel Rea as Vice President, Mr. Charles J. Canda as Treasurer, and Mr. Thomas B. Rea as Secretary. According to the plans prepared by Lindenthal the bridge would cross the Hudson River in the vicinity of 23rd Street and would carry 14 railway tracks.

In 1897, the promoters of the North River Bridge Co. refused to provide any information on their plans for the bridge. However, it was known that the length of the suspension span and the total length of the bridge proper would be 3,100 ft. (944.9 m) and 7,340 ft. (2,237.2 m), respectively. No definite traffic arrangements were made with any of the railway companies operating on the New Jersey side. The railway companies did not show their willingness to cooperate with each other in the discussion of construction of the necessary connections and terminals. Without the support from the railroads the bridge project was dead (Engineering News 1897).

There was no activity reported by the promoters of Lindenthal's bridge during 1898 and 1899. In 1900, the U.S. Senate passed Senator Sewell's bill extending the time for completion of Lindenthal's bridge to January 1, 1902 (Railroad Gazette, 1900). In 1901, the North River Bridge Co. entered into discussions with two or three railroads to back its plan without any success (Railroad Gazette 1901).

For a two-year period, 1902–1903, Lindenthal was appointed as Commissioner of Bridges by Mayor Low of New York City; and no activities took place for the construction of the Hudson River Bridge.

The biggest setback to the Hudson River Bridge project came when the PennRR in 1904 decided to go it alone and build two tunnels under the Hudson, a major passenger terminal in Manhattan near 34th Street, and a yard for its trains in Queens by building tunnels under the East River. There were no major activities from 1905 to 1911.

10 ACTIVITIES FROM 1912 TO 1920

In an article published in the New York Times which was reprinted in Engineering News, Gustav Lindenthal (1912) acknowledged that the real deterrent to the construction of the Hudson River Bridge was its cost ranging from $70 million to $100 million including the cost of the right-of-way. He presented a case in his article for building the bridge by connecting it with the handling of freight in Manhattan.

The editorial analysis of the history of the Hudson River Bridge (Engineering News 1912) agreed with Lindenthal that time was ripe for the construction of the Hudson River Suspension Bridge.

One of the readers, who wrote a letter after reading Lindenthal's article and Engineering New's editorial comments, asked Lindenthal to reduce the cost of his bridge by putting one or two piers in the river (Smith 1913). In his reply, Lindenthal (1913) explained why one or two piers in the Hudson River would not reduce the cost of the bridge, and cautioned against underestimating the total cost of the project by not taking into account contingencies, delays, interest payment during construction, cost of administration, right-of-way, and connections and approaches to the bridge.

In 1920, Lindenthal (1920) discussed the proposed Hudson River bridge, its capacity and cost compared with tunnels, and the bridge's influence on New York's transportation problems at a meeting of the American Society of Mechanical Engineers held in New York on December 9, 1920.

11   ACTIVITIES FROM 1921–1930 AND PARTICIPATION OF AMMANN

In 1921, the Hudson River Bridge and Terminal Association was formed to obtain public support to build the bridge designed by Lindenthal, with terminal facilities on both sides of the river for passengers and freight (Engineering News-Record 1921a). One of the members of the committee was Samuel Rae, president of the PennRR.

Another organization, the Hudson River Bridge Corporation, was formed also in 1921. Under its umbrella, a large engineering organization was built up which was at work under the direction of Gustav Lindenthal, chief engineer, making structural and traffic connection studies and developing plans. Cooperation of all railroads entering New York was secured in the effort to adapt the project as efficiently as possible to New York terminal traffic requirements.

Othmar H. Ammann, one of the best known and most respected long span bridge builders of the 20th century, worked for Lindenthal on the Hell Gate Bridge in New York City between 1912 and 1917. He left Lindenthal in 1917 due to lack of work, and rejoined in December 1920 to work on the Hudson River Suspension Bridge. During these 3 years, between 1917 and 1920, Lindenthal was planning the bridge on an even grander scale.

Lindenthal's new design published in the April 1921 issue of Scientific American connected West 57th Street in Manhattan with 50th Street in Weehawken in New Jersey. The proposed bridge had a main span of 3,240 ft. (987.6 m) and two decks. The upper deck was for pedestrians and 16 lanes of vehicular traffic, whereas the lower deck supported 12 railroad tracks. The granite-clad steel towers were 840 ft. (256 m) tall. Lindenthal even planned an office tower to generate income to pay for the $100 million bridge project (Figure 3). The elevation and cross-section of the bridge are shown in Figures 4 and 5, respectively.

Lindenthal often asked Ammann to present his scheme at public and professional meetings. The negative sentiments and loud opposition to the bridge project expressed at these meetings convinced Ammann that there was a real need for a vehicular bridge and not for a bridge with 12 railroad tracks. He pleaded with Lindenthal to reduce the scale of the project and relocate the site. Lindenthal criticized Ammann for this timidity and short-sightedness and for not looking far enough ahead (Rastorfer 2000). In March of 1923, Ammann left Lindenthal, and opened his own consulting engineering firm. He developed a proposal for a vehicular bridge across the Hudson River with a price tag of $40 million, and presented it at a meeting of the Connecticut Society of Engineers on February 19, 1924 (Rastorfer 2000). Ammann's design was approved by the Port of New York Authority, and the first vehicular bridge across the Hudson River was opened to traffic in 1931. It is known today as the George Washington Bridge.

The application submitted by the North River Bridge Co. to the War Department for a permit to build the bridge showed a suspension bridge with minimum vertical clearance above the mean high water at center of 175 ft. (53.3 m) for a width of 500 ft. (152.4 m) and 166 ft. (50.6 m) for a width of 1,500 ft. (457.2 m) and an unobstructed channel between the established pierhead lines. The War Department scheduled a hearing in the Army Building at 39 Whitehall Street in New York City at 10:00 AM on September 9, 1927 (Engineering News-Record 1927a).

The hearing was headed by Colonel F.C. Boggs, and the project was presented by Lindenthal and supported by Francis Lee Stuart and Samuel Rae, the former president of the PennRR. Opposition to the application was voiced by shipping interests, and they asked for a clear height of 215 ft. (65.5 m) Others opposed to the bridge questioned the need for Lindenthal's bridge in view of the traffic facilities provided by the Fort Lee bridge and the Holland Tunnel under construction (Engineering New-Record 1927b). No decision was made by the army engineers.

Meanwhile, it was reported in Railway Age (1929a) that the Baltimore & Ohio RR acting with the North River Bridge Co. had filed an application with the Secretary of War for approval of Lindenthal's plans. The cost estimate submitted with the plans was $180 million. Secretary of War James W. Good stated on May 2, 1929 that General Edgar Jadwin, Chief of Engineers, would report on the examination of the general plans shortly. The Baltimore & Ohio RR made a statement that it was an "interested observer" of the bridge project, and flatly denied that it planned to use the bridge as a means of getting access to Manhattan for a railroad terminal (Engineering News-Record 1929a).

Figure 3.   Rendering of Lindenthal's proposed bridge with office tower (Scientific American, 1921b).

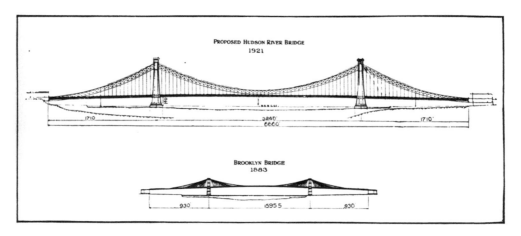

Figure 4.   Elevation of proposed Hudson River Suspension Bridge compared with that of Brooklyn Bridge
(Scientific American, 1921a).

On May 29, 1929, Secretary of War Good made an announcement that he was rejecting the
application of the North River Bridge Co. for permission to bridge the Hudson River at 57th Street
in New York City unless the plans were revised to provide a 200 ft. (60.96 m) clearance above the
mean high water at the center of the bridge, and 185 ft. (56.4 m) at the pierhead lines. The plans as
submitted by the North River Bridge Co. provided only 175 ft. (53.3 m) of clearance at the center
of the spans. The decision was concurred by Major General Edgar Jadwin, Chief of Engineers
(Engineering News-Record 1929b).

In support of his decision, the report of the Secretary of War stated in part that "nowhere is the
protection of navigation more important than at New York, which is preeminent among the ports

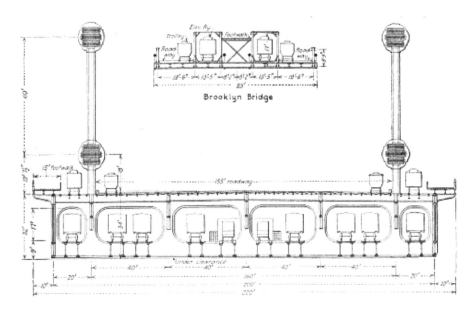

Figure 5.   Cross section of proposed bridge (Scientific American, 1921a).

of the United States and of the world as well." And, "that no unnecessary bar should be put in the way of the great vessels coming to this port laden with freight for Manhattan, or for the interior of the country" (Railway Age 1929c). The editorial in Railway Age (1929b) described the decision by the Secretary of War as "a judgement by the opponent's attorney".

In May of 1930, the new Secretary of War Hurly, accompanied by Chief of Engineers Major General Lytle Brown, inspected the proposed bridge site, and as a result a new hearing on the construction of a combined railroad and vehicular bridge across the Hudson River at 57 Street was scheduled on June 4, 1930 in New York City before a special board of officers headed by Col. G.M. Hoffman of the Corps of Engineers.

The new application was made by the North River Bridge Co. of Jersey City, NJ, and showed a maximum vertical clearance of 184.5 ft. (56.2 m) for a width of 300 ft. (91.44 m) at the center of the span, and the lowest 175.8 ft. (53.6 m) for the same width (Engineering News-Record 1930a).

At the hearing held in New York City on June 4, 1930, a special board of army engineers consisting of Colonels G.M. Hoffman, G.B. Pillsbury, and Harley B. Ferguson heard the testimony of a number of witnesses for and against the bridge (Engineering News-Record 1930b). The North River Bridge Co.'s arguments were presented by Francis Lee Stuart, a consulting engineer who stated that raising the vertical clearance to 200 ft. (60.96 m) would create too steep a grade for most locomotives to climb. Stuart also offered $3,000 to $8,000 per ship for telescoping of the upper masts and funnels of ships requiring more than 184 ft. (56.1 m) of vertical clearance. David B. Steinman, consulting engineer for the bridge company, argued that the trend was to reduce the mast heights of ships, and that the decision would have created a precedent to be followed in the future.

Arguments in opposition to the bridge included the effect on traffic congestion, lowering of the real estate values, and public sentiment that cross-river traffic should be cared for by tunnels than by overhead structures.

12   ACTIVITIES FROM 1931 TO 1933

The special board of army engineers submitted an adverse report on the application of the North River Bridge Co., stating that the clearances for the projected structure were inadequate (Engineering News-Record 1931a).

The Secretary of War disapproved the revised plans of the North River Bridge Co. on June 9, 1931. The revised plans had a center clearance of 179.8 ft. (54.8 m) and the pierhead clearance of 160.6 ft. (48.9 m), as opposed to the minimum center clearance of 200 ft. (60.96 m) as required by the Army Engineers Corps (Engineering News-Record 1931b).

The letter from the Secretary of War, Patrick J. Hurley, addressed to the President of the North River Bridge Co. read in part that "there are ships in operation in the harbor now which cannot maneuver under your proposed bridge without alterations in them more serious than the mere lowering of the masts" (Engineering News-Record 1931c). The investigation by the Engineering News-Record indicated that of the ships entering the port of New York there were none with a funnel height of 175 ft. (53.3 m), and only one ship had a funnel height of over 150 ft. (45.7 m).

Lindenthal at the age of 83 filed one more application to obtain a permit from the Secretary of War to build his bridge at 57th Street in Manhattan; and one more time the application was declined. The Secretary of War stated that it was his interpretation of the law that when there is serious question as to the necessity of the bridge, any interference with navigation by it is to be considered unreasonable. The Secretary cited the opposition to the bridge, especially by the Port of New York Authority which had begun construction of a vehicle tunnel under the river at 38th Street (Engineering News-Record 1933), and the bridge would come into direct and powerful competition with the tunnel, for which the Authority had borrowed a large sum from the federal government (Railway Age 1933).

It is appropriate to include a short biography of this remarkable person, Gustav Lindenthal.

13   GUSTAV LINDENTHAL, 1850-1936 (FRANKLAND, 1940)

Gustav Lindenthal was born in Brunn, Austria on May 21, 1850 (Figure 6). He was educated at Politechnicum (Dresden, Germany) college, and received practical training from 1866 to 1870. He came to the U.S. in 1874. He worked at the Centennial Exposition as a laborer and then as a designer. In 1878 he joined Atlantic Great Western Railroad as a bridge engineer. In 1881, Lindenthal started his own engineering practice in Pittsburgh, and found assignments in the design and construction of important bridges for railroads and bridge companies.

He received recognition and very good publicity for replacing the Smithfield Street suspension bridge, originally built by John A. Roebling, with a 350 ft. (106.7 m) span double-elliptical steel truss in 1882. This and several other projects in Pittsburgh brought him to the attention of Samuel Rae of the Pennsylvania Railroad who later supported Lindenthal in the proposal for the Hudson River bridge. It was Lindenthal's paper in 1888 presented at the ASCE convention, "The North River Bridge Problem, with a Discussion on Long Span Bridges," that excited not only the engineering community but the general public. The real estate industry sold parcels of farmland at inflated prices to buyers who were speculating that the price of their properties would increase once the bridge was built.

Lindenthal spent his entire career in private practice except for a two year period (1902–1903) when he was appointed the Bridge Commissioner of New York City by Mayor Low. His activities and performance as New York City's Bridge Commissioner are covered in detail by Gandhi (2013). The most notable project of his professional practice was the Hell Gate Arch Bridge in New York. Originally the firm of Boller and Hodge was selected by the PennRR to design this bridge, but with the help of Samuel Rae the project was given to Lindenthal. On the Hell Gate arch project, Lindenthal hired Othmar H. Ammann, David B. Steinman, and Charles S. Whitney, who would later become world-renowned engineers in their own right.

Before the Hell Gate Bridge was completed, the Norfolk and Western Railroad retained Lindenthal to design a bridge across the Ohio River at a site a few miles east of Sciotoville, Ohio with a river width of more than 2,000 ft. (609.6 m) and requirements of two navigation channels. Between 1914 and 1917 Lindenthal designed a two-span continuous bridge over a central pier with a record-breaking combined span-length of 1,550 ft. (457.2 m).

Figure 6.   Gustav Lindenthal (Wildman 1921).

It was said of Lindenthal that he never designed two bridges alike because he looked to the design of each bridge as a unique problem and created a customized solution. His love of beauty in engineering works simultaneously with his search for the structural solution. In his search for aesthetic design, he did not hesitate to consult architects whenever he had to deal with an important bridge project.

Lindenthal's last professional assignment was design and construction of bridges in Portland,

Oregon to carry the city's major thoroughfares across the Willamette River, which he completed in 1928. Lindenthal died on July 31, 1935, in his 86th year.

14   CONCLUSIONS

The conclusions as to why Lindenthal did not succeed in his lifelong quest for the Hudson River Railroad Suspension Bridge are summarized below.

This is a sad story with a sad ending as far as Lindenthal is concerned. He depended on railroads, specifically the PennRR, to raise the money to build his ambitious project. In the early 1890s, if the railroads agreed to contribute towards the building of the bridge, the space to build the station yards and connections was available in Manhattan at a reasonable price. However, the smaller railroads mistrusted the PennRR, and its overbearing influence in decision making in the overall project.

In the early 1900s when the economy once again started growing, the PennRR knew that the other railroads were not interested in teaming up; and in 1904 decided to go alone and build two tunnels under the Hudson, and Penn Station at 34th Street in Manhattan. The development of electric locomotives for use in the tunnels was one major reason to go with the tunnels by the PennRR. This caused Lindenthal to lose his biggest supporter of his bridge project.

It was evident, as early as 1903 that any bridge that ignored the growth of population in 40 or so towns in New Jersey, and about 500,000 commuters bravely crossing the Hudson daily, would not pay for itself. Lindenthal and his railroad friends ignored this trend, and kept adding the number of railroad tracks to the bridge, thereby increasing the cost of the project to an unreachable level.

In the early 1920s when the use of automobiles was on the rise and the railroads were declining, Ammann urged his boss Lindenthal to scale down his bridge, and move the bridge location to an undeveloped area in the north to reduce the cost of the project. Instead, Lindenthal ridiculed Ammann for his lack of imagination and not looking 1,000 years further down the road. This forced Ammann to leave Lindenthal, start his own business, and plan his own bridge to cross the

Hudson, which was accepted by the newly formed bi-state agency Port of New York Authority in the mid-1920s.

Even after the opening of the George Washington Bridge crossing the Hudson River in 1931, Lindenthal deluded himself into believing that a railroad suspension bridge was still a viable option, and submitted his scheme as late as 1933 to the Army Corps of Engineers. This time it was the Port of New York Authority which filed objection to Lindenthal's bridge on the grounds that the Port Authority had borrowed funds from the U.S. Government for construction of the Lincoln Tunnel, and if the bridge were permitted to be built it would siphon off the toll from the Lincoln Tunnel. This was the last time, three years before his death, that Lindenthal faced his final defeat.

ACKNOWLEDGEMENTS

The author thanks Dr. Bojidar Yanev, P.E. of NYCDOT for his careful review of this paper and constructive suggestions; and Brenda Hill and Jaclyn Rabinowitz of Gandhi Engineering for their patience and expert assistance in preparing and editing of this paper.

REFERENCES

Bixby, W.H. 1895. Wind Pressure in Engineering Construction. *Engineering News* 33(11): 175–184.

Ende, Max Am, 1889a. A Criticism of Mr. Lindenthal's Project of a Bridge over the North River, New York and a Suggestion. *The Engineer* 67(20): 411–413.

Ende, Max Am, 1889b. The Proposed North River Suspension Bridge at New York City. *Engineering and Building Record* 20(18): 49.

Engineering and Building Record 1888a. Proposed Bridge over the Hudson River at New York City. 17(7): 102–103.

Engineering and Building Record 1888b. Mr. Gustav Lindenthal of Pittsburg, PA. 18(1): 7.

Engineering News and American Contract Journal 1884a. A Project for Bridging the Hudson River at New York. 11(8).

Engineering News and American Contract Journal 1884b. Bridge Project Over the Hudson. 11(10): 116.

Engineering News 1888a. Society Proceedings. 19(1): 12.

Engineering News 1888b. The North River Bridge with a discussion of Long Span Bridges. 19(4): 57–59.

Engineering News 1888c. Editorial 19(4): 62.

Engineering News 1890a. The North River Bridge Bill. 23(8): 170.

Engineering News 1890b. The North River Bridge Bill. 23(14): 313–314.

Engineering News 1890c. Editorial on rights of Congress and the States 24(4): 80–81.

Engineering News 1891a. The Plans for the North River Suspension Bridge. 25(18): 415.

Engineering News 1891b. The North River Bridge Co. of New York. 26(18): 415.

Engineering News 1892a. The Rival Hudson River Bridge Companies. 27(1): 415.

Engineering News 1892b. Breaking Grounds for the New York and New Jersey Bridge. 27(1): 19–20.

Engineering News 1892c. The North River Bridge Co. 27(15): 341.

Engineering News 1894a. New York and New Jersey Bridge Bill. 31(5): 81.

Engineering News 1894b. Experts' Report up on the Proposed Hudson River Bridge at New York. 32(10): 187–190.

Engineering News 1894c. The Maximum Practicable Length for Suspension Bridges (Part I). 23(21): 423–425.

Engineering News 1894d. The Maximum Practicable Length for Suspension Bridges (Part 2). 32(22): 444–446.

Engineering News 1894e. The Maximum Practicable Length for Suspension Bridges (Part 3). 32(23): 463–465.

Engineering News 1897. Plans for extending transportation facilities in and about New York-North River transit. 38(22): 344.

Engineering News-Record 1921a. Hudson River Bridge Association to be incorporated. 86(15): 656.

Engineering News-Record 1921b. Planning for Hudson River Bridge at New York City. 86(21): 900–901.

Engineering News-Record 1927a. Army Hearing on 57th St. Bridge over Hudson Set for September 9. 99(7): 282.

Engineering News-Record 1927b. Hearing on Bridge project at 57th Street, Manhattan is held by Army Engineers. 99(11): 446.

Engineering News-Record 1929a. Baltimore and Ohio not interested in Hudson River Bridge. 102(19): 767.

Engineering News-Record 1929b. Clearance of 200ft. Required for Hudson River Bridge. 102(23): 920.

Engineering News-Record 1930a. War Department Reopens hearing on Hudson River Bridge. 104(20): 821–822.

Engineering News-Record 1930b. Army Engineers Hear Renewed Pleas for Hudson River Bridge. 104(24): 989–990.

Engineering News-Record 1931a. Adverse Report on New York's Proposed 57th Street Bridge. 106(14): 576.

Engineering News-Record 1931b. Bridge Plans at New York Again Disapproved. 106(24): 985.

Engineering News-Record 1931c. Text of War Department Refusal to Application for New Hudson River Bridge. 106(21): 635.

Engineering Record 1894a. Letter from Charles Macdonald of Union Bridge Co. 30(20): 322–323.

Engineering Record 1894b. Theory of Continuous Stiffening Truss. 30(21): 339–340.

Engineering Record 1894c. Letters from W. Hildenbrand. 30(22): 357–359.

Engineering Record 1894d. Modified plan and Estimate from Hildenbrand. 30(23): 357–358.

Engineering Record 1894e. Letter from New York State Chamber of Commerce. 30(23): 376.

Frankland, F.H. & Schmitt, F.E. 1940. Memoir of Gustav Lindenthal. *Transactions of the American Society of Civil Engineers* 105: 1790–1794.

Gandhi, K. 2013. Lindenthal and the Manhattan Bridge eyebar chain controversy. In Khaled M. Mahmoud (ed.), *Durability of Bridge Structures*. CRC Press 2013: 285–300.

Lindenthal, G. 1893. Bridging the Hudson at New York. *Engineering Magazine* 6: 213–222.

Lindenthal, G. 1895. Will a Bridge Across the Hudson River Bay? *Engineering Magazine* 10: 261–268.

Lindenthal, G. 1889. A Reply to Certain Criticism on the Proposed North River Suspension Bridge at New York City. *Engineering News* 22(3): 58–59.

Lindenthal, G. 1894. The North River Bridge. *Engineering Record* 30(24): 390–392.

Lindenthal, G. 1912. An Opportunity for a Bridge Across the North River at New York City. *Engineering News* 68(25): 1143–1144.

Lindenthal, G. 1913. Piers for a North River Bridge at New York City Would Effect no Savings in Cost over a Single Span. *Engineering News* 69(3): 126.

Lindenthal, G. 1920. Proposed Hudson River Bridge. *Engineering News-Record* 85(26): 1246–1247.

Pope, Thomas, 1811. A Treatise on Bridge Architecture. *Printed for the author by Alexander Niven* (New York): 288 and 13 plates.

Railroad Gazette 1893. Editorial-untitled. 25: 930.

Railroad Gazette 1900. The Hudson River Bridge. 32: 175.

Railroad Gazette 1901. The North River Bridge. 33: 865.

Railway Age 1929a. New York-New Jersey Railway Bridge Proposed. 86(18): 1065–1066.

Railway Age 1929b. A Judgment by the Opponent's Attorney. 86(23): 1314–1315.

Railway Age 1929c. Plans for Hudson River Bridge not Approved. 86(23): 1343.

Railway Age 1933. Hudson River Bridge not Approved. 95(22): 773.

Rastorfer, Darl 2000. *Six bridges, the Legacy of Othmar H. Ammann*. New Haven: Yale University Press.

Scientific American 1921a. The Hudson River Bridge. 124(17): 324.

Scientific American 1921b. The New York Approach to the Hudson River Bridge Showing the Massive Anchorage, With Office Building Superimposed. 124(26): 508.

Smith, A. 1913. Should a Pier be Permitted in the North River for a Hudson River Bridge? *Engineering News* 69(1): 37.

Wildman, Edwin 1921. Famous Leaders of Industry 2nd Series. *The Page Company* (Boston): 339.

Chapter 28

Rehabilitation of the West Broadway Bridge over the Passaic River, Paterson, New Jersey

G.M. Zamiskie & J.G. Chiara
TranSystems Corporation, Newark, NJ, USA

ABSTRACT: The National Register of Historic Places-eligible West Broadway Bridge is an unique 325-foot long three span structure dating to 1898 located at the base of the Great Falls in the City of Paterson, New Jersey. The bridge is technologically significant as an early example of the patented Melan-type reinforcing system that advanced the use of steel and concrete to achieve longer and more efficient arch spans. Growing traffic demands, poor physical condition, insufficient capacity, scour vulnerability, inadequate safety features, and lost/altered architectural elements were the driving needs for this historic rehabilitation and preservation project. The purpose and goals are to increase the load carrying capacity of the bridge, correct and arrest the causes of moisture-penetration and related material failure, provide for long-term conservation of deteriorated fabric, restore well documented but lost original detailing, and improve inadequate and aesthetically inappropriate safety features.

1 INTRODUCTION

1.1 *Project scope*

The West Broadway Bridge in Paterson New Jersey not only spans the Passaic River, it has spanned the twentieth century providing vehicles and pedestrians a main throughway to the center of Paterson. Built in 1897-1898, the three-span concrete and steel Melan arch bridge was technologically innovative in its early years. Passaic County completed this important rehabilitation project in 2008. It improved the crossing in order to continue its service into the twenty first century and preserving this unique structure that has merited listing on the National Register of Historic Places.

Construction of the $5.2 million project started in May 2006 and was completed in 2008 (Figure 1.). However, the planning, design and approval process began in spring 1999 and followed the National Environmental Policy Act (NEPA) protocol for projects receiving federal funds under the jurisdiction of the NJDOT Local Aid Program.

TranSystems Corporation (TranSystems) was retained by Passaic County to undertake the scoping study of the West Broadway Bridge to document existing conditions, investigate environmental impacts and evaluate alternative solutions including replacement and rehabilitation for the historic bridge. Upon completion of the scoping program that included cultural resource identification, alternatives analysis and environmental screening, TranSystems began the design phase for the historic rehabilitation in 2004.

Passaic County's commitment to the bridge rehabilitation extended beyond upgrading to current design standards and achieving greater structural capacity while retaining the bridge's appearance to include conservation and repair of existing materials, reconstruction of deteriorated and missing masonry details, replication of the original "lost" ornamental pedestrian railings, and installation of a period appropriate modern lighting system.

1.2   *Project purpose and need*

In cooperation with the North Jersey Transportation Planning Authority and the New Jersey Department of Transportation, Passaic County identified the need to improve and upgrade the existing West Broadway crossing of the Passaic River to successfully preserve and extend the service life of this historically significant transportation structure. The existing bridge served as a crossing since 1898. Long-term moisture penetration, stream forces, and the demands of modern day traffic usage were accelerating its deterioration. Aesthetic details, particularly the metalwork, were removed and replaced with unsuitable modifications over time. A vehicular railing system, installed along the curb lines was not only structurally inadequate, but also aesthetically inappropriate for the bridge and its setting. The bridge had reached a state of physical deterioration that necessitated intervention through rehabilitation. The proposed project goals included:

- Measures to correct the cause of long-term moisture-penetration related material failure
- Provision for conservation of deteriorated historic fabric
- Increase in structural capacity to carry current traffic loads and meet current design standards
- Installation of scour countermeasures to eliminate the possibility of scour induced failure
- Upgrades to the ornamental pedestrian railings to current requirements

2   HISTORY OF THE WEST BROADWAY BRIDGE

2.1   *Historic overview*

The West Broadway Bridge is a concrete-steel Melan arch bridge built in 1898 for the City of Paterson. Consultation with the New Jersey Historic Preservation Office identified the West Broadway Bridge both as a contributing resource to the Great Falls/Society for Establishing Useful Manufacturers Historic District and as an individually eligible resource (Guzzo, 2002). The bridge is technologically distinguished as an early example of the patented Melan-type reinforcing system. The Melan patent marked a technological advancement in construction in which steel reinforcement was used in combination with concrete to facilitate lighter, stronger, and more efficient arch construction (Melan, 1893). An Engineering News article: *Three-Span Melan Arch Bridge Across the Passaic River, Paterson, NJ, March 16, 1899*, describes the merits of the design that was developed by the Melan Arch Construction Company and Mr. Edwin Thatcher, ASCE. By achieving a low rise to span ratio, the design reduced the number of piers in the river and maintained acceptable grades at the approaches. The geometry and the efficiency resulting from the concrete-steel combination are apparent when one considers the original shallow depth of the arch at its crown, a mere 15 inches.

When completed in 1898, the handsome elliptical arch bridge was the second longest Melan-type bridge in the country, ranking after a similar span in Topeka, Kansas completed in 1896-97 (Figure 2.). Industrial Historian Donald C. Jackson states that the Topeka Bridge was the "first major reinforced concrete arch bridge in the United States". The West Broadway bridge is one of the most important concrete-steel spans in the Northeast based on its date, structural type, and remarkably complete state of preservation. It is a nationally significant example of the Melan arch bridge technology, and it was designed by noted civil engineer Edwin Thacher. Thacher was a proponent of the Melan arch which is a series of parallel iron or steel I-beams curved to the profile of the soffit and encased in plain concrete. Joseph Melan, a Viennese engineer, was granted an American patent for his design in 1894. Fritz von Emperger, a German-born engineer, built the first Melan arch in the United States at Rock Rapids, Iowa, and he is credited with popularizing the design in this country. Emperger made additions to the Melan system, adding a beam to the deck and joining the deck and arch beams by means of bars set on radial lines. He was granted a patent in 1897 for the changes to Melan's patent.

Edwin Thacher and William Mueser formed the Concrete-Steel Engineering Company in New York City in 1901, and the firm was responsible for many important Melan-type bridges in the

Figure 1. Completed Rehabilitation of the West Broadway Bridge, 2008.

country, including the 8-span Grand Avenue Viaduct, Milwaukee, Wisconsin, built in 1907; a 7-span Melan arch built at Wichita, Kansas, built in 1911; the 6-span Hudson River bridge at Glens Falls, New York, built in 1914-15; the bridge over the Mississippi River at Minneapolis composed of five 231' spans built in 1907. Thacher was a versatile engineer who received many patents including one for the "Thacher Cylindrical Slide-Rule," the "Thacher Steel Bridge Truss," the "System of Concrete Steel Arches," and the "Thacher Combination Bridge Truss" among others. He held positions of chief engineer for the Decatur Bridge Company of Decatur, Alabama, and the Keystone Bridge Company of Pittsburgh before opening his own consulting firm in Louisville, Kentucky. He was responsible for the 1891 Walnut Street Bridge across the Mississippi River at Chattanooga and the 1892 Costilla Crossing Bridge over the Rio Grande in Colorado which was an example of the Thacher truss patented in 1884. Thacher formed a partnership with W.H. Keepers in 1894 at Detroit, and the partnership lasted as Thacher and Keepers until October of 1899. It was this firm that designed the West Broadway Bridge at Paterson. Thacher remained with the Concrete-Steel Engineering Company until his retirement in 1912. The Melan arch was replaced in popularity in this country by the reinforced concrete arch span (McCahon, 2001).

The West Broadway Bridge is one of two Melan-type arch bridges in the state designed by Edwin Thacher. The other is located in Branch Brook Park in Newark, Essex County. It was designed by the Concrete-Steel Engineering Company in 1905. Both are technologically and historically important as examples of the experimentation associated with the introduction of concrete and steel arch bridges in this country.

The West Broadway Bridge is evaluated as individually distinguished. The surrounding area has been redeveloped and has lost most of its historic character. Consequently, the historic setting for the technologically distinguished span has been lost and the boundary limited to the span itself and the westerly side of West Broadway at the Passaic River crossing.

2.2 Historic setting

The project is located in the City of Paterson, New Jersey in a formerly industrial area along the Passaic River characterized by a high degree of urban renewal and limited redevelopment. West Broadway is a two-lane, arterial city street that has long been an important thoroughfare through Paterson. The setting is now dominated by land cleared as part of urban renewal programs. Most of the 19th and early-20th century buildings in the area have been removed as a result of 1960s and

West Street Bridge, Paterson, N. J.

Figure 2.   Postcard Image West Broadway Bridge, c.1900.

1970s urban renewal efforts thus stripping the area of its once-impressive and historically important character.

There has been some redevelopment including a modern, fast-food building built at the southwest corner of West Broadway and Memorial Drive intersection and stands as the only building on the south side of the bridge. A modern, high-rise, subsidized housing complex built in 1963 by the Housing Authority of Paterson is north of Presidential Boulevard. Late-19th and early-20th century brick buildings built to the lot line are beyond the southwest quadrant of the bridge are highly altered. The island west of the bridge was significantly developed by the Alfano family starting about 1905 to be the local wholesale fruit and produce market, a function it served until 1931 and named the Island Market. The 19th century buildings on the island that were associated with its use as an amusement park were washed away in the devastating 1903 flood that also destroyed portions of the upstream side of the bridge.

The project is located at the north end of the Society for Establishing Useful Manufactures (SUM) district, this country's first attempt to harness the entire power of a major river to power factories. Chartered in 1791, the society was successful over nearly 60 years in developing a power canal system to capture the power from the Great Falls of the Passaic River to meet ever-expanding industrial usage. An impressive assemblage of textile and industrial factories were built along their series of canals, and the area flourished as one of the premier antebellum industrial districts in the state. Most of the development was on the south side of the river with only the Addy Silk mill, started in 1852, on the north side. The great historical importance of the SUM was recognized about 1975 when it was designated a National Historic Landmark (McCahon, 2001).

The island in the middle of the river just west of the West Broadway Bridge is designated on period maps as a wooded island in the river with natural banks. It served as a park and the 1887 Sanborn maps depicts it as Philion's Little Coney Island complete with a theater, a billiard parlor, and a rifle range. There was an entrance gatehouse accessed from the West Broadway Bridge. The island driveway remains comprised of a concrete span supported on the upstream spandrel wall. None of the buildings associated with Philion's park survived the 1903 flood. By 1915, the island had been developed by Pasqual Alfano and his business associates into the city's wholesale fruit and produce market. The Island Market Company flourished until 1931.

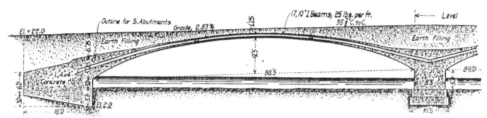

FIG. 2.—LONGITUDINAL SECTION OF 88.5-FT. MELAN ARCH BRIDGE FOR WEST ST. BRIDGE, PATERSON, N. J.

Figure 3.   Engineering News, March 16, 1899 Schematic of West Broadway Bridge.

3   REHABILITATION CONSIDERATIONS AND ANALYSES

3.1   *Existing bridge description*

The West Broadway Bridge is a well-proportioned 3-span, steel and concrete elliptical arch bridge of the patented Melan reinforcing system. The concrete spandrel walls and arches are faced with ashlar brownstone masonry. The structure is 290' long with each of the cinder-filled arch spans approximately 89' clear. The rise of the arch is 9-6". The overall bridge width is 54' and carries a 2-lane, 35' wide roadway with 8'-3" wide sidewalks to each side. The roadway is flanked by 2'-3" high beam guide rails and the sidewalks enclosed with 3'-10" high metal railings with lattice panels along the bottom half (Figure 4).

Brownstone comprises the masonry spandrel walls, cutwaters and voussoirs. Granite capstones and coping complete the masonry detailing. The bridge was originally appointed with ornate pedestrian railings and lamp posts. Over the years, the bridge's condition had suffered. The most severe damage occurred in the 1903 flood of Paterson which destroyed portions of two of the spans.

The Melan reinforcing system is comprised of 10" I-beams set 3' on center and embedded in concrete of variable thickness. The depth of the concrete section varies from 15"± at the crown to 66"± at the skewbacks. Expansion joints are built in the spandrel walls over the piers to provide for thermal changes.

The foundations for the arch bridge, as documented in the Engineering News Article of 1899, consist of 18' wide concrete abutments embedded approximately 5' below the channel bed (Figure 3). The concrete cutwater piers are 10.5' wide at the base and 8.5' wide at the stem. The referenced article indicates that the southernmost pier is founded on piles (Unknown, ENR, 1899). Since their original construction, steel sheeting has been installed around the piers and a concrete apron poured above the pier footings between the pier wall and the steel sheeting to provide scour protection.

3.2   *Existing bridge condition*

The intrados of the arch spans were in poor condition and typically exhibited numerous large spalls exposing the structural steel reinforcing beams. The steel beams, where exposed, were severely corroded (Figure 5). The arch spandrel walls had minor areas of loose and missing mortar and several cracked stones throughout. Water infiltration through the cracked and deteriorated roadway and sidewalk surfaces had occurred, as evidenced by extensive wet and damp areas throughout the arch intrados at the time of project scoping (Chiara, 2005). Results of non-destructive testing indicated that the existing cinder fill on the bridge exhibited moisture entrapment at the concrete interface and along the longitudinal construction joints. Continued deterioration of the primary structural components, if not arrested, would further reduce the load carrying capacity of the bridge.

The abutments were in fair condition with areas of missing mortar throughout. The west end of the north abutment footing was exposed adjacent to the 66" diameter outlet pipe. There were several

Figure 4.   Roadway View West Broadway Bridge, 2008.

Figure 5.   Existing Condition – Arch Intrados, 2000.

missing stones at this location resulting in a large void. The stone masonry retaining walls had several missing stones and areas throughout with missing mortar. The piers were in fair condition with areas of loose and missing mortar throughout and moderate to severe efflorescence below the springline at each face. There were numerous missing and loose stones at the pier noses (Figure 6). The exposed concrete core was severely scaled at these locations and the east end of the piers had small trees growing through the mortar joints Pickering, Corts and Summerson, 1997).

The ornamental pedestrian railing had random minor to moderate rust and some locations of loose or bent members. The railings were misaligned at several locations. The existing vehicular railing exhibited impact damage at many locations.

Figure 6.   Existing Condition – Pier Nose, 2000.

Figure 7.   Existing Condition – Railing Safety Features, 2000.

3.3   *Existing structural capacity*

Prior to the rehabilitation, the bridge was load posted for 10, 10, and 14 Tons for the legal load vehicle types 3 (25 Tons), 3S2 (40 Tons), and 3-3 (40 Tons), respectively. As such, the bridge was categorized as "structurally deficient" according to the NBIS inspection report at the time of the alternatives analysis development. Rating calculations performed resulted in an Inventory rating of 13 Tons and an Operating rating of 26 Tons for the desired HS25 (45 Ton) design vehicle. In spite of the load posting, it was evident that the bridge was used on a regular basis by trucks and busses due to its central urban location. Based on this usage, an increase in the structural capacity of the arch was required to safely carry current traffic loadings and to bring the bridge elements to current design standards.

3.4   *Safety features*

The existing ornamental pedestrian railings provided adequate height (3'-10") for pedestrian traffic. However the railings did not meet present day standards for pedestrian loading. The geometric configuration of the pedestrian railings was consistent with AASHTO standards for minimum spacing between elements but numerous segments had been retrofitted and modified with substandard elements. Several posts and rails were loose and completely disconnected posing a hazard to the traveling public. As the pedestrian railing was not from original construction, replacement of the substandard railing was recommended.

Due to the historic nature of the bridge, the proposed pedestrian railings are ornamental and are not capable of meeting traffic impact requirements. Vehicular railings are required at the curb lines to protect the pedestrian railings and were desired by public opinion and the owner. The existing steel vehicular beam guide rails were structurally inadequate and in poor physical condition (Figure 7). The existing post spacing was too great and the rail element stiffening on the bridge and in the approach transition zones was substandard. Guide rail end treatment at the bridge approach corners and at the island access driveway was also substandard. Due to the above deficiencies, the existing vehicular railing system required replacement.

3.5   *Aesthetic and architectural features*

As described above, the bridge's aesthetics have suffered from physical deterioration of materials primarily as a result of long-term moisture infiltration and deferred maintenance. In addition, there has been a loss of original architectural features on the bridge. The existing pedestrian railing was installed, according to county records, after 1936 and has been subjected to unsuitable modifications and repairs. These modifications and repairs consist of strengthening measures that incorporate insensitively installed material and connections. A vehicular railing system at the bridge curb lines is not only structurally inadequate and in poor physical condition, but also an aesthetically inappropriate modern intrusion to the bridge. The luminaires and light standards that were once installed atop the concrete pylons at the piers are missing and only the electrical conduits that once fed them remain. This rehabilitation project proposes to reproduce lost or insensitively altered architectural features.

4   REHABILITATION WORK

Due to the bridge's state of physical deterioration, insufficient structural capacity, and inadequate safety features it was not considered prudent or feasible to leave the noted conditions uncorrected. Several alternatives were investigated during the scoping process, including rehabilitation; rehabilitation with strengthening; a new bridge on existing alignment (loss of existing arch); and a new bridge on an alternate alignment with the adaptive reuse of the existing arch as a pedestrian/bicycle facility. Since replacement of the bridge was not regarded as necessary due to its overall sound physical condition and geometric adequacy, and in consideration of its historic significance, rehabilitation with strengthening was determined to be the only "prudent and feasible" means of remedying the outstanding deficiencies and meeting the project needs. This option was realized with no historic properties adversely affected. The proposed project was developed as sensitive bridge rehabilitation performed in accordance with *The Secretary of the Interior's Standards for the Treatment of Historic Properties*.

4.1   *Physical condition*

4.1.1   *Arch*

The existing bituminous concrete roadway pavement, concrete sidewalks, steel vehicular railing system and fill material were removed to expose the arch extrados and interior faces of the spandrel walls. All hollow-sounding, spalled and cracked cast-in-place concrete was carefully removed at theses exposed surfaces. Exposed structural steel was appropriately cleaned and primed prior to patching with concrete. The arch extrados and interior faces of the spandrel walls were waterproofed with a membrane system.

A new drainage system consisting 1½" diameter PVC pipe weepholes were core drilled through the existing concrete arch installed inconspicuously at 4' spaces across the width of the bridge adjacent to the springlines at both ends of each arch span. The drainage system will aid in preventing any moisture that should infiltrate the new roadway pavement and concrete sidewalks from contacting the arch extrados.

After installation of the waterproofing membrane and drainage system construction, a variable thickness cast-in-place, concrete relief arch was constructed atop the existing arch within the limits of the roadway. New concrete sidewalks, backfill, and bituminous concrete roadway pavement were installed after construction of the concrete relief arch.

All hollow sounding, spalled, scaled, and cracked cast-in-place concrete was carefully removed from the arch intrados. Exposed structural steel was cleaned and primed prior to installing concrete repairs. Deteriorated cast-in-place concrete areas were repaired in accordance with generally accepted concrete conservation methodologies. The new concrete material design was consistent in composition (ratios of lime/cement and aggregate based on laboratory analysis of the existing), color, finish/texture, and detailing to the existing.

Areas of missing and deteriorated pointing at the spandrel walls were appropriately repointed in accordance with generally accepted conservation practices to match the existing work in composition (ratios of lime/cement and aggregate based on laboratory analysis of the existing), color, finish/texture, and joint style. Detrimental vegetation was removed from the arch stone masonry surfaces prior to repointing.

4.1.2 *Abutments*

Replacement stones were sought to match the type, appearance, durability, and strength of the existing stones used in the original construction. As with the spandrel walls, all repointing at the abutments was performed in accordance with generally accepted conservation practices to match the existing work in composition, color, finish/texture and joint style.

4.1.3 *Piers*

The lost and damaged portions of the pier cutwaters were rebuilt using in-kind material and replacement brownstone as needed to match the original architectural features. The existing stone facing at the cutwaters was first documented, numbered, measured, photographed, dismantled, and salvaged for reuse in creating alike architectural features that are authentic restorations. Replacement stones match the type, appearance, durability, and strength of the existing stones used in the original construction. Stones were retrieved from the waterway and reused to the fullest extent. The underlying concrete cores of the pier cutwaters, noted to be unsound material where exposed, were completely removed and reconstructed. As with the spandrel walls and abutments, all repointing and new pointing at the piers, including cutwater reconstruction, was performed in accordance with generally accepted conservation practices to match the existing work in composition color, finish/texture and joint style. The cracks in the concrete aprons surrounding the piers were sealed with a pressure-injected grout. Detrimental vegetation was removed from the pier surfaces prior to reconstruction and repointing.

4.2 *Structural capacity*

The structural capacity of the bridge was increased to current load requirements, permitting the removal of the existing load posting by placing a relief arch on top of the existing arches. Design analyses included 1) a composite relief arch option whereby the Melan arch and the new cast-in-place relieving arch would be made integral and 2) an independent relief arch design option.

Figure 8.   Setting Precast Concrete Panel Formwork and Jacking Insert Detail, 2006.

There were several considerations that favored an independent arch in preference to a composite design.

The West Broadway Bridge, as an early example of the Melan reinforcing system, utilizes a fixed-depth I-beam arch rib. In the absence of detailed original plans, an accurate determination of the depth of concrete above the steel beam along the span length could not be made. Non-destructive methods were employed; however, it remained unclear as to the shape of the arch and location of the single steel reinforcing beam within various sections. Further, based on the advanced deterioration of the Melan arch, a definitive load-share determination could not be made as it is difficult to quantify how much load would be carried by the original arch if constructed compositely with the new.

From a preservation perspective, the composite construction could potentially result in an accelerated rate of deterioration of the Melan arch under live load. The constancy of dead load in carrying its own self-weight was viewed as beneficial to extending the life of the structure and preserving original historic fabric. Another consideration was the capacity of the new relief arch. If deterioration of the Melan arch advances to a critical level, then the weight of the Melan arch that served as the formwork for constructing the relief arch would be carried by the new relief arch.

The optimal solution was the use of precast concrete panel forms to serve first, as formwork for construction of the cast-in-place relief arch and second, to act as a composite section with the relief arch once cured to design strength. The precast forms were intentionally roughened and provided with hook reinforcement bars as shear stirrups allowing for a minimal depth section. Similar to the composite section between the Melan and new relief arch; the Melan arch served as the shoring support for the precast panels and new relief pour. The design employed use of a series of threaded rods and sacrificial bearing plates set directly on the Melan arch extrados. The precast panel forms were supported from the threaded rods at planned elevations above the Melan arch extrados (Figure 8). The goal was to form a deflection gap between the new arch intrados or bottom and the Melan arch extrados. The reinforcement steel was installed cast-in-place relief arch was poured and allowed to adequately cure (Figure 9).

After the new concrete arch reached the specified design strength, the temporary threaded rods were extracted from the panels allowing the new relief arches to be self-supporting. All live load and new construction dead loads were now independent from the historic Melan arches. The new relief arches were imperceptible once the final roadway pavement was installed.

Once the new arch cured and threaded rods extracted, the Melan arch was subject to its own dead load and relieved of the dead load and live load now fully supported on the new arch. A 3" deflection gap remained between the extrados of the Melan and the intrados of the new arch that is the bottom of the precast panels (Figure 10). The gap could be seen at the stage line assuring that

Figure 9.   Setting Reinforcement Steel at Relief Arch, 2007.

Figure 10.   Deflection Gap between the Melan and Relief Arch, 2007.

the elevated precast platform performed as planned. To tie the new relief arches vertical dowel bars were drilled and anchored into the existing arch supports behind the abutment springlines and pier supports.

To facilitate construction of the relief arch, the existing large granite capstones on top of the spandrel walls were documented, marked, and removed. Adding a matching granite stone band atop the existing granite fillet at each spandrel wall and subsequently resetting the existing capstones accommodated the additional superstructure depth resulting from the relief slab. The placement of the new material was performed in a manner that left the original arch and spandrel walls unaltered and undisturbed and expresses where new fabric has been placed. The shape of the new band conforms to the profile of the relief arch. The band is a new element, but it is a discreet, small detail executed in matching material. It is located under large overhanging capstones that will cast a shadow line. It does not affect the significance or aspects of integrity of the bridge. It is a new visual element that is in scale and character with the bridge and district (Figure 11).

Figure 11.   New Stone Coursing – delineation of original and raised profile, 2006.

4.3   *Scour vulnerability*

To reduce the potential of damage from scour, riprap approximately 4' deep was placed as a stream liner around the steel sheeting at the piers and at the abutments. The scour protection placed in the stream within the historic district will be primarily subaqueous. The channel was also cleared of accumulated rubble stone, debris, and sediment.

4.4   *Safety features*

According to limited available documentation, the existing pedestrian railing was not original to the structure completed in 1898. Much of the existing railing were insensitively repaired and retrofitted with components that did not match original material and architectural features. A new ornamental pedestrian railing meeting current AASHTO loading and geometric requirements for pedestrian usage was installed on the bridge atop the reset capstones. Detailed records of the original pedestrian railing do not exist. Therefore, replication of the original railing was completed through photographic documentation dated 1899. The rails were finished with the color of the original railing. Connections between railing components was made with rivets and square-headed bolts to match the appearance of the original connections. A matching sliding gate was installed at the island access driveway. The gate is kept in a locked-closed position and will be only opened if needed for emergency evacuation of the island. As part of the pedestrian railing replication, the concrete pylons located over the piers were recast based upon details shown in the 1899 photograph (Figure 12). The bridge plaques were reset on these pylons. An original cast iron end post dating to the 1898 railing was utilized to recast four new cast iron end posts for the replacement railing system.

In consideration of the non-crashworthy nature of the proposed pedestrian railings, a vehicular railing system continues to be warranted along the curb lines, evidenced by recent impact damage to the existing vehicular rails. The existing vehicular railing system on the bridge and approaches was replaced with a crash tested system per *NCHRP Report 350 – Recommended Procedures for the Safety Performance Evaluation of Highway Features* commensurate to the roadway classification and design speed. The Illinois Curb-Mounted Tubular Bridge Rail, a metal tube bridge railing system tested to the TL-4 level, was installed that lends itself to this historic bridge due to its open appearance and simple, clean shape. Removable crash-worthy bollards were installed at the driveway for continuity in the vehicular barrier (Figure 13).

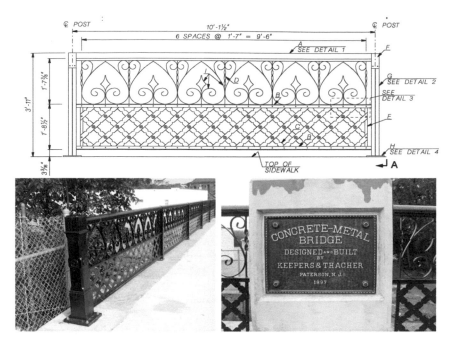

Figure 12.   Pedestrian Railing Panel replicated from 1898 Postcard and Original Bridge Plaque, 2008.

Figure 13.   Vehicular Railing TL-4 Level; Removable Bollards, 2008.

4.5 *Aesthetics and architectural features*

According to limited available documentation, luminaires and light standards were once installed atop the concrete pylons as evidenced by remaining standard anchor bolt circles and center wiring conduit. During design consultation, the Paterson Historic Preservation Commission indicated a desire to include installation of ornamental bridge lighting in the proposed project scope.

As the original lighting luminaries and standards are a lost architectural feature, there is limited photographic documentation of detail available for replication. However, through close coordination with the Paterson Historic Preservation Commission, new lighting assemblies included the

Figure 14.   Masonry Field Mock-Up and SHPO Consultation, 2006.

luminaire style name "Franklin Square" and light post style name "Galtier" as manufactured by Spring City Electrical Manufacturing Company. The posts and luminaires were painted to match the color of the ornamental pedestrian railing. Bridge lighting assemblies were fabricated to be vandal proof units to assure serviceability and longevity. Close coordination with the State Historic Preservation Office and the Executive Director of the Paterson Historic Preservation Commission was a continuous effort in the final development and completion of the bridge lighting plan, aesthetics and preservation mitigation (Figure 14).

5   CONCLUSIONS

The proposed rehabilitation of the bridge provided much benefit to the public using the bridge and the Great Falls of Paterson/Society for Useful Manufactures (S.U.M.) Historic District by returning the structure to full utility and conserving/restoring its original aesthetic appearance to the greatest extent possible. Benefits to the public include much needed safety improvements to the vehicular and pedestrian railings and a positive means to close the private driveway access from mid-span of the bridge for the protection of pedestrians. By increasing the structural capacity all weight limit postings were removed allowing for unrestricted vehicle passage critical to the downtown area of Paterson.

REFERENCES

Chiara, J. 2005. *Application for Project Authorization Under New Jersey Register of Historic Places Act.*
Guzzo, D. 2002. *Section 106 Consultation Letter of Opinion*; State of New Jersey Dept. of Environmental Protection, Historic Preservation Office.
McCahon, M. 2001. *Identification of Cultural Resources and Determination of Effects Report.*
Melan, J. 1893. U.S. Patent Office, *Vault for Ceilings, Bridge, &c., Letters Patent No. 505054.*
Pickering, Corts and Summerson, 1997. *Cycle 7 Re-Evaluation Bridge Survey Report, Sructure No. 1600-017.*
Unknown, 1899. Engineering News Record, *Three-Span Melan Arch Bridge across the Passaic River, Paterson, N.J.*

Author index

T - #0497 - 071024 - C344 - 246/174/15 - PB - 9780367737931 - Gloss Lamination